Nitric Oxide Synthase:
Characterization and Functional Analysis

Nitric Oxide Synthase: Characterization and Functional Analysis

Edited by

Mahin D. Maines

Department of Biophysics
University of Rochester Medical Center
Rochester, New York

ACADEMIC PRESS

San Diego   London   Boston   New York   Sydney   Tokyo   Toronto

Front cover photograph: Cortical neurons.

This book is printed on acid-free paper. ∞

Academic Press, Inc.
525 B Street, Suite 1900, San Diego, California 92101-4495, USA
http://www.apnet.com

Academic Press Limited
24-28 Oval Road, London NW1 7DX, UK
http://www.hbuk.co.uk/ap/

International Standard Serial Number:

International Standard Book Number: 0-12-185301-2

PRINTED IN THE UNITED STATES OF AMERICA
96 97 98 99 00 01 EB 9 8 7 6 5 4 3 2 1

Table of Contents

Section III Assessment of Nitric Oxide-Mediated Functions at Cell and Organ Levels

Contributors

Article numbers are in parentheses following the names of contributors. Affiliations listed are current.

TIMOTHY R. BILLIAR (14), Department of Surgery, A-1010 Presbyterian University Hospital, University of Pittsburgh Medical Center, Pittsburgh, Pennsylvania 15213

BERNHARD BRÜNE (24), Department of Medicine IV/Experimental Medicine, Faculty of Medicine, University of Erlangen-Nürnberg, 91054 Erlangen, Germany

HANS R. BRUNNER (28), Division of Hypertension, Centre Hospitalier Universitaire Vaudois, 1011 Lausanne, Switzerland

MICHEL BURNIER (28), Division of Hypertension, Centre Hospitalier Universitaire Vaudois, 1011 Lausanne, Switzerland

GABRIEL CENTENO (28), Division of Hypertension, Centre Hospitalier Universitaire Vaudois, 1011 Lausanne, Switzerland

ANTHONY H. CHEUNG (16), Department of Medicine, Medical Sciences Building, Room 7358, University of Toronto, Toronto, ON M5S 1A8, Canada

FRANCESCO COSENTINO (19), Division of Cardiology, Cardiovascular Research, University Hospital, CH-3010 Bern, Switzerland

TED M. DAWSON (6, 20), Department of Neurology, Neuroscience, and Physiology, John Hopkins University School of Medicine, Baltimore, Maryland 21287

VALINA L. DAWSON (6, 20), Department of Neurology, Neuroscience, and Physiology, John Hopkins University School of Medicine, Baltimore, Maryland 21287

JAN DE VENTE (7), Department of Psychiatry and Neuropsychology, European Graduate School for Neuroscience, University of Limburg, Maastricht, Netherlands

JAMES F. EWING (9), Department of Biochemistry and Molecular Biology, NitroMed Research Laboratories, NitroMed, Inc., Boston, Massachusetts 02118

FRANK M. FARACI (23), Departments of Internal Medicine and Pharmacology, Cardiovascular Center, University of Iowa College of Medicine, Iowa City, Iowa 55242

JOHN GARTHWAITE (21), The Cruciform Project, London W1P 9LN, United Kingdom

MICHAEL F. GOY (8), Department of Physiology, Center for Gastrointestinal Biology and Disease, and Curriculum in Neurobiology, University of North Carolina, Chapel Hill, North Carolina 27599

CHRISTIAN HARTENECK (11), Institut für Pharmakologie, Freie Universität Berlin, Berlin, Germany

YUKITOSHI IZUMI (25), Departments of Psychiatry, Anatomy, and Neurobiology, Washington University School of Medicine, St. Louis, Missouri 63110

FREDERICK KIECHLE (2), Department of Clinical Pathology, William Beaumont Hospital, Royal Oak, Michigan 48073

WON-KI KIM (27), Laboratory of Cellular and Molecular Neuroscience, Children's Hospital and Program in Neuroscience, Harvard Medical School, Boston, Massachusetts, 02115; and Ewha Medical School, Yangchun-hu, Seoul 158-750, Republic of Korea

PETER KLATT (11), Institute für Pharmakologie und Toxikologie, Karl-Franzens-Universität Graz, A-8010 Graz, Austria

HARRIET KRUSZYNA (1), Department of Pharmacology and Toxicology, Dartmouth Hitchcock Medical Center, Dartmouth Medical School, Hanover, New Hampshire 03755

EUGENIUSZ KUBASZEWSKI (2), Department of Chemistry and Institute of Biotechnology, Oakland University, Rochester, Michigan 48309

EDUARDO G. LAPETINA (24), School of Medicine, Case Western Reserve University, Cleveland, Ohio 44106

ZHIPING LI (8), Department of Physiology, Center for Gastrointestinal Biology and Disease, and Curriculum in Neurobiology, University of North Carolina, Chapel Hill, North Carolina 27599

STUART A. LIPTON (27), Laboratory of Cellular and Molecular Neuroscience, Children's Hospital & Program in Neuroscience, Harvard Medical School, and Beth Israel Hospital, Brigham and Women's Hospital, Massachusetts General Hospital, Boston, Massachusetts 02115

BARBARA M. LIST (11), Institut für Pharmakologie und Toxikologie, Karl-Franzens-Universität Graz, A-8010 Graz, Austria

JIANWEI LIU (18), Molecular Cardiobiology Program and Department of Pharmacology, Boyer Center for Molecular Medicine, Yale University, New Haven, Connecticut 06536

JOSEPH LOSCALZO (3, 4), Evans Department of Medicine and Department of Biochemistry, Whitaker Cardiovascular Institute, Boston University School of Medicine, Boston, Massachusetts 02118

THOMAS F. LÜSCHER (19), Division of Cardiology, Cardiovascular Research, University Hospital, CH-3010 Bern, Switzerland

TADEUSZ MALINSKI (2), Department of Chemistry and Institute of Biotechnology, Oakland University, Rochester, Michigan 48309

PHILIP A. MARSDEN (16), Department of Medicine, Medical Sciences Building, University of Toronto, Toronto, ON M5S 1A8, Canada

PAVEL MARTASEK (12), Department of Biochemistry, University of Texas Health Science Center, San Antonio, Texas 78284

BETTIE SUE SILER MASTERS (12), Department of Biochemistry, University of Texas Health Science Center, San Antonio, Texas 78284

BERND MAYER (11), Institute fur Pharmakologie und Toxikologie, Karl-Franzens-Universität Graz, A-8010 Graz, Austria

KIRK McMILLAN (12), Pharmacopeia, Inc., Princeton, New Jersey 08540

STEVEN MENNERICK (25), Departments of Psychiatry, Anatomy, and Neurobiology, Washington University School of Medicine, St. Louis, Missouri 63110

DAHLIA MINC-GOLOMB (15), Institute of Human Genetics, The Chaim Sheba Medical Center, Tel Hashomer 52621, Israel

MASAKI NAKANE (10, 13), Pharmaceutical Discovery, Abbott Laboratories, Abbott Park, Illinois 60064

PAUL K. NAKANE (5), Department of Anatomy III, Nagasaki University School of Medicine, Nagasaki 852, Japan

JONATHAN NISHIMURA (12), Department of Biochemistry, University of Texas Health Science Center, San Antonio, Texas 78284

ULRICH POHL (22), Institute of Physiology and Pathophysiology, Johannes Gutenberg-University Mainz, D-55 099 Mainz, Germany

JENNIFER S. POLLOCK (10, 13), Pharmaceutical Discovery, Abbott Laboratories, Abbott Park, Illinois 60064

PREM PONKA (29), Lady Davis Institute for Medical Research, Sir Mortimer B. Davis—Jewish General Hospital, McGill University, Montreal, Quebec H3T 1E2, Canada

DAVID R. REPASKE (8), Division of Endocrinology, Children's Hospital Medical Center, Cincinnati, Ohio 45229

DES R. RICHARDSON (29), Lady Davis Institute for Medical Research, Sir Mortimer B. Davis—Jewish General Hospital, McGill University, Montreal, Quebec H3T 1E2, Canada

LORI G. ROCHELLE (1), Department of Anatomy, Physiological Sciences, and Radiology, College of Veterinary Medicine, North Carolina State University, Raleigh, North Carolina 27606

LINDA J. ROMAN (12), Department of Biochemistry, University of Texas Health Science Center, San Antonio, Texas 78284

MARK SALTER (21), Wellcome Research Laboratories, Beckenham, Kent BR3 3BS, United Kingdom

KURT SCHMIDT (11), Institut für Pharmakologie und Toxikologie, Karl-Franzens-Universität Graz, A-8010 Graz, Austria

ERIN M. SCHUMAN (26), Division of Biology, 216-76, California Institute of Technology, Pasadena, California 91125

JOAN P. SCHWARTZ (15), Molecular Genetics Section, Clinical Neuroscience Branch, National Institute of Neurological Disorders and Stroke, National Institutes of Health, Bethesda, Maryland 20892

WILLIAM C. SESSA (17, 18), Molecular Cardiobiology Program and Department of Pharmacology, Boyer Center for Molecular Medicine, Yale University, New Haven, Connecticut 06536

ROGER P. SMITH (1), Department of Pharmacology and Toxicology, Dartmouth Hitchcock Medical Center, Dartmouth Medical School, Hanover, New Hampshire 03755

ERIC SOUTHAM (21), Wellcome Research Laboratories, Beckenham, Kent BR3 3BS, United Kingdom

JONATHAN S. STAMLER (27), Department of Respiratory and Cardiovascular Medicine, Duke University Medical Center, Durham, North Carolina 27710

HARRY W. M. STEINBUSCH (7), Department of Psychiatry and Neuropsychology, European Graduate School for Neuroscience, University of Limburg, Maastricht, Netherlands

EDITH TZENG (14), Department of Surgery, University of Pittsburgh, Pittsburgh, Pennsylvania 15261

GILBERT R. UPCHURCH, JR. (3), Evans Department of Medicine and Department of Biochemistry, Whitaker Cardiovascular Institute, Boston University School of Medicine, Boston, Massachusetts 02118

GEORGE N. WELCH (3), Evans Department of Medicine and Department of Biochemistry, Whitaker Cardiovascular Institute, Boston University School of Medicine, Boston, Massachusetts 02118

ERNST R. WERNER (11), Institut für Medizinische Chemie und Biochemie, Universität Innsbruck, A-6020 Innsbruck, Austria

HEATHER M. YOUNG (7), Department of Anatomy and Cell Biology, University of Melbourne, Parkville 3052, Victoria, Australia

YING-YI ZHANG (4), Evans Department of Medicine and Department of Biochemistry, Whitaker Cardiovascular Institute, Boston University School of Medicine, Boston, Massachusetts 02118

ELLEN M. ZIMMERMAN (8), Department of Internal Medicine, University of Michigan School of Medicine, Ann Arbor, Michigan 48109

CHARLES F. ZORUMSKI (25), Departments of Psychiatry, Anatomy, and Neurobiology, Washington University School of Medicine, St. Louis, Missouri 63110

Preface

Because of the immense interest in the biological function of nitric oxide (NO), this volume is devoted to this molecule and related components. This simple molecule, which acts in a fairly specific manner, is now recognized as an integral part of various homeostatic mechanisms, including the central and peripheral nervous system and cardiovascular system, as well as host–defense interactions, where it can directly alter vital functions. The impact of the biological actions of NO is further augmented by its activity that affects gene expression. The unwavering interest in NO relates to the seemingly unending and astonishing diversity of cellular and physiological functions that are attributed to it. The role of NO in the nervous system is of particular interest because, unlike neurotransmitters such as GABA, dopamine, and glutamate, which have highly restricted regional distribution, the ability to produce NO is found in many cells in the brain. And, because NO can diffuse across membranes, its range of activity far extends its local effects. As the significance of NO activity in the brain becomes more apparent, the methods that allow investigators to identify its role in cellular and physiological functions become invaluable.

In assembling this volume, I have attempted, in cooperation with a distinguished group of contributors, to produce a useful tool for neuroscientists and other investigators in the field of NO and those entering it. I am grateful to the contributors for their cooperation and thankful to the staff of Academic Press, particularly Shirley Light, for their support and assistance.

MAHIN D. MAINES

Methods in Neurosciences

Editor-in-Chief

P. Michael Conn

Section I

Enzyme Activity Measurements, Nitric Oxide Detection, and Tissue Localization

[1] Electron Paramagnetic Resonance Detection of Nitric Oxide in Biological Samples

Lori G. Rochelle, Harriet Kruszyna, and Roger P. Smith

Introduction

Nitric oxide (NO), a small inorganic gas molecule, can freely diffuse in and out of cells and tissues. The physicochemical properties of NO including high diffusibility and reactivity result in a short biological half-life and make detection and quantification of NO difficult under physiological conditions. Electron paramagnetic resonance (EPR) is a method that can be used to detect molecules with unpaired electrons (paramagnetic molecules) such as NO. Free NO is not detected by EPR because of its rapid relaxation time, but it is possible with this technique to take advantage of spontaneous endogenous reactions for binding and stabilizing NO in extracellular or intracellular locations. Nitric oxide interacts with the constituent sulfhydryls and transition metals of a variety of endogenous molecules. Electron paramagnetic resonance has been a particularly useful technique for detecting NO bound to iron proteins, which constitute some of the predominant cellular NO complexes. Binding to iron-containing molecules such as hemoglobin permits not only detection, but also stabilization, characterization, and quantification by EPR. Nitric oxide interacts with endogenous molecules integral to the exterior or interior of the cell or with exogenous materials that may or may not be taken up by the cells.

Electron paramagnetic resonance is an especially attractive method for detecting NO in biological samples because under some conditions, the sample can be directly introduced without further preparation into the EPR tube for immediate analysis or snap frozen in liquid nitrogen for future analysis. Most other components of cells and tissues are invisible to this technique, which provides the necessary specificity for the NO-bound iron complexes discussed in this chapter.

Electron Paramagnetic Resonance Spectroscopy

Electron paramagnetic resonance spectra are produced by energy absorbances analogous to UV/Vis or IR spectroscopy. In this technique, the energy source emits microwave frequencies that are absorbed by unpaired electrons in the sample. When unpaired electrons of paramagnetic molecules are placed in a strong magnetic field, incident electromagnetic radiation (applied perpendicular to the field) is absorbed as the electrons go from one electron spin state to another, that is, the magnetic moments of the electrons change from one discrete value to another (1). Energy absorption that permits transition between energy states (i.e., resonance), occurs at microwave frequencies ($\sim 10^{10}$ Hz). This energy can be described by the basic equation of quantum physics:

$$E = h\nu = hc/\lambda \tag{1}$$

where h is Planck's constant; ν, the frequency; c, the speed of light; and λ, the wavelength. For various practical reasons such as equipment, geometry, and detection sensitivity, the magnetic field is varied rather than the frequency of the incident radiation, which is varied in other spectroscopic techniques. Equation (1) can be rewritten in a form that contains parameters relevant to EPR:

$$h\nu = g\beta H \quad \text{or} \quad g = h\nu/\beta H$$

where g is the electron g factor; β, the Bohr magneton (constant); and H, magnetic field strength at resonance (in gauss, G).

The electron g factor is unitless and identifies specific transitions between spin states. It serves as a fingerprint for a particular paramagnetic substance (it locates the position of resonance) and it can be determined experimentally from the absorption spectrum of the substance. When the unpaired electron is free, as in a radical with delocalized unpaired electrons, the g value is 2.0023. Molecules with localized unpaired electrons will exhibit g values higher or lower than 2.0023 so that even small shifts in the EPR spectra are significant and specific.

Figure 1 is an idealized spectrum of a free unpaired electron expressed as absorbance (Fig. 1a) and as a first derivative (Fig. 1b). Electron paramagnetic resonance spectra are presented as a first derivative of the absorbance of microwave energy (dA/dH) where A is the absorbance and H is the magnetic field (Fig. 1b). These spectra can be integrated once to obtain the absorbance and twice to calculate the area under the absorbance curve. The area under the curve is proportional to the concentration of unpaired electrons in the sample. Most EPR instruments are interfaced with a computer and contain software capable of doubly integrating spectra. To quantify a signal, a series of standards is prepared and scanned with the same EPR parameters used for the unknowns. A calibration curve that relates concentration to area under the curve (signal intensity) can then be constructed.

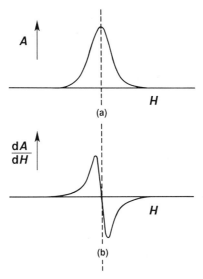

Fig. 1   Idealized absorbance spectra. Spectrum (a) is an absorbance mode as presented in UV/Vis and IR spectroscopy. Spectrum (b) is a first derivative as presented in EPR spectroscopy. For EPR, the x axis is expressed in units of gauss (magnetic field) or g factor. [Modified from H. M. Swartz and S. M. Swartz, *Methods Biochem. Anal.* **29,** 207 (1983). Copyright © 1983 John Wiley & Sons, Inc. Reprinted by permission of John Wiley & Sons, Inc.]

We use the following EPR parameters to analyze iron protein-bound NO: All spectra are obtained on a Bruker ESP300 X-band spectrometer (Karlsruhe, Germany) operating at 20 mW microwave power, 100 kHz modulation frequency, and 10 G modulation amplitude. The samples in EPR tubes are scanned with a 500 G sweep width (3350 center field) nine times at −196°C in a Wilmad EPR Dewar flask with a 5 msec time constant at a rate of 33 G/sec. We construct a calibration curve relating the concentration of nitrosylhemoglobin [HbNO] in 2 mm (i.d.) tubes to the area under the curve that fits the relationship

$$[\text{HbNO}],\ \text{m}M = 2.75 \times 10^{-6}(\text{area}) - 1.19 \times 10^{-3}$$

HbNO was prepared by fully reducing and nitrosylating isolated hemoglobin tetramers (Fig. 2B,C). With these parameters, we have been able to quantify as little as 2 μM and up to 1 M concentrations of HbNO. Other current reviews by Wilcox and Smith (2) and Kalyanaraman (3) provide extensive technical details of EPR parameters for detecting NO in chemical and biological reactions.

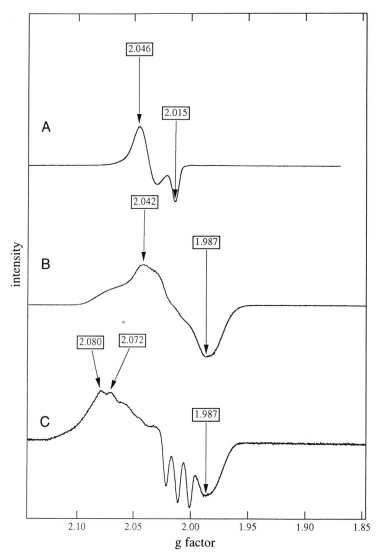

FIG. 2   Electron paramagnetic resonance spectra of NO-bound iron proteins from samples at $-196°C$. Values in boxes are experimentally derived g factors for each maximum or minimum. (A) Spectra of FeNOSR complex from endothelial cells incubated with sodium nitroprusside. (B) Spectra of HbNO comprised of nitrosylated β subunits. (C) Spectra of HbNO comprised of nitrosylated α subunits. Spectra (B) and (C) were produced by nitrosylating subunits isolated by an isoelectric focusing technique [H. Kruszyna, R. Kruszyna, R. P. Smith, and D. E. Wilcox, *Toxicol. Appl. Pharmacol.* **91**, 429 (1987)].

Samples

Our typical samples for analysis by EPR range from 0.25 to 0.50 ml. These are introduced into an EPR tube with an internal diameter of either 2 or 4 mm. Samples can be liquid, minced tissue, or suspended or pelleted cells. The sample is introduced into the tube by Pasteur pipette and frozen in liquid nitrogen. The tube is then placed in the spectrometer cavity inside a thin finger Dewar which retains a small volume of liquid nitrogen and holds the tube at a specified height in the cavity. The sample must fill the tube to a height (4 cm in the case of the Bruker 300 instruments) that is sufficient in order to be seen through the window of the EPR spectrometer. If the sample is smaller, it can be positioned within the EPR window as a frozen pellet in the tube. This allows analysis of smaller samples (100 μl or less).

Trapping by Hemoglobin

Nitric oxide avidly binds to iron proteins. These molecules can serve as spin traps that stabilize NO for detection by EPR. Hemoglobin (Hb) is an attractive spin trap because it is an endogenous molecule that can be sampled by routine procedures in relatively large quantities, and its affinity for NO is more than 1000-fold greater than for CO or O_2. Samples of Hb from blood have been analyzed to infer NO production in blood and tissue *in vivo*. *In vitro* NO production can be assessed from NO trapped by deoxyhemoglobin or suspended erythrocytes. Hemoglobin NO is characterized by two distinct EPR spectra. The predominance of either spectra depends on nitrosylation of the α or β subunits of Hb (Fig. 2B,C). The triplet hyperfine feature upfield (to the left of) of $g = 2.00$ in the α-nitrosylated HbNO spectrum is the most easily distinguished signature of samples with low concentrations of HbNO. This feature is intensified by allosteric effectors such as inositol hexaphosphate (4), low pH (5), and low oxygen tension (6).

Hemoglobin: In Vivo

Production of NO by the vasculature *in vivo* has been measured indirectly from the erythrocytic HbNO concentration in shed blood. Separating erythrocytes from plasma removes the plasma copper-containing metalloprotein, ceruloplasmin, from the sample. Ceruloplasmin displays an EPR signal adjacent to that of HbNO and confounds interpretation of the signal for HbNO.

With this protocol, HbNO has been detected in blood samples from patients receiving nitroglycerin (7) or inhaling NO (8), from mice treated with nitrova-sodilators (9), from rats treated with endotoxin or in hemorrhagic shock (10), from mice and rats treated with bacterial endotoxin (11), and from pregnant rats (12).

Hemoglobin: In Vitro

The constituent iron of hemoglobin must be reduced (Fe^{2+}) to serve as a ligand for NO. Hemoglobin iron is maintained in the necessary reduced state by an endogenous reductase within the erythrocytes, but once red cells are lysed the iron undergoes oxidation. Consequently, chemical reduction is usually necessary for isolated Hb or hemolysed erythrocytes. Deoxygenation of Hb is also necessary because the oxygen ligand of oxyhemoglobin readily reacts with NO (13), producing nitrogen oxides that are not detectable by EPR. For the same reason, HbNO cannot be produced with hemoglobin solutions in aerobic conditions (>5% oxygen atmosphere) due to preferential reaction of NO with O_2 and oxidation of Hb to metHb (14). Under anaerobic conditions and <1% oxygen, deoxyHb readily forms HbNO by binding free NO without producing metHb (14, 15).

Methemoglobin is the oxidized species formed on aerobic incubation of Hb with NO. The amount of metHb in solutions or in fresh blood can be quantified by a spectrophotometric method (16). It also can be detected by EPR, but it is not easily quantified with this technique due to the broadness of the peak produced at $g = 6.00$. Methemoglobin is not only produced on oxidation of Hb by NO but also by other nitrogen oxides, such as nitrite, and much more slowly by O_2 as discussed above.

Suspensions of erythrocytes for *in vitro* experiments are usually prepared by washing fresh blood (treated with anticoagulant) three times with isotonic saline. Erythrocyte suspensions can be used for up to 1 week if they are stored at 4°C in a Krebs–Ringer buffer with glucose. Deoxygenated Hb solutions or erythrocyte suspensions are prepared by equilibrating a slowly stirred thin layer of Hb or erythrocytes in a sealed container with a flow-through stream of N_2 gas for 20 min. Commercially available Hb (which can contain as much as 50–95% metHb) requires chemical reduction with sodium dithionite or sodium borohydride.

Erythrocytes cocultured with alveolar macrophages trapped greater amounts of NO on activation of the macrophages with bleomycin (17). We have been able to compare the amounts of NO produced from nitrovasodilat-ing drugs by reacting them with deoxyhemoglobin and trapping the NO on reduced heme groups (15). Hemoglobin has been used in numerous other studies as an NO sink and as an inhibitor of the biological effects of NO, but, as discussed above, reactions with NO under aerobic conditions can produce a variety of forms of Hb and nitrogen oxides, precluding accurate quantification of NO.

Hemoglobin Agarose

Another expedient method that controls oxidation of the Hb spin trap and collects free NO is to pass the reaction solution through Hb covalently attached to a solid agarose support in closed columns. Hemoglobin–agarose (Hb–agarose) bead columns are assembled as depicted in Fig. 3. In this column, the Hb can be reduced and deoxygenated just prior to introduction of the sample that contains NO.

The slurry of Hb–agarose is added to the EPR tube assembly (F through I, Fig. 3). The Hb gel is rinsed with 10 ml H_2O through the column. The Hb is reduced by rinsing and retaining 10 mg of sodium dithionite/ml 0.1 M phosphate buffer (pH 7.4) on the column. At this step, the column is capped (assembly A through E, Fig. 3) and deaerated by a nitrogen gas stream introduced through A. Then, 4 ml of deoxygenated buffer is injected on top of the column and rinsed through to leave an aqueous miniscus above the top of the agarose column. The sample can then be introduced on top of the column (through A) and pushed through with positive pressure from a stream of nitrogen or argon gas (through A). When introducing a sample or rinse, it is necessary to vent the apparatus (through B) only during the introduction to provide space for the extra volume, while preventing any inflow of air into the apparatus. The sample should also be deoxygenated to prevent oxidation of NO or Hb.

Columns have been used to trap nitrite chemically reduced to NO in human plasma (18) and to measure NO released into the gas phase from a solution of the nitrovasodilating drug SIN-1 (19).

Trapping with Other Nonheme Iron Proteins

Endogenous NO–iron complexes have been detected by EPR not only in erythrocytes, but also in endothelial cells (20) and smooth muscle cells (21), in macrophages activated with lipopolysaccharide (LPS) (22), in macrophages (23), hepatocytes (24, 25), and pancreatic islet cells (26) exposed to cytokines, and in tumor cells exposed to activated macrophages (27) without addition of trapping molecules. Whereas the complex in the case of erythrocytes and one of the complexes in smooth muscle cells is HbNO, the other complexes are classified as nonheme iron proteins that have not been fully

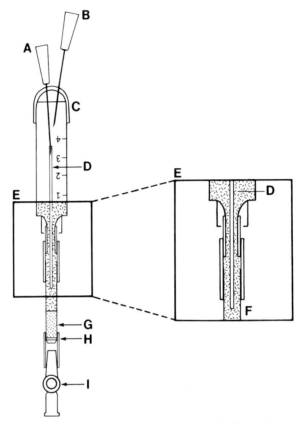

FIG. 3 Apparatus for perfusing, deaerating, and reducing a hemoglobin–agarose bead column. A and B, 20-gauge needles, $1\frac{1}{2}$ inches; C, septum; D, PE 100 polyethylene tubing; E, 5-ml syringe barrel with the flange removed; F, 12 cm, 4-mm i.d. quartz EPR tube tapered at outlet; G, 3 cm column of Hb–agarose; H, fritted disk; I, stopcock. Assemblies A–E and F–I are connected with interlocking pieces of Tygon tubing.

characterized. These proteins display EPR characteristics of NO covalently attached to an iron nucleus bound to a sulfhydryl moiety (FeNOSR complex). Large numbers of cells (0.5×10^8 to 10^9) per sample are required to produce concentrations of FeNOSR complexes detectable by EPR. Accordingly, some investigators add exogenous iron complexes to maximally trap and stabilize any NO produced by the cells (28–31).

In all cases, it is necessary to maximize the concentration of cells/unit volume in the EPR tube. This can be accomplished by pelleting cells by centrifugation or by culturing cells on microbeads. Cell pellets are transferred to an EPR tube where the dilution of small samples is minimized by freezing a plug of buffer in the bottom of the EPR tube before introducing the sample (24). All collection, centrifugation, and transfer procedures should be done in closed containers with the smallest possible headspace in order to retain NO in solution.

Polarized cells, such as endothelial cells, that require attachment for growth and survival may be cultured on microbeads and then introduced into EPR tubes, perfused, and analyzed directly in the tubes. We have used an EPR tube perfusion column to examine the cellular reduction mechanism that produces NO from the vasodilating drug sodium nitroprusside (32). Aliquots of cells cultured on beads are pipetted into a modified EPR tube (Fig. 4) and allowed to settle on the frit to a height of 3 cm (~2 × 10^6 endothelial cells). The EPR tube is then converted to a perfusion column by filling it with buffer, connecting a stopcock with tubing to the fritted disk end, and placing a septum over the other end. The beads are dispersed by tapping the tube. The bead slurry in the tube is then perfused by injecting buffer or argon or nitrogen gas at a slow rate through a needle puncturing the septum. With this method, supernatant can be collected separately from the cells, or the entire tube minus the stopcock can be snap frozen for EPR analysis.

Conclusions and Future Directions

Electron paramagnetic resonance detection and quantification of NO is possible when NO is bound to iron proteins. The NO-binding iron proteins are ubiquitous in cells and tissues and have been characterized by EPR as heme (e.g., hemoglobin, NO synthase, guanylate cyclase) or nonheme (e.g., ferritin, ferredoxin, transferrin). Hemoglobin has a very high affinity for NO and

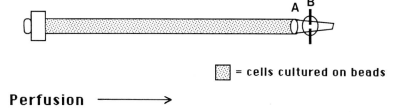

= cells cultured on beads

Perfusion ⟶

FIG. 4 Electron paramagnetic resonance tube cell perfusion column. A, Fritted disk; B, stopcock.

has been a useful trap for detecting NO released into the bloodstream and for quantifying NO generation *in vitro*. Nitrosylated nonheme iron proteins also appear to be important for stabilizing intracellular or extracellular NO. The NO-binding proteins most likely serve in several capacities including trafficking of NO from a site of release to a site of action or serving as sinks for inactivation of NO and as targets for activation or deactivation by NO. Electron paramagnetic resonance detection of protein-bound NO in conjunction with free NO analyses could permit important studies on the spatial and temporal distribution of NO in biological systems.

Acknowledgments

Some of the research reviewed here was supported by National Institutes of Health Grant 14127 from the National Heart Lung and Blood Institute. The Bruker ESP-300 spectrometer was purchased with National Science Foundation Grant CHE-8701406. We are grateful for the long-standing collaboration with our colleague Dean E. Wilcox of the Department of Chemistry, Dartmouth College.

References

1. I. D. Campbell and R. A. Dwek, "Biological Spectroscopy." Benjamin/Cummings, London, 1984.
2. D. E. Wilcox and R. P. Smith, *Methods* (*San Diego*) **7,** 59 (1995).
3. B. Kalyanaraman, *Meth. Enzymol.* in press (1995).
4. R. Hille, J. S. Olson, and G. Palmer, *J. Biol. Chem.* **254,** 12110 (1979).
5. E. Trittlevitz, H. Sick, and K. Gersonde, *Eur. J. Biochem.* **31,** 578 (1972).
6. H. Kosaka, Y. Sawai, H. Sakaguchi, E. Kumura, N. Harada, M. Watanabe, and T. Shiga, *Am. J. Physiol.* **266,** C1400 (1994).
7. L. R. Cantilena, R. P. Smith, S. Frasur, H. Kruszyna, R. Kruszyna, and D. E. Wilcox, *J. Lab. Clin. Med.* **120,** 902 (1992).
8. A. Wennmalm, G. Benthin, A. Edlund, L. Jungersten, N. Kieler-Jensen, S. Lundin, U. N. Westfelt, A. S. Petersson, and H. Waagstein, *Circ. Res.* **73,** 1121 (1993).
9. R. Kruszyna, H. Kruszyna, R. P. Smith, and D. E. Wilcox, *Tox. Appl. Pharm.* **94,** 458 (1988).
10. U. Westenberger, S. Thanner, H. H. Ruf, K. Gersonde, G. Sutter, and O. Trentz, *Free Radical Res. Commun.* **11,** 167 (1990).
11. Q. Wang, J. Jacobs, J. DeLeo, H. Kruszyna, R. Kruszyna, R. P. Smith, and D. E. Wilcox, *Life Sci.* **49,** PL-55 (1991).
12. K. P. Conrad, G. M. Joffe, H. Kruszyna, R. Kruszyna, L. G. Rochelle, R. P. Smith, J. E. Chavez, and M. D. Mosher, *FASEB J.* **7,** 566 (1993).
13. M. P. Doyle and J. W. Hoekstra, *J. Inorg. Biochem.* **14,** 351 (1981).

14. H. Kruszyna, R. Kruszyna, L. G. Rochelle, R. P. Smith, and D. E. Wilcox, *Biochem. Pharmacol.* **46,** 95 (1993).
15. H. Kruszyna, R. Kruszyna, R. P. Smith, and D. E. Wilcox, *Toxicol. Appl. Pharmacol.* **91,** 429 (1987).
16. R. P. Smith, *Clin. Toxicol.* **4,** 273 (1971).
17. A. E. Huot, H. Kruszyna, R. Kruszyna, R. P. Smith, and M. P. Hacker, *Biochem. Biophys. Res. Commun.* **182,** 151 (1992).
18. A. Wennmalm, B. Lanne, and A.-S. Petersson, *Anal. Biochem.* **187,** 359 (1990).
19. Y. Wang, L. G. Rochelle, H. Kruszyna, R. Kruszyna, R. P. Smith, and D. E. Wilcox, *Toxicology* **88,** 165 (1994).
20. A. Mülsch, P. I. Mordvintcev, A. F. Vanin, and R. Busse, *Biochem. Biophys. Res. Commun.* **196,** 1303 (1993).
21. Y.-J. Geng, A.-S. Petersson, A. Wennmalm, and G. K. Hansson, *Exp. Cell Res.* **214,** 418 (1994).
22. J. R. Lancaster, Jr., and J. B. Hibbs, Jr., *Proc. Natl. Acad. Sci. U.S.A.* **87,** 1223 (1990).
23. C. Pellat, Y. Henry, and J.-C. Drapier, *Biochem. Biophys. Res. Commun.* **166,** 119 (1990).
24. A. K. Nussler, D. A. Geller, M. A. Sweetland, M. Di Silvio, T. R. Billiar, J. B. Madriaga, R. L. Simmons, and J. R. Lancaster, Jr., *Biochem. Biophys. Res. Commun.* **194,** 826 (1993).
25. J. Stadler, H. A. Bergonia, M. Di Silvio, M. A. Sweetland, T. R. Billiar, R. L. Simmons, and J. R. Lancaster, Jr., *Arch. Biochem. Biophys.* **302,** 4 (1993).
26. J. A. Corbett, J. R. Lancaster, Jr., M. A. Sweetland, and M. L. McDaniel, *J. Biol. Chem.* **266,** 21351 (1991).
27. J.-C. Drapier, C. Pellat, and Y. Henry, *J. Biol. Chem.* **266,** 10162 (1991).
28. L. A. Shinobu, S. G. Jones, and M. M. Jones, *Acta Pharmacol. Toxicol.* **54,** 189 (1984).
29. P. Mordvintcev, A. Mülsch, R. Busse, and A. Vanin, *Anal. Biochem.* **199,** 142 (1991).
30. A. Mülsch, A. Vanin, P. Mordvintcev, S. Hauschildt, and R. Busse, *Biochem. J.* **288,** 597 (1992).
31. A. Komarov, D. Mattson, M. M. Jones, P. K. Singh, and C.-S. Lai, *Biochem. Biophys. Res. Commun.* **195,** 1191 (1993).
32. L. G. Rochelle, H. Kruszyna, R. Kruszyna, A. Barchowsky, D. E. Wilcox, and R. P. Smith, *Toxicol. Appl. Pharmacol.* **128,** 123 (1994).

[2] Electrochemical and Spectroscopic Methods of Nitric Oxide Detection

Tadeusz Malinski, Eugeniusz Kubaszewski, and Frederick Kiechle

Introduction

In this chapter spectroscopic, spectrometric, and electrochemical methods are described for the measurement of nitric oxide (NO) in single cells, tissues, and biological fluids. The measurement of the concentration of NO is a challenging problem due to the short half-life of NO in biological systems (1, 2).

The currently used instrumental techniques for NO measurements are spectroscopic and electrochemical methods. Mass spectrometry and gas chromatography have been used occasionally for NO detection but are much less sensitive. Spectroscopic methods include UV–visible spectroscopy, electron spin resonance spectroscopy (described elsewhere in this volume), and chemiluminescence spectrometry. Electrochemical methods include voltammetry, amperometry, and coulometry. Electrochemical methods offer several features that are not available from analytical spectroscopic methods. Most important is the capability afforded by the use of ultramicroelectrodes for direct *in situ* measurements of NO in single cells near the source of NO synthesis. Electrochemical methods detect NO directly and are based on electron exchange between NO and an electrode. All spectroscopic methods detect NO indirectly and generate a product following a reaction with NO or one of its oxidation products, NO_2^- or NO_3^-. Therefore, electrochemical methods are more suited for *in situ, in vitro,* and *in vivo* monitoring of kinetics of NO release (minute concentration changes, moles/liter with time) while spectroscopic methods are better for the determination of the total amount of NO (moles) produced by a certain number of cells or milligrams of tissue.

Ultraviolet–Visible Spectroscopy with Use of Azo Dyes

The typical application of azo dyes to detect NO_2^- is the Griess reaction (3, 4). The two-step assay is based on the observation that nitrite reacts with *p*-aminibenzenesulfonamide (diazotization reaction). In the second step a

Methods in Neurosciences, Volume 31

product of diazotization undergoes reaction with *N*-(1-naphthyl)ethylenedia-mine (diazonium coupling) to form an azo derivative dye that is readily monitored by UV–visible spectroscopy. The diazotization reaction is also called the Sandlmeyer reaction. It is compatible with the presence of a wide variety of substituents on the aromatic ring (Ar):

$$Ar—NH_2 + HNO_2 + 2H^+ + X^- \rightarrow Ar—N^+{=}NX^- + H_2O \quad (1)$$

The diazonium coupling reactions are typical electrophilic aromatic substitution processes in which the positively charged diazonium ion is the electrophilic substrate that reacts with the electron-rich ring of phenol or arylamine. The reactions occurring between the Griess reagent and nitrite may consequently be described by Eqs. (2) and (3):

Diazotization:

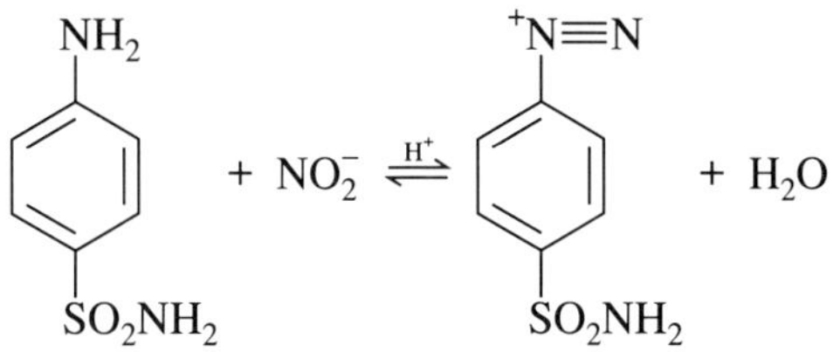

$$(2)$$

Diazonium Coupling:

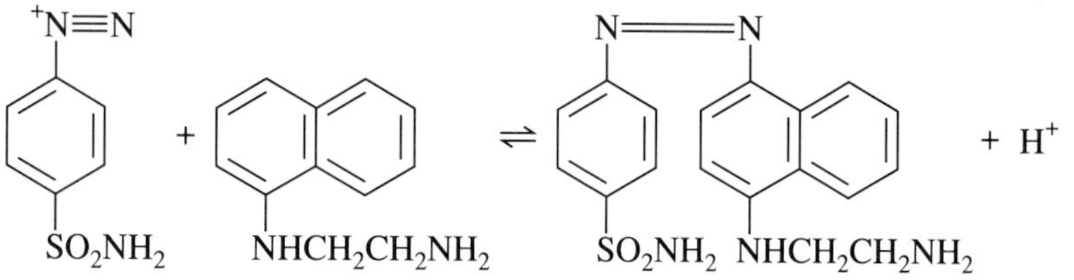

$$(3)$$

The azo dye formed from these Griess-type reactions absorbs in the visible region. The bluish color is observed after reaction of nitrite ions with Griess reagents. A typical absorption spectrum is shown in Fig. 1. A maximum of absorbance is observed at 548 nm at pH 2.0. The yield of the diazotization reaction is highest in acidic media. Therefore, the acidity of the media will affect the reactions. Consequently, pH is an important parameter in the optimization of this method as illustrated in Fig. 2.

The wavelength of maximum absorbance is shifted from 548 to 488 nm with increasing pH from 2.0 to 5.2, respectively. Reaction of NO_2^- with the Griess reagent has its maximum absorbance in a relatively narrow range of

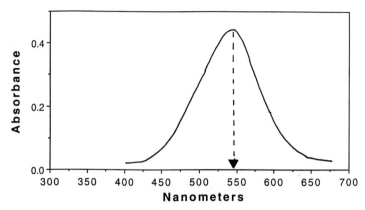

FIG. 1   Ultraviolet–visible spectrum obtained after reaction of NO_2^- (10 μM) with Griess reagent.

pH (pH 1.8–2.0). At higher pH a dramatic decrease of absorbance is observed, and at pH 6.0 the solution becomes optically transparent. At a constant concentration of N-(1-naphthyl)ethylenediamine dichloride and NO_2^- the absorbance changes with the change of concentration of p-aminobenzenesulfonamide. The maximum absorbance is observed at a concentration range of 0.2–0.3 mM p-aminobenzenesulfonamide (Fig. 3). The diazotization takes place between nitrite and p-aminobenzenesulfonamide to form a diazonium ion which later is coupled with N-(1-naphthyl)ethylenediamine dichloride. Therefore, an adequate concentration of p-aminobenzenesulfonamide is required to react with all of NO_2^- and maintain the high sensitivity of the Griess method. A linear relationship between the concentration and absorbance (linear correlation coefficient r better than 0.997) is observed up to about 5 μM of NO_2^-. A detection limit for NO_2^- (defined as the lowest concentration which produces a signal three times higher than noise) is about 3×10^{-7} M. Precision (expressed as a percentage of the relative standard deviation) is better than 5%. The incubation time (time needed to develop the color) is about 20–30 min.

Ultraviolet–Visible Spectroscopy with Use of Hemoglobin

A reduced form of deoxyhemoglobin ($HbFe^{2+}$) or oxyhemoglobin [(Hb-$Fe^{2+})O_2$] can react with NO to produce methemoglobin [$Hb(Fe^{3+})$] (5, 6):

$$NO + (HbFe^{2+})O_2 \rightarrow Hb(Fe^{3+}) + NO_3^- \qquad (4)$$

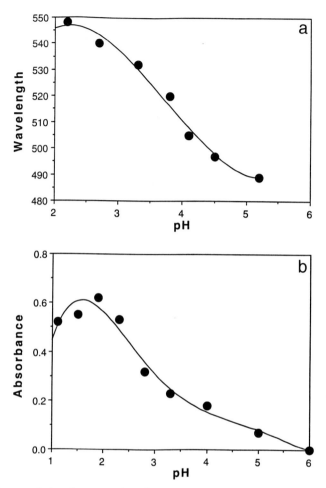

FIG. 2   Change of absorbance and peak wavelength of Griess reagent with pH. Data were obtained after reaction of NO_2^- (15 μM) with Griess reagent.

Reaction (4) is fast and can be used for kinetic studies of NO release. The UV–visible spectrum of hemoglobin shows high molar absorptivity ($\varepsilon = 131$ and 133 mM^{-1} cm^{-1} for oxyhemoglobin and deoxyhemoglobin, respectively) of the Soret band (Table I). During oxidation of the reduced form of hemoglobin, the Soret band is shifted to a shorter wavelength and absorptivity is slightly increased (Fig. 4). The theoretical detection limit for the hemoglobin method can be as low as $7 \times 10^{-8} M$ due to high molar absorptivity of the Soret band. However, this detection limit can be achieved only at stoichiometric

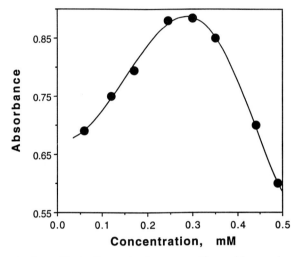

FIG. 3   Concentration effect of *p*-aminobenzenesulfonamide on absorbance of the Griess reagent [concentration of *N*-(1-naphthyl)ethylenediamine 0.02 mM; 22 μM NO$_2^-$; pH 2.0].

concentrations of NO and reduced hemoglobin. This ratio of reactants will not be found during the analysis of biological samples, where there is an excess of reduced hemoglobin compared to the concentration of NO. In real analysis the methemoglobin peak will overlap with the peak for reduced hemoglobin. This overlap will affect absorbance readings, and the detection limit ($\sim 1 \times 10^{-8}$ to $5 \times 10^{-8} M$) is only slightly better than the Griess method. A linear relationship between absorbance and concentration of NO will be observed over two orders of magnitude with a correlation coefficient better

TABLE I   Molar Absorptivity Coefficients (ε) and Band
Wavelengths (λ_{max}) of Reduced and Oxidized
Forms of Hemoglobin

Compound	λ_{max} (nm)	ε (mM^{-1} cm^{-1})
I. Oxyhemoglobin, (HbFe^{2+})O$_2$	415	131
II. Deoxyhemoglobin, HbFe^{2+}	433	133
III. Methemoglobin, HbFe^{3+}	406	161
Isosbestic point		
IV. (HbFe^{2+})O$_2$ $\rightleftharpoons$ HbFe^{3+}	411	124
V. HbFe^{2+} $\rightleftharpoons$ HbFe^{3+}	420	126

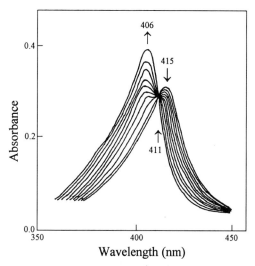

FIG. 4   Time-resolved (resolution 8 msec) spectra (absorbance versus wavelength, nm) obtained during reaction of deoxyhemoglobin with nitric oxide.

than 0.996. Precision of the method should be better than 5%. The incubation time for the hemoglobin–NO reaction is about 100–200 msec.

Chemiluminescence Spectrometry

The chemiluminescence technique is based on the fact that during a chemical reaction a significant fraction of intermediates or products are generated in excited electronic states. The chemical energy released in the reaction is used to produce excited species. The emission of photons from the excited molecules is measured. For detection of NO with the chemiluminescence method a chemical reaction of NO and O_3 is used (7, 8):

$$NO + O_3 \rightarrow (NO_2^* + NO_2) + O_2 \qquad (5)$$
$$\text{mixture}$$

A significant fraction of nitrogen dioxide produced in reaction (5) is in a metastable excited state (NO_2^*). Molecules of nitrogen dioxide in the excited state (half-life about 10^{-6} sec) are photon-emitting species:

$$NO_2^* \rightarrow NO_2 + \text{photon} \tag{6}$$

At constant O_3 concentration, the number of photons emitted is proportional to the concentration of NO.

Major components of the instrument for NO measurement with chemiluminescence include an ozone generator, reactor flow cell, and photomultiplier (Fig. 5). The continuous flow method yields a signal that is easy to measure. The detection limit is about 10^{-8} M and depends on the yield of photons generated in the reactor chamber as well as the sensitivity of the photomultiplier. A plot of signal (usually expressed in arbitrary units or as a current or voltage generated by the photomultiplier) versus NO concentration shows two linear regions with different slopes for concentrations between 2×10^{-8} M to 2×10^{-7} M and 3×10^{-7} M to 3×10^{-6} M.

Chemiluminescence is a sensitive method but requires at least two steps in sample preparation: chemical reduction of NO_2^- and/or NO_3^- back to NO and transfer of NO from the liquid phase to the gas phase. Both steps may introduce significant error. In the first step, exogenous nitrite, allyl nitrite ions, or nitroso compounds also can generate NO. In the second step, the removal of NO from the solution can be different, especially at low concentra-

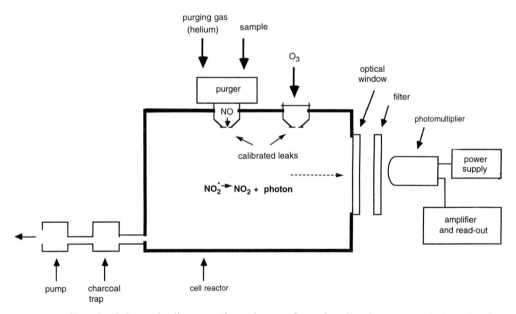

FIG. 5   Schematic diagram of continuous flow chemiluminescence photometer for measurement of nitric oxide.

tions. For example, removal of NO under normal atmospheric pressure from a solution by bubbling an inert gas, such as argon or helium, is effective only for NO concentrations greater than 10^{-7} M. Also NO has been known to be absorbed strongly on many solid materials, including glass, artifactually lowering the concentration. Ammonia, olefins, and sulfur gases produce chemiluminescence with ozone and may interfere with NO measurement.

Electrochemical Methods

Electrochemical methods currently available for NO detection are based on the electrochemical oxidation of NO on solid electrodes. If the current generated during NO oxidation is linearly proportional to the concentration, the oxidation current can be used as an analytical signal. This current can be measured in either an amperometric or a voltammetric mode, both methods providing a quantitative signal. In the amperometric mode, current is measured while the potential at which NO is oxidized is applied and kept constant. In the voltammetric mode, the current is measured while the potential is linearly scanned through the region that includes NO oxidation. Although the amperometric method is faster than the voltammetric method, voltammetry provides not only quantitative but also qualitative information that can prove that the current measured is in fact due to NO oxidation.

Generally, the oxidation of NO on solid electrodes proceeds via a two-step mechanism. The first electrochemical step is a one-electron transfer from an NO molecule to the electrode resulting in the formation of a nitrosonium cation:

$$NO - e^- \rightarrow NO^+ \tag{7}$$

NO^+ is a relatively strong Lewis acid and in the presence of OH^- is converted to nitrite (NO_2^-):

$$NO^+ + OH^- \rightarrow HNO_2 \tag{8}$$

The rate of the chemical reaction (8) increases with increasing pH. Since the oxidation potential of nitrite in aqueous solution is only 60–80 mV more positive than that of NO, oxidation of NO on solid electrodes with scanned potential results in the transfer of two additional electrons. Thus nitrite is ultimately oxidized to nitrate, NO_3^-, the final product of electrochemical oxidation of NO.

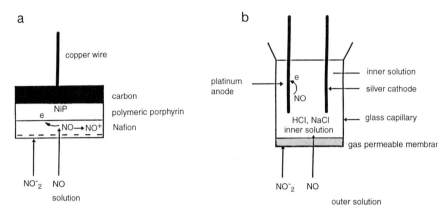

FIG. 6   Schematic diagram of (a) porphyrinic sensor (auxiliary and reference electrode are not shown) and (b) Clark type probe for determination of nitric oxide.

Figure 6 shows sensors that have been developed for the electrochemical measurement of NO. One is based on the electrochemical oxidation of NO on a conductive polymeric porphyrin (porphyrinic sensor) (9). The other is based on an oxygen probe (Clark electrode) and operates in the amperometric mode (10). In the porphyrinic sensor, NO is oxidized on a polymeric metalloporphyrin (n-type semiconductor) on which the oxidation reaction occurs at 630 mV [versus SCE (standard calomel electrode)], 270 mV lower than the potential required for the comparable metal or carbon electrodes. The current efficiency (analytical signal) for the reaction is high, even at the physiologic pH of pH 7.4.

At an operational potential of 0.63–0.65 V, the sensor does not respond to other gases such as oxygen, carbon dioxide, and carbon monoxide. The sensor is too slow to respond to superoxide, which in the biological environment is scavenged in fast reactions with other molecules including NO ($k = 3 \times 10^9 \, M^{-1} \, sec^{-1}$). The sensor is interference-free from the following readily small oxidizable secretory products at concentrations of at least two orders of magnitude greater than their expected physiological concentrations: epinephrine, norepinephrine, serotonin, dopamine, ascorbate, acetylcholine, glutamate, glucose, the NO-decay product NO_2^-, and peptides containing tryptophan, tyrosine, and cystine. The porphyrinic sensor has a response time of 0.1 msec at micromolar NO concentrations and 10 msec at the detection limit of 5 nM. A linear relationship between current and concentration is observed up to about $10^{-5} \, M$.

The Clark probe consists of platinum wire as a working electrode (anode) and silver wire as the counter electrode (cathode). The electrodes are mounted in a capillary tube filled with a sodium chloride–hydrochloric acid solution separated from the analytic solution by a gas-permeable membrane. A constant potential of 0.9 V is applied, and current (analytical signal) is measured from the electrochemical oxidation of nitric oxide on the platinum anode.

The detection limit of the Clark probe is about 10^{-8} M and can be achieved by measuring NO homogeneous synthetic solutions where a constant NO concentration can be established on both sides of the gas-permeable membrane. However, when the probe is used for measurements of NO concentrations in biological materials with heterogeneous environments, several limitations affect the utility of the Clark probe. The electrode response is slow, being 1.4 to 3.2 sec for 50% rise time and 2.2–3.5 sec for 75% rise time at an NO concentration of 1 μM. This response time is about 1000 times slower than a porphyrinic sensor and results from not only the long distance NO must diffuse to the electrode, but also the significant consumption of NO by redox reaction outside as well as inside the probe. In addition, because of its larger size, the electrode cannot be placed exactly at the site of high NO concentration, the membrane of the cell. Nitric oxide, which is supplied to the probe by diffusion, is consumed not only in fast reactions such as with superoxide but also in much slower reactions with oxygen and protein. The fraction of the NO that finally reaches the capillary probe chamber can be oxidized by oxygen, which diffuses easily through the membrane and is present in the capillary at concentrations much higher than those of NO. The Clark probe has a very high temperature coefficient (10 times higher than the porphyrinic sensor) due to the change of the NO diffusion rate through the membrane with temperature. The temperature coefficient of the Clark probe is 14–18% per degree Celsius, which may be viewed as a significant change of current (baseline drift) equivalent to changes of NO concentration of 65–80 nM per degree Celsius. Therefore, the temperature during NO measurement has to be carefully controlled (within ±0.1°C) to avoid small fluctuations of temperature that can lead to serious errors. The Clark probe is sensitive not only to O_2 and NO but also to N_2O and CO. Significant changes in the baseline will also be observed due to fluctuations of CO_2 concentration in the inner solution of the probe. This effect is more noticeable at lower pH where the electrode is most sensitive to NO.

Methods

Griess Method

In this method 0.172 g of p-aminobenzenesulfaniamide is dissolved in 10 ml of 0.5 M hydrochloric acid. The final concentration of p-aminobenzenesulfonamide is 0.1 M. The solution is stable for at least 1 month. A 10 mM

stock solution of N-(1-naphthyl)ethylenediamine in 1 M hydrochloric acid is prepared. For a reaction medium 5.59 g of KCl is dissolved in 19.5 ml of 1.0 M HCl and diluted with distilled water to 500 ml. The pH is adjusted with 1.0 M HCl and 1.0 M NaOH to pH 2.00. A sample containing NO and/or NO_2^- is dissolved in reaction medium, and p-aminobenzenesulfonamide and N-(1-naphthyl)ethylenediamine are added such that the final concentration is 0.2 and 20 mM, respectively. After 30 min of incubation, 2 ml of solution is transferred to a cuvette, and a full spectrum from 350 to 700 nm is recorded or absorbance is measured at constant wavelength of 548 nm. The concentrations of NO in the medium are calculated after comparison of absorbance values with those of a standard curve constructed with various concentrations of authentic NO.

Griess reagent will react with NO_2^- but will not react with nitrate (NO_3^-). To determine the total amount of NO released from biological system, NO_3^- has to be reduced back to NO_2^-. Nitrate ion can be reduced by metallic zinc in an ammoniacal solution of pH 10.8 in the presence of manganous hydroxide. This reduction is quantitative only in ammoniacal solution of pH 10.2 to 11.2. One milliliter of analyte (solution containing NO_3^-) and 2 ml of 0.1 M $Mn(OH)_2$ is added to 5 ml of buffer (NH_4Cl/NH_4OH; pH 10.8) containing 5 g of zinc wire. The temperature of the reacting mixture has to be maintained at 10°–12°C (water–ice bath). The rate of conversion of nitrate to nitrite with this reduction procedure is 98.3 ± 0.3%.

Hemoglobin Method

Twenty milligrams of bovine hemoglobin (methemoglobin, myoglobin, or metmyoglobin) is dissolved in 1–2 ml of buffer (pH 7–7.4). After the hemoglobin is completely dissolved, 1–2.5 mg of $Na_2S_2O_4$ (sodium hydrosulfite) is added, and the solution is stirred by bubbling oxygen (10–20 min). The reduced form of oxyhemoglobin (bright, light red color) is purified on a Sephadex G-25 column (diameter 10–20 mm, length 10–20 cm) using a buffer of pH 7.4 as the mobile phase. At a flow rate of the buffer of 1 ml/min, the retention time for reduced oxyhemoglobin is about 10 min. Deoxyhemoglobin can be prepared from hemoglobin also by reduction with $Na_2S_2O_4$ in deoxygenated Krebs–bicarbonate solution.

The purity and concentration of reduced hemoglobin can be established by UV–visible spectroscopy. Spectroscopic parameters are listed in Table I. The stock solution (millimolar concentration) is added to buffer in a quartz cuvette. The final concentration of reduced hemoglobin in the cuvette should be higher than the expected NO concentration (usually in the range of 1–10

μM). The spectrum of hemoglobin has to be monitored in the range of 350–500 nm before and after addition of the NO sample.

After addition of the sample (2–100 μl) containing NO, a decrease in the peak at 415 nm [for $(HbFe^{2+})O_2$ or 433 nm (for $HbFe^{2+}$)] with the simultaneous appearance and increase of a peak at 406 nm should be observed. A calibration curve can be made by plotting (versus NO concentration) the difference of absorbance between the methemoglobin peak at 406 nm and the isosbestic point at 411 nm for $(HbFe^{2+})O_2$ or 420 nm for $HbFe^{2+}$.

Chemiluminescence Method

For liquid biological samples, NO has to be driven into the gas phase in order to be determined by chemiluminescence methods. This can be achieved by bubbling the solution with an inert gas under vacuum conditions (headspace method). The specimen can also be injected into a glass purge chamber previously evacuated to 2–3 mm Hg to remove O_2. A glass purge chamber has a frit in its floor permitting NO to be removed from the liquid sample by bubbling it with helium (10–15 ml/min for 40–70 sec). The mixture of NO in helium is drawn into a reaction cell chamber filled under low pressure with a mixture of ozone and helium. A response to NO will be monitored as a current or potential generated by the photomultiplier in the form of a sharp peak. The best method of calibration is to plot an integrated area under the peaks versus concentration. The chemiluminescence method measures only the NO present. Therefore, for most biological samples conversion of NO_2^- and NO_3^- to NO is required. Nitric oxide may be generated using the acidification and reduction process described previously for the Griess method. Sodium or potassium iodide can also be used as the reduction medium.

Porphyrinic Sensor

Microsensors are produced by threading a carbon fiber (diameter 7 μm, Amoco) through the pulled end of a capillary tube with about 1 cm left protruding. Nonconductive epoxy is put at the glass–fiber interface. After the epoxy cement drawn into the tip of the capillary has cured, the carbon fiber is sealed in place. The carbon fiber is then sharpened by gradual burning (propane–air microburner, 1300–1400°C). The sharpened fiber is immersed in melted wax–rosin (5 : 1 w/w) at a controlled temperature for 5–15 sec and, after cooling, is sharpened again. The flame temperature and the distance of the fiber from the flame needs to be carefully controlled. The resulting electrode is a slim cylinder with a small diameter (0.5–2 μm) rather than a short taper, a geometry that aids in implantation and increases the active surface area. The tip (length 2–6 μm) is the only part of the carbon fiber where electrochemical processes can occur. For the sensor to be implanted into a cell, this length must be less than the cell thickness. The unsharpened end of the fiber is attached to a copper wire lead with silver epoxy cement.

A polymeric film of nickel(II) tetrakis(3-methoxy-4-hydroxyphenyl)porphyrin [Ni(II) TMHPP] is synthesized according to a procedure described previously (11). The polymeric film of Ni(II) TMHPP is deposited on a single carbon fiber electrode from a solution of 5×10^{-4} M monomeric Ni(II) TMHPP using cyclic scanning of potential between 0.2 and 1.0 V (versus a saturated calomel electrode) with a scan rate of 100 mV/sec for 10–15 scans. Dip coating the dried polymeric porphyrin/carbon fiber tip (3 times for 5 sec) in 1% Nafion in ethanol (Aldrich, Milwaukee, WI) produces a thin anionic film that repels or retards charged ionic species while allowing the small neutral and hydrophobic NO molecule access to the underlying catalytic surface.

For *in vivo* and *in vitro* measurements of NO in tissue and blood, a catheter-protected porphyrinic NO sensor is constructed from the needle of a 22-gauge 1-inch-long intravenous catheter–needle unit (Angiocath, Becton Dickinson, Lincoln Park, NJ) truncated and polished flat so that it was 5 mm shorter than a 20-gauge catheter (12–14). A bundle of seven carbon fibers (each 7 μm in diameter, protruding 5 mm, 12 Ω^{-1} cm^{-1}, Amoco) is mounted inside the hollow truncated 22-gauge needle with conducting epoxy cement. After curing, the exterior of the truncated needle is coated with nonconductive epoxy (2-TON, Devcon, Danvers, MA) and allowed to cure again. The protruding 5-mm carbon fiber bundle is made more sensitive and selective for NO by covering it with polymeric porphyrin and Nafion, before calibration with NO. To implant the porphyrinic NO sensor, ventricular tissue is pierced with a standard 20-gauge angiocatheter needle (clad with its catheter containing 4×400 μm ventilation holes near the tip), which is then advanced to the desired location. After intracavitary contact, the catheter tip is withdrawn 2–4 mm. The position of the catheter is secured, and the placement needle is removed and quickly replaced with the truncated 22-gauge porphyrinic NO sensor.

The quality of the fine structure of the conductive polymeric film is related to the selectivity and sensitivity of the porphyrinic sensor (15). The film must exhibit a large catalytic effect on the oxidation of nitric oxide as well as sufficient electronic (metallic) conductivity. Both of these features depend on an organization of porphyrinic molecules in the polymeric film. The current efficiency depends on the central metal (iron $\geq$ nickel $>$ cobalt $\gg$ zinc, copper). If the current efficiency of a given polymeric porphyrin is not affected by changing the central metal, the film probably does not have the desired properties for NO oxidation. The desirable properties of the film depend not only on the structure of monomeric porphyrins, but also on the amount of impurities, method of polymerization, supporting electrolyte concentration, range of potential scanned, properties of the solid support surface on which the film is plated, distribution of the electrical field (orientation of solid electrode versus counter electrode), shape of the solid electrode, and thickness and structural homogeneity of the porphyrinic film.

Response and Calibration of Porphyrinic Sensor

Differential pulse voltammetry (DPV) and differential pulse amperometry (DPA) can be used to monitor analytical signals, where current is linearly proportional to NO concentration. In DPV, a potential modulated with 40 mV rectangular pulses is linearly scanned from 0.4 to 0.8 V. The resulting voltammogram (alternating current versus voltage plot) contains a peak due to NO oxidation. The maximum current of the peak should be observed at a potential of 0.63 to 0.67 V (at 40 mV pulse amplitude), the characteristic potential for NO oxidation on the porphyrinic/Nafion sensor. Differential pulse voltammetry is used primarily to verify that the current measured is due to NO oxidation. In differential pulse amperometric measurements, a potential of 0.63 to 0.67 V modulated with 40 mV pulses is kept constant, and a plot of alternating current versus time is recorded. The amperometric method (with a response time better than 10 msec) provides rapid quantitative measurement of minute changes of NO concentration. Differential pulse voltammetry also provides quantitative information but requires approximately 5–40 sec for the voltammogram to be recorded; therefore, it is used mainly for qualitative analysis.

Many other electroanalytical techniques including normal pulse voltammetry, square wave voltammetry, fast scan voltammetry, and coulometry can be used to measure nitric oxide with a porphyrinic sensor. Amperograms and voltammograms can be recorded with two- or three-electrode systems. However, differential pulse voltammograms should always be recorded in three-electrode systems in order to obtain accurate and reproducible values of peak potentials. Three-electrode systems consist of an NO sensor working electrode, a platinum wire (0.25 mm) counter electrode, and a silver|silver chloride or SCE as the reference electrode. The reference electrode is omitted from the two-electrode system. The porphyrinic sensor can be connected to any fast response potentiostat for amperometric or coulometric measurements or to a voltammetric analyzer (a potentiostat and waveform generator) for voltammetric measurements. An instrumental current sensitivity of the 100 pA/inch will be sufficient for most measurements using a multifiber sensor. For single-fiber experiments, at least 10 times more sensitivity is required; this can be achieved by adding a low noise current-sensitive preamplifier to the potentiostat. The sensor can be calibrated with saturated NO solution at 2 mM as a standard, or by preparing a calibration curve using the standard addition method.

A typical amperometric response of the sensor under different flow conditions is shown in Fig. 7. In a flowing or stirred homogeneous NO solution, a rapid increase in current reaching a plateau occurs after addition of NO. In heterogeneous flowing solutions with high localized concentrations of NO, such as blood, the amperometric signal is observed in the form of a peak. In static solutions where mass transport of NO to the electrode is due only to a diffusion process, a linear increase in current reaching a plateau is followed by a decrease in current due to NO oxidation.

Clark Probe for Nitric Oxide Detection

The Clark probe, which was originally designed for the detection of oxygen, is a glass pipette whose opening is sealed with a gas-permeable membrane of thick rubber. Only a low molecular weight gas can readily diffuse into the glass pipette through the membrane and be oxidized or reduced at the surface

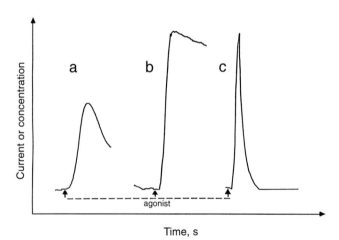

Fig. 7 Amperograms (current–time curves) showing porphyrinic sensor response for nitric oxide in a homogeneous static solution (a), homogeneous laminar flow solution (b), and heterogeneous laminar flow solution (c).

of the metal electrode (working electrode). In the Clark probe for oxygen detection, a working electrode (platinum) is polarized with a potential -0.8 V (versus silver counter/reference electrode), and current due to the reduction of oxygen is observed. For the detection of NO with the Clark probe, only the polarization of the electrodes must be changed. To oxidize nitric oxide on the platinum electrode, polarization of the electrodes has to be reversed to 0.8 V (versus the silver electrode) instead of -0.8 V for oxygen detection. As the electrolyte in the Clark probe for NO detection, a mixture of NaCl and HCl (1 M) can be used. The probe does not respond to NO in a basic solution.

The Clark probe operates in the amperometric mode, with the direct current being measured using simple electronic potentiostat (current–voltage converter). As an analytical signal, current is measured at a constant potential of 0.8–0.9 V. For a Clark probe made with a glass pipette diameter of 300 μm, 50 μm platinum anode and 200 μm silver cathode, the current developed is between 3 and 100 pA/μM. The relatively low current is due to several factors of which the major one is a low net NO concentration in the glass pipette as compared with the concentration at the source of NO release.

Most commercially available miniaturized oxygen probes (Clark electrodes, Clark probes) can be used for the detection of NO. A low cost probe can be relatively easily prepared as follows, using a glass pipette, silver and platinum wires, and a gas-permeable membrane. The tip of a pipette (1–3 cm length) with a diameter of 150–350 μm is flame polished. The pipette is then filled with an acidic solution of electrolyte to provide sufficient conductivity and the proper environment for NO oxidation; usually, 30 mM NaCl and 0.13 mM HCl are used. The optimal pH is about pH 3.5; above pH 3.5 the electrode sensitivity will be significantly reduced, and in basic solution the probe shows high noise levels and is not sensitive to NO. A soft wax (beeswax), melted or dissolved in turpentine, is painted on the outer surface of the pipette, and the tip is then sealed with a gas-permeable membrane. Several different materials can be used as gas-permeable membranes: Teflon, polyethylene, silicon rubber, or chloroprene rubber, with the chloroprene rubber membrane being easiest to prepare. Common rubber cement can also be used as a material for simple preparation of the gas-permeable membrane. The electrode tip can be placed on the surface of an air bubble of rubber cement or a solution of chloroprene rubber. The sealing procedure has to be repeated two to three times to obtain sufficient membrane strength. The membrane thickness will vary depending on the material used for membrane preparation. Teflon membranes will usually have a thickness of 6–8 μm, and the chloroprene rubber membrane will be 1–5 μm.

A Teflon-coated platinum wire (diameter 50–100 μm) is then inserted into the pipette. The Teflon coating is removed by gradual burning from the wire tip. About 100–200 μm of bare platinum wire must be exposed to the solution

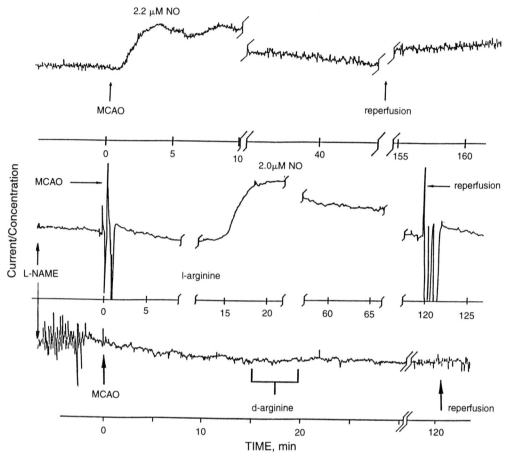

FIG. 8   Continuous amperometric monitoring of nitric oxide production in rat brain during middle cerebral artery occlusion and reperfusion (a) in the absence of L-NAME, (b) in the presence of L-NAME followed by injection of L-arginine, and (c) in the presence of L-NAME followed by injection of D-arginine.

and closely positioned to the membrane at the orifice. As a counter/reference electrode, a silver wire (diameter 100–200 μm) is inserted into the pipette. The Clark probe can be connected to any commercially available potentiostat with the capability to measure relatively small (picoampere) currents. The platinum electrode must have a positive potential of 0.9–0.8 V relative to the silver electrode. A current change can be continuously monitored with a chart recorder or storage oscilloscope. The Clark probe can also be connected to a simple and inexpensive current–voltage converter made of an operational amplifier.

Measurements in Brain with Porphyrinic Sensor

Figure 8 presents a typical amperometric curve obtained from the *in situ* measurements of NO concentrations in rat brain before, during, and after middle cerebral artery occlusion (MCAO) (16). The porphyrinic sensor was stereotaxically implanted perpendicularly into the ipsilateral parietal cortex at coordinates 0.8 mm posterior and 4.5 mm lateral to the bregma and 3 mm below the dura. An increase in NO concentration to a semiplateau of 2.2 μM was observed after 4 min. L-NAME (N^w-nitro-L-arginine methyl ester), an inhibitor of NO synthase, injected before MCAO diminished production of nitric oxide. However, after injection of L-arginine the concentration of NO was restored to 2.0 μM, the level observed in the first experiment. The injection of D-arginine under similar conditions did not restore production of NO.

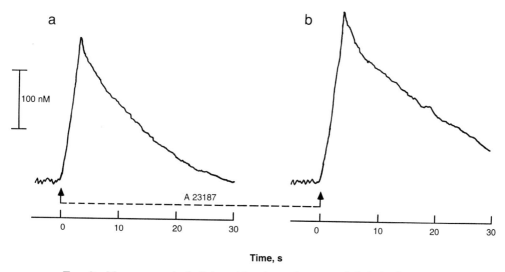

FIG. 9   Measurement of nitric oxide release from arteriole in brain. The release was stimulated by the calcium ionophore A23187 in an absence (a) and presence (b) of SOD.

Measurement of Nitric Oxide in Arteriole

Figure 9a shows a typical amperogram of nitric oxide released from an arteriole in rat brain. A multifiber (3 fibers) 12-μm diameter porphyrinic sensor was placed on the surface of an arteriole. Immediately after injection of a solution of the calcium ionophore A23187, a rapid increase of NO concentration was observed. The rate of concentration increase was 120 nM/sec, and a peak concentration was reached 2.1 sec after injection of the calcium ionophore. In the presence of superoxide dismutase (SOD, 100 U/ml), a 9% increase of NO concentration is observed (peak concentration 290 nM) (Fig. 9b). This indicates that a portion of the NO release in the absence of SOD is consumed in fast reactions with superoxide generated simultaneously in the system. This portion of NO concentration cannot be measured with the porphyrinic sensor, which even though it has a relatively fast response time (milliseconds), is still too slow to compete with fast (faster than 1 msec) chemical reaction(s).

Acknowledgment

This work was supported in part by a grant from William Beaumont Hospital Research Institute.

References

1. F. Kiechle and T. Malinski, *Am. J. Clin. Pathol.* **100**, 567 (1993).
2. S. Archer, *FASEB J.* **1**, 349 (1993).
3. F. K. Bell, J. J. O'Neill, and R. M. Burgison, *J. Pharm. Sci.* **52**, 637 (1963).
4. L. J. Ignarro, G. M. Buga, K. S. Wood, R. E. Byrns, and G. Chaudhuri, *Proc. Natl. Acad. Sci. U.S.A.* **84**, 9265 (1987).
5. M. P. Doyle and J. W. Hoekstra, *J. Inorg. Biochem.* **14**, 351 (1981).
6. M. Feelisch and E. Noack, *Eur. J. Pharmacol.* **139**, 19 (1987).
7. T. Hoki, *Biomed. Chromatogr.* **4**, 128 (1990).
8. S. L. Archer and N. J. Cowan, *Circ. Res.* **68**, 1569 (1991).
9. T. Malinski and Z. Taha, *Nature* (*London*) **358**, 676 (1992).
10. K. Shibuki, *Neurosci. Res.* **9**, 69 (1990).
11. T. Malinski, B. Ciszewski, J. Bennett, J. R. Fish and L. Czuchajowski, *J. Electrochem. Soc.* **138**, 2008 (1991).
12. T. Malinski, S. Patton, B. Pierchala, E. Kubaszewski, S. Grunfeld, and K. V. S. Rao, *in* "Frontiers of Reactive Oxygen Species in Biology and Medicine" (K. Asada and T. Yoshikawa, eds.), p. 207. Elsevier, Amsterdam, 1994.
13. D. J. Pinsky, M. C. Oz, S. Koga, Z. Taha, M. J. Broekman, A. J. Marcus, H.

Liao, Y. Naka, J. Brett, P. J. Cannon, R. Nowygrod, T. Malinski, and D. M. Stern, *J. Clin. Invest.* **93,** 2291 (1994).

14. D. J. Pinsky, N. Yoshifumi, N. C. Chowdhury, H. Liao, M. C. Oz, R. E. Michler, E. Kubaszewski, T. Malinski, and D. M. Stern, *Proc. Natl. Acad. Sci. U.S.A.* **91,** 12086 (1994).

15. T. Malinski, Z. Taha, S. Grunfeld, A. Burewicz, P. Tomboulian, and F. Kiechle, *Anal. Chim. Acta* **279,** 135 (1993).

16. T. Malinski, F. Bailey, Z. G. Zhang, and M. Chopp, *J. Cereb. Blood Flow. Metab.* **13,** 335 (1993).

[3] Detection of S-Nitrosothiols

George N. Welch, Gilbert R. Upchurch, Jr.,
and Joseph Loscalzo

Introduction

Nitric oxide (NO) is an extremely labile, heterodiatomic molecule that is produced by a variety of cell types including the vascular endothelium, platelets, macrophages, and neuronal cells. In neuronal tissue, nitric oxide is produced rapidly and transiently by a constitutive form of the enzyme NO synthase (nNOS). Evidence suggests a putative role for NO as a neurotransmitter (1), and it has also been demonstrated that NO is involved both in modulating cholinergic neurotransmission (2) and in mediating neuronal ischemia–reperfusion injury (3). Nitric oxide is rapidly inactivated by a number of biochemical species in the physiological milieu, including heme iron (4), superoxide anion (5), and oxygen (6).

Low molecular weight thiols react with NO in the presence of O_2 (by way of NO^+ or N_2O_3) under physiological conditions to form thionitrites or S-nitrosothiols (RSNOs) (7). The RSNOs confer stability to the NO radical and minimize its interaction with reactive oxygen species in the biological milieu. S-Nitrosothiol formation may also attenuate NO toxicity because S-nitrosothiols are much less reactive than NO and, therefore, are less likely to react with oxygen to form toxic oxidizing species (8). S-Nitrosothiols appear to play an integral role in a variety of different NO-dependent physiological processes, particularly in the vasculature. The demonstration that the predominant redox form of NO in mammalian plasma is S-nitroso serum albumin (9) has confirmed the physiological relevance of S-nitrosothiols. The total concentration of RSNOs in plasma is approximately 1 μM, and S-nitroso serum albumin accounts for greater than 85% of this plasma RSNO pool. In contrast, the plasma concentration of free NO is approximately 3 nM (9). These observations have reinforced the need for developing new analytic techniques capable of separating and detecting S-nitrosothiols in plasma and other biological fluids.

Methods in Neurosciences, Volume 31

High-Performance Liquid Chromatography with Electrochemical Detection

In 1977, Rubenstein and Saetre developed a novel method for detecting biological thiols using an electrochemical technique based on the catalytic oxidation of thiols by mercury (10). The detection process used cation-exchange chromatography to separate thiols which are then identified on the basis of relative redox potential and affinity for a mercury electrode (10). Slivka and colleagues subsequently modified this methodology using reversed-phase high-performance liquid chromatography (RP-HPLC) coupled to a thin film Au|Hg electrode to detect S-nitrosothiols, as well (11). The optimal reductive potential for the low molecular weight S-nitrosothiols, including S-nitrosoglutathione (GSNO), S-nitroso-L-cysteine (SNO-Cys), and S-nitrosohomocysteine (SNO-homocysteine) was found to be −0.15 V using hydrodynamic voltammetry. Chromatography was performed with a C_{18} reversed-phase HPLC column, and the electrochemical detector consisted of a dual Au|Hg electrode set in series at both oxidizing (+0.15 V) and reducing (−0.15 V) potentials versus an Au|AgCl reference electrode. Using this novel analytical device, S-nitrosothiols are rapidly and sensitively separated and detected in the nanomolar concentration range (12).

Electrochemical Principles

S-Nitrosothiols are separated initially during elution from the C_{18} chromatography column and then subjected to electrochemical detection (ECD). The electrochemical detector is arranged with the dual electrode set in series; the reducing (upstream) electrode is set at −0.15 V and the oxidizing (downstream) electrode is set at +0.15 V. After the eluate leaves the column, S-nitrosothiols are detected at the upstream (reducing) electrode during a strong reducing reaction that produces a thiol end product. The current generated (peak height) at this electrode is used to identify and quantify specific S-nitrosothiols. The thiol produced at the upstream electrode is incidentally detected at the downstream oxidizing electrode. The half-reaction that occurs at the reducing electrode (13) is

$$2RSNO + 2H^+ + 2e^- \rightarrow 2RSH + 2NO$$

The half-reaction that occurs at the oxidizing electrode (14) is

$$2RSH + Hg \rightarrow Hg(SR)_2 + 2H^+ + 2e^-$$

Chemicals

Citrate–phosphate–dextrose (CPD) anticoagulant (15.6 mM citric acid, 90 mM sodium citrate, 16 mM NaH$_2$PO$_4$ · H$_2$O, and 142 mM dextrose, pH 7.35) is used in acquiring plasma samples. Diethylenetriaminepentaacetic acid (DTPA) is used to chelate free metal ions and reduce metal ion-induced redox reactions. Chelex-treated phosphate-buffered saline (PBS) is used for the same purpose.

The mobile phase consists of 0.1 M monochloroacetic acid, 0.125 mM disodium ethylenediaminetetraacetic acid (EDTA), 1.25 mM sodium octyl sulfate, and 1% (v/v) acetonitrile, pH 2.8. After preparation, the mobile phase should be filtered through a 0.2-μm filter. All reagents should be of the highest analytical grade possible, and the water used for the preparation of the mobile phase should be deionized, HPLC-grade water.

Instrumentation

The HPLC–ECD system consists of several commercially available components. The HPLC column used is a C$_{18}$ (octadecylsilane) column [Bioanalytical Systems (BAS), West Lafayette, IN], and the electrochemical detector comprises a dual Au|Hg electrode (BAS) and an Ag|AgCl reference electrode (BAS). The system also requires a Rheodyne 20-μl sample injector (Model 712S, Berkeley, CA) and a BAS Chromatographic Analyzer (Model 200A). The chromatographic analyzer is linked to a computer supplied with chromatographic analysis software (BAS). The mobile phase is stored in a 1-liter glass flask connected to the chromatographic analyzer. Dissolved oxygen rapidly oxidizes both thiols and S-nitrosothiols during elution; therefore, the mobile phase must be thoroughly deoxygenated during operation by bubbling a constant stream of helium through it. Samples should also be individually deoxygenated with helium for 3–5 min prior to injection. The column flow rate should be set between 0.5 and 0.8 ml/min.

The electrochemical detector consists of a dual electrode with two circular gold electrodes. The mercury–gold amalgam is prepared by applying a drop of triply distilled mercury to each electrode surface (14). Each surface should be entirely covered with mercury. After 3–5 min, the mercury is removed with the edge of a thin plastic card, and the electrode is wiped with tissue to remove reflective areas that are indicative of excess mercury. A correctly prepared electrode should have a dull sheen. The electrode may then be installed in the chromatographic analyzer and a potential applied. The electrode requires an equilibration period of approximately 1 hr before it may be used.

Sample Preparation and Chromatographic Analysis

The detection of S-nitrosothiols is challenging owing to their high degree of reactivity with molecular oxygen and contaminant metals. Dissolved oxygen and trace amounts of metal ion contaminants rapidly oxidize both thiols and S-nitrosothiols during sample preparation and elution, leading to poor reproducibility of measurement. S-Nitrosothiol stability is also both pH- and temperature-dependent (15). It is therefore crucial to deoxygenate both the mobile phase and individual samples prior to analysis. The mobile phase should be deoxygenated with helium for at least 1 hr before initiating chromatographic analysis. Each sample should be deoxygenated with helium for 3–5 min immediately before injection. Authentic standards should be prepared in Chelex-treated buffer, and the metal ion chelator diethylenetri-aminepentaacetic acid (DTPA) should be added to all samples (final concentration 2 mM) during preparation to minimize metal ion-induced redox reactions.

Plasma samples should be prepared after drawing whole blood into tubes containing 1 ml of CPD anticoagulant (containing 20 mM DTPA) for every 9 ml of blood. The blood specimen should then be immediately centrifuged at 1000 g for 10 min in a centrifuge set at 4°C. The plasma (upper layer) is ultrafiltered to remove proteins using a Centrifree micropartition system (Amicon, Beverly, MA) with a molecular weight cutoff of 30,000 at 1500 g (12). The final filtrate is gently deoxygenated with a stream of helium gas for 3–5 min. A 30-μl aliquot is then injected into the Rheodyne injector port (20-μl injection loop) and chromatographic analysis is initiated. Overnight freezing of some samples at −70°C may result in attenuation of signal; therefore, all samples should be analyzed as soon as possible after acquisition. During processing, all samples should be stored on ice.

Standard Curves and Recoveries

S-Nitrosothiols are identified and quantified by comparison with authentic standards. The calibration curve is linear over a range of 50 nM to 50 μM ($r = 0.99$) (11). Intra- and interassay variabilities are less than 10%. The typical immediate recovery of SNO-Cys from plasma is 93%.

Photolysis–Chemiluminescence

Principle

Chemiluminescence relies on the reaction of NO with ozone. The chemical oxidation of NO by ozone yields nitrogen dioxide in an excited state. Relaxation from this excited state produces distinctive light emission (chemiluminescence) that is directly proportional to NO concentration. In a standard chemiluminescence device, a pyrolyzer is used to release the nitrosyl radical (NO·). In this modification, the pyrolyzer is replaced wth a photolysis cell (Nitrolyte, Thermedics Detection, Chelmsford, MA) and linked to a conventional chemiluminescence device [Model 543 Thermal Energy Analyzer (TEA), Thermedics] (9).

Instrumentation and Assay Procedure

The photolysis cell is composed of a narrow-bore borosilicate glass coil (15 loops each of 8.8 cm external diameter with a total height of 12.0 cm and a total length of 414.6 cm) with a 200-W high-pressure mercury vapor lamp located in its center (9). Biological samples (100 μl) are injected directly into the photolysis cell via a HPLC pump with water as the mobile phase at a flow rate of 1 ml/min. The sample is next irradiated with UV light (300–400 nm bandwidth), resulting in complete homolytic photolysis of the S–N bond of RSNOs and thereby allowing measurement of total NO content (free and bound NO). When the device is operated with the light off, only free NO is measured. After a total exposure time of 5 sec, the sample is transported from the photolysis cell by a carrier stream of helium or argon (4 min) into two cold traps where the aqueous phase is removed. The gas stream then carries free NO into the chemiluminescence spectrophotometer where NO is detected by its reaction with ozone. The distinctive chemiluminescence signal that is produced is proportional to free NO concentration and is recorded with an integrator (Model 3393A, Hewlett-Packard).

Standard Curves and Limits of Detection

Standard curves for S-nitrosothiols, including SNO-Cys, GSNO, and S-nitrosoalbumin, were determined and the correlation coefficients were ≥0.98 for all standards (9). The limit of detection for S-nitrosothiols is approximately 0.01 μM (9), and the intra- and interassay variabilities are less than 2%.

Capillary Zone Electrophoresis

Capillary zone electrophoresis (CZE) is a well-established analytical technique used to separate and identify different compounds reliably on the basis of molecular mass and charge. However, standard CZE is unsuitable for *S*-nitrosothiol detection because it involves alkaline conditions under which RSNOs are extremely unstable, decomposing rapidly to yield nitric oxide (16). Our group, therefore, modified standard CZE by using acidic rather than basic electrophoretic conditions for detecting RSNOs. Electrophoretic mobility [$cm^2/(sec/V)$] can be related to migration time, t_m, by the equation $\mu_e = L^2/tV = 5.39 \times 10^{-4}/t_m$ substituting $L_d L_t$ for L^2, where L_d is capillary length from injector site to detector and L_t is the entire length of the capillary (17).

Instrumentation

The capillary electrophoresis system is a Bio-Rad HPC-100 system (Bio-Rad, Richmond, CA), fitted with a silica-coated capillary (20 cm × 25 μm). The single-pen strip-chart recorder used to record electrophoretic separations is a Model 1321 (Bio-Rad) with a chart speed of 1.0 cm/min.

Sample Preparation

Immediately prior to use samples should be diluted in electrophoresis buffer (0.01 *N* HCl, 0.01 *M* sodium phosphate, pH 2.3) (17). Stock solution concentrations are determined by the Saville reaction (18).

Assay Procedure

A 10-μl aliquot of sample is loaded for 9 sec at 11 kV. At the end of each run, the capillary is rinsed with the electrophoresis buffer. A positive polarity is set on the power supply to ensure that cations migrate toward the detector. Optimal sensitivity is assured by monitoring the eluted sample at 200 nm initially and *S*-nitrosothiol is subsequently confirmed at 320 nm.

Standard Curves and Limits of Detection

The relationship between concentration and peak height is linear with a correlation coefficient of 0.99; the intra- and interassay variability is less than 2%. The major shortcoming of the technique is a lack of sensitivity; the limit of detection of the technique is in the micromolar range (17).

References

1. T. M. Dawson and S. H. Snyder, *J. Neurosci.* **14,** 5147 (1994).
2. M. C. Baccari, F. Calanai, and G. Staderini, *Neuroreport* **5,** 905 (1994).
3. J. W. Kuluz, R. J. Prado, and D. Dietrich, *Stroke* **24,** 2023 (1993).
4. R. M. J. Palmer, A. G. Ferrige, and S. Moncada, *Nature (London)* **327,** 524 (1987).
5. L. J. Ignarro, *Circ. Res.* **65,** 1 (1989).
6. L. J. Ignarro, *FASEB J.* **3,** 31 (1989).
7. J. S. Stamler, D. I. Simon, J. A. Osborne, M. E. Mullins, O. Jaraki, T. Michel, D. J. Singel, and J. Loscalzo, *Proc. Natl. Acad. Sci. U.S.A.* **89,** 444 (1992).
8. J. S. Stamler, D. J. Singel, and J. Loscalzo, *Science* **258,** 1898 (1992).
9. J. S. Stamler, O. Jaraki, J. A. Osborne, D. I. Simon, J. F. Keaney, Jr., J. A. Vita, D. J. Singel, C. R. Valeri, and J. Loscalzo, *Proc. Natl. Acad. Sci. U.S.A.* **89,** 7674 (1992).
10. D. L. Rubenstein and R. Saetre, *Anal. Chem.* **49,** 1036 (1977).
11. A. Slivka, J. S. Scharfstein, C. Duda, J. S. Stamler, and J. Loscalzo, *Circulation* **88,** I-523 (1993).
12. J. S. Scharfstein, J. F. Keaney, Jr., A. Slivka, G. N. Welch, J. A. Vita, J. S. Stamler, and J. Loscalzo, *J. Clin. Invest.* **94,** 1432 (1994).
13. G. N. Welch, G. R. Upchurch, Jr., J. Scharfstein, A. Slivka, and J. Loscalzo, *Meth. Enzymol.* in press (1996).
14. L. A. Allison and R. E. Shoup, *Anal. Chem.* **55,** 12 (1982).
15. P. Ramdev, J. Loscalzo, M. Feelisch, and J. S. Stamler, *Circulation* **88,** I-523 (1993).
16. L. J. Ignarro, H. Lipton, J. C. Edwards, W. H. Baricos, A. L. Hyman, P. J. Kadowtiz, and C. A. Gruetter, *J. Pharmacol. Exp. Ther.* **218,** 739 (1981).
17. J. S. Stamler, and J. Loscalzo, *Anal. Chem.* **64,** 779 (1992).
18. B. Saville, *Analyst* **83,** 670 (1958).

[4]   Nitrosation of Proteins

Ying-Yi Zhang and Joseph Loscalzo

Introduction

Nitrosation reactions introduce the NO group into an organic molecule. When covalently bound to a carbon, nitrogen, oxygen, or sulfur atom, the incorporated NO is referred to as a nitroso group; when coordinated to a transition metal, it is referred to as a nitrosyl group. Under physiological conditions, nitrosation reactions lead to the formation of three types of X–NO bonds: S-nitroso, such as that in S-nitrosocysteine; N-nitroso, such as that in N-nitrosotryptophan; and iron-nitrosyl complexes.

Methods of Protein Nitrosation

Although the nitrosation of low molecular weight bioorganic molecules has been studied extensively over the past 30 years, protein nitrosation has only relatively recently been the subject of investigation. With the identification of nitric oxide (NO) as a product of many cell types and the well-known reactivity of this radical and its activated redox forms (1), the possibility that nitrosation may occur under physiological conditions has been entertained. Considerable experimental evidence is mounting that protein nitrosation occurs with NO generation *in vivo*. Nitrosation of proteins provides a means by which to store NO and also represents a form of posttranslational modification that modulates protein function and cell phenotype. In this overview, we provide a concise summary of the methods for nitrosating proteins *in vitro* and the identification of the resulting products.

Nitrosation with Nitric Oxide Gas

Nitric oxide gas is a poor nitrosating agent. Although NO is able to react with heme-bound Fe(II), there is no evidence that NO can directly nitrosate other nonradical molecules. However, NO reacts with molecular oxygen rapidly, and the resulting nitrogen oxides such as N_2O_3 ($ON^{\delta+}-^{\delta-}ONO$) are

Methods in Neurosciences, Volume 31

efficient nitrosating agents. Thus, nitrosation of proteins using gaseous solutions of NO requires that the reactions proceed in the presence of oxygen.

The mechanism by which NO in the presence of O_2 supports nitrosation of thiols to form S-nitrosothiols is not well understood. It is clear, however, that S-nitrosothiols form under physiological conditions, and that their rate of formation is comparable to that for oxidation of NO to nitrite (2).

Three methods have been used to nitrosate proteins with gaseous NO. In the first method, the protein solution is incubated with a NO-saturated buffer. The latter is prepared by deoxygenating the buffer with argon or nitrogen gas for 15 min. NO gas is then bubbled into the solution for 5 min. The container is sealed with a septum and the buffer saturated with NO is transferred to the protein solution with an air-tight syringe. (Note that NO is a potentially toxic gas owing to the generation of N_2O_3 and N_2O_4 in the presence of oxygen, and all reactions described herein should therefore be performed in a well-ventilated hood.) The saturating concentration of NO gas in water is 3.28 μM at 0°C and 1.93 μM at 25°C. The precise concentration of NO in the buffered reagent solution should be verified by chemiluminescence spectroscopy, oxygen consumption (3), or spectroscopic detection of nitrosylhemoglobin (4).

The second method for nitrosating proteins with NO gas is to expose the protein solution to NO that slowly diffuses from a Silastic tube through which the gas flows at a controlled rate (5). The nitrosation efficiency of this method is considerably improved over that observed with simply mixing gaseous solutions of NO with protein solutions. Experimental conditions for this procedure involve placing 3–10 ml of a buffered protein solution in a 10- to 20-ml glass vial containing a small magnetic stir bar. The vial is sealed with a rubber stopper, and two 21-gauge needles are inserted through the stopper into the vial. Inside the vial, the tips of the needles are connected with a piece of Silastic tubing (0.025 inch i.d., 0.047 inch o.d., Dow-Corning, Flushing, NY). One needle is connected to an NO source [10% (v/v) in argon], while the other is open to air in the hood. The NO gas flow is adjusted to 2–5 ml/min using a sensitive flow meter. Nitric oxide diffuses through the wall of the Silastic tubing into the protein solution where mixing is facilitated by stirring. The final concentration of NO in the protein solution is determined by the Griess reaction (6).

A third method used to nitrosate proteins involves placing the buffered protein solution in dialysis tubing and dialyzing the solution against a buffer saturated (i.e., constantly bubbled) with NO gas. This method is somewhat less consistent than the two previously described methods owing to the poorly controlled delivery of NO and the inefficiency of mixing NO in the reaction solution.

Nitrosation with Low Molecular Weight NO Donors: Transnitrosation

Various NO donors have also been used to nitrosate proteins, including *S*-nitroso-*N*-acetyl-penicillamine (SNAP), *S*-nitroso-L-cysteine (SNO-Cys), *S*-nitrosoglutathione (GSNO), sodium nitroprusside [$Na_2(FeCN)_5NO)$] (SNP), 3-morpholinosydnonimine (SIN-1), and diethylamine dinitric oxide [$(C_2H_5)_2N(N[O]NO)^-Na^+$] (DEANO). The first three molecules are *S*-nitroso compounds, and the last two contain *N*-nitroso groups. Sodium nitroprusside has a metal-coordinated NO, but in the presence of hydroxyl anion, this nitrosyl complex is converted to the nitro complex [$Fe(CN)_5NO_2^{4-}$]. Glyceryl trinitrate (GTN) has also been used to nitrosate proteins; however, this NO donor must undergo cellular metabolism to effect nitrosation. All of these nitrosating agents are not stable at neutral pH and undergo spontaneous denitrosation in aqueous solution. This observation has led to the assumption that these compounds support nitrosation by releasing the NO radical. However, direct transnitrosation from these compounds to protein functional groups could also account for their ability to nitrosate proteins. Detailed mechanistic studies of this process have not, as yet, been published.

Nitrosation of proteins by these nitrosating species is commonly performed by incubating the protein with the NO donor at pH 7–8 for 5–15 min at room temperature. The nitrosating agent is removed from the protein solution by passing the reaction solution over a desalting column. The reagents SNP, GTN, SIN-1, and DEANO are commercially available. The reagents SNAP, SNO-Cys, and GSNO are prepared by mixing solutions of *N*-acetyl-penicillamine, L-cysteine, and glutathione, respectively, with equimolar $NaNO_2$ in 0.5 *N* HCl. The reaction is typically complete within 1 min; for reasons of stability, the solution is neutralized by addition of NaOH immediately before addition to the protein solution.

Transnitrosation appears to be an important mechanism *in vivo* for distributing NO among different pools, especially between low molecular weight thiols (L-cysteine and glutathione) and serum albumin (7–9). In addition, transnitrosation reactions may lead to *S*-nitrosation of other cysteinyl thiol groups in proteins that evoke unique neuronal responses characteristic of the bioactivity of NO in the nervous system (10).

Nitrosation with $NaNO_2/HCl$

The reaction of $NaNO_2$ and HCl yields one of the oldest and best-studied nitrosating reagents, HNO_2. We use the denotation $NaNO_2/HCl$ to define the nitrosating reagent rather than HNO_2 because under conditions of protein nitrosation other equilibria exist that contribute to the formation of other nitrosating species, albeit at much lower concentrations than HNO_2 itself. In this solution, the following equilibria exist:

$$3HNO_2 \rightleftharpoons 2NO + HNO_3 + H_2O$$
$$2HNO_2 \rightleftharpoons N_2O_3 + H_2O$$
$$HNO_2 + H_3O^+ \rightleftharpoons NO^+ + 2H_2O$$
$$HNO_2 + H_3O^+ + Cl^- \rightleftharpoons NOCl + 2H_2O$$

For protein nitrosation, we typically use 0.5 N HCl. Serum albumin, the protein with which we have the greatest experience in this regard, is stable under these conditions and can be reversibly renatured; however, other proteins may require higher pH values to ensure stability and renaturation. Since the pK_a of HNO_2 is 3.35, nitrosating proteins with this reagent requires that the reaction proceed below pH 3.0.

Nitrosation is typically performed by incubating the protein solution with $NaNO_2$ in 0.5 N HCl at room temperature. The protein solution and acidified nitrite solution are freshly prepared immediately prior to use. For bovine serum albumin, the incubation time for efficient nitrosation is approximately 30 min; longer incubation times lead to precipitation of the protein, especially when conducted at high concentrations (50–100 mg/ml). S-Nitrosation of the single free cysteine, Cys-34, in albumin proceeds to completion under these conditions within 5 min; longer incubation times may be required to N-nitrosate the indole nitrogen of one of the two tryptophan residues in albumin. Importantly, S-nitrosation is preferred over N-nitrosation, even under these rather extreme conditions of acidic pH.

Nitric Oxide Assays of Nitrosated Proteins

The NO groups incorporated in proteins can be measured by one of two methods: the Saville assay (11) and photolysis–chemiluminescence analysis (12). The advantage of the Saville assay is that it does not require special instrumentation; however, its sensitivity (detection limit, 1 μM) is approximately two orders of magnitude lower than that of the photolysis–chemiluminescence method (detection limit, 0.01 μM).

Saville Assay

The Saville assay is carried out by first removing free nitrite from the protein solution. This is achieved by mixing 0.1% ammonium sulfamate with the protein solution in 0.4 N HCl for 1–2 min (total volume, 0.5 ml). After this initial step, 0.5 ml of a solution containing 0.25% $HgCl_2$ and 3% sulfanilamide in 0.4 N HCl is added. This step leads to the displacement of NO from S-NO groups in proteins by the mercuric ion and to diazotization of sulfanilamide. The resulting diazonium ion is detected by addition of 0.5 ml of 0.1% N-1-(naphthyl)ethylenediamine dihydrochloride in 0.4 N HCl. The reaction is allowed to proceed for 10 min, and the absorbance at 540 nm recorded. The S-nitroso content of a protein solution is determined by comparison with a standard curve constructed using a 1–20 μM range of nitrite concentrations assayed similarly except that the ammonium sulfamate step is omitted.

The Saville assay was originally designed to detect the S-nitroso group. With simple modification, however, N-nitroso groups can also be detected. Owing to the instability of these latter groups at acidic pH and to the inability of mercuric ion to displace NO from N-NO functional groups, one can omit the $HgCl_2$ from the reaction and compare the nitrite content in the presence and absence of mercuric ion. This procedure allows one to differentiate S-NO from N-NO groups. As the assay is limited by the rate of spontaneous denitrosation, longer incubation times are required and should be determined for each specific molecule studied. For example, a development time of 30 min is required for N-nitroso-L-glycyl-L-tryptophan. It is important to note that there will be incomplete recovery of measurable N-NO groups when the ammonium sulfamate step is used in this reaction; thus, free nitrite may be better removed using a rapid desalting column step.

Photolysis–Chemiluminescence Spectroscopy

Nitric oxide derived from S-NO or N-NO functional groups in proteins can also be detected by first photolyzing the S–NO or N–NO bond and subsequently detecting the homolytically released NO by its chemiluminescent reaction with ozone. Photolysis–chemiluminescence analysis is performed using a thermal energy analyzer (TEA Model 543), without a pyrolyzer, linked to a photolytic interface (Nitrolyte, Thermedics Detection, Inc., Chelmsford, MA) (12, 13). The protein sample (100 μl) is introduced into the system using a high-performance liquid chromatography pump with water as the mobile phase at a flow rate of 1 ml/min. The sample enters a narrow borosilicate glass tube coiled around a 200-W mercury vapor lamp that emits light between 300 and 400 nm. Nitric oxide is homolytically cleaved from the S-NO or N-NO functional group by photolysis and is then carried by argon or helium through two cold traps which separate it from the aqueous phase. The NO then enters the thermal energy analyzer under vacuum where it reacts with ozone to produce a chemiluminescent product. The chemiluminescence signal is detected and recorded, and the concentration of NO is determined from a standard curve derived from GSNO over a range of concentrations from 10 to 1500 nM. This method of detection can be used to measure S-NO, N-NO, and metal-coordinated NO. Nitrite also yields a weak signal that is approximately 500-fold lower than that of the nitroso- derivatives.

References

1. J. S. Stamler, D. S. Singel, and J. Loscalzo, *Science* **258,** 1898 (1992).
2. D. A. Wink, R. W. Nims, J. F. Derbyshire, D. Christodoulou, I. Hanbauer, G. S. Cox, F. Laval, J. Laval, J. A. Cook, M. C. Krisna, W. G. DeGraff, and J. B. Mitchell, *Chem. Res. Toxicol.* **7,** 519 (1994).
3. R. M. Clancy, Y. Miyazaki, and P. C. Cannon, *Anal. Biochem.* **191,** 138 (1990).
4. L. J. Ignarro, G. M. Buga, K. S. Wood, R. E. Byrns, and G. Chaudhuri, *Proc. Natl. Acad. Sci. U.S.A.* **84,** 9265 (1987).
5. S. Tamir, R. W. Lewis, T. R. Walker, W. M. Deen, J. S. Wishnok, and S. R. Tannenbaum, *Chem. Res. Toxicol.* **6,** 895 (1993).
6. P. Griess, *Chem. Ber.* **12,** 426 (1879).
7. J. F. Keaney, Jr., D. I. Simon, J. S. Stamler, O. Jaraki, J. Scharfstein, J. A. Vita, and J. Loscalzo, *J. Clin. Invest.* **91,** 1582 (1993).
8. J. Scharfstein, J. F. Keaney, Jr., A. Slivka, G. N. Welch, J. A. Vita, J. S. Stamler, and J. Loscalzo, *J. Clin. Invest.* **94,** 1432 (1992).
9. M. A. Kaufman, I. Castelli, H. Pargger, and L. J. Drop, *J. Pharmacol. Exp. Ther.* **273,** 855 (1995).
10. D. T. Hess, L.-H. Lin, J. A. Freeman, and J. J. Norden, *Neuropharmacology* **33,** 1283 (1994).
11. B. Saville, *Analyst* **83,** 670 (1958).
12. J. J. Conboy and J. J. Hotchkiss, *Analyst* **114,** 155 (1989).
13. D. H. Fine, F. Rufeh, D. Lieb, and D. P. Rounbehler, *Anal. Chem.* **47,** 1188 (1975).

[5] Immunohistochemical Methods for Nitric Oxide Synthase

Paul K. Nakane

Introduction

The usual prerequisites for immunohistochemistry are applicable also for successful immunohistochemical localization of nitric oxide synthase (NOS). The prerequisites include an availability of specific antisera to a desired antigen and a method to fix the antigen at its native sites while retaining its antigenicity and tissue morphology. To date there is no method that retains antigenicity totally in well-preserved tissues, and one is forced to compromise between sensitivity and morphology depending on the answer sought. For this reason various immunohistochemical methods and permutations are provided in this chapter.

Basic Methods

In performing immunohistochemistry, there are basically three approaches in tissue handling. The first is to perform the procedure on sections obtained from tissues that have been frozen fresh, the second on sections from tissues which were fixed and then frozen, and the third on sections from tissues which were fixed, dehydrated, and embedded in paraffins or plastics. The first approach permits histological observation only at the light microscopic level, and tissue morphology is somewhat poor; the second and third approaches allow localization of antigens both at the light and electron microscopic levels with good morphology. On the other hand, retention of antigenicity and accessibility of an antibody to antigens is usually better in the fresh frozen sections than in fixed tissue sections.

Fixation

The differences in retention of antigenicity result from the manner by which proteinaceous antigens are made insoluble in an aqueous solution. Organic solvents such as acetone and ethanol that are often used to fix frozen sections of fresh tissue denature the antigens via unfolding and form little or no cross-linking between the precipitated proteins. Fixatives used for pieces of tissue or for perfusion fixation are mostly aldehyde derivatives and form complexes with basic hydrophilic groups such as amino and hydroxy groups of the antigens. Such fixatives form cross-links between neighboring proteins and insolubilize the antigens. In other words, fixation by organic solvents mainly alters the tertiary but not the primary structure of the antigens, whereas the other types of fixatives modify proteinaceous antigens at both levels.

The accessibility of antibodies to epitopes depends on the porosity of the tissues, the presence or absence of interfering materials associated with the epitopes, and the size of the antibodies. The solvent fixatives make tissues porous by extracting lipid and carbohydrates during fixation, and most of the epitopes remain exposed. The aldehyde-type fixatives, however, retain most of the lipid and carbohydrates, react with epitopes which are usually hydrophilic, and modify or mask the epitopes. The smaller the size of the antibody the faster and better it diffuses into tissue sections (1). This is particularly evident in sections of tissues fixed with the aldehyde-type fixatives.

The speed of fixation of antigens, particularly soluble antigens, should also be considered. For soluble antigens, it is important that the antigens do not diffuse away from their native site before fixation and become fixed on other sites, thereby leading to an incorrect conclusion. The nitric oxide synthases are complicated if all three major isozymes in a given tissue are to be localized because bNOS (brain NOS) is quite soluble, whereas iNOS (inducible NOS) is less soluble than bNOS but more soluble than eNOS (endothelial NOS) which is present in a particulate form.

Antibodies

There are basically four types of antisera: (1) polyclonal antibodies made against isolated proteins, (2) monoclonal antibodies made against isolated proteins, (3) polyclonal antibodies made against synthetic peptides, and (4) monoclonal antibodies made against synthetic peptides. All types are available for the localization of NOS. On the basis of nucleotide sequences of bNOS (2), eNOS (3), and iNOS (4), some antibodies will recognize all three isozymes (5) and some will only recognize one isozyme (see product information from vendors such as Affinity BioReagents, Neshanic Station, NJ; Euro-Diagnostica AB, Malmo, Sweden; Incstar, Stillwater, MN; Labosev, Giessen, Germany; Transduction Laboratories, Lexington, KY). Because the sequence of a given isozyme is similar among species (6), an antibody against an isozyme of one species may cross-react with the isozyme of another species. It has been our experience that, for routine immunohistochemical localization of NOS, polyclonal antibodies against synthetic peptides usually produce the best reproducible results. Monoclonal antibodies against isolated proteins are also useful; however, some have a low avidity, and modifications of washing conditions may be necessary. Monoclonal antibodies against synthetic peptides have not been useful for some of the materials we have studied, but others have used them successfully (7).

Method I: Sections of Fresh Frozen Tissues

Step 1. Cut tissues into pieces 10 mm (width) × 10 mm (width) × 3 mm (thickness) or smaller. The size of the pieces is important. If the thickness exceeds 3 mm, the tissue will not freeze evenly; if the widths of the tissue sections are larger than 10 mm, it will be difficult to pick up frozen sections without distortion. Place the tissues on the bottom of an aluminum boat [about 10–12 mm (width) × 10–12 mm (width) × 25–30 mm (depth)] and fill the boat with Tissue Tek optimal cutting temperature (OCT) compound (Miles, Elkhart, IN). Carefully lower the boat onto a mixture of ethanol and dry ice; avoid spilling the mixture into the boat. Do not use the quick-freezing stage that comes with a cryostat as it often compresses tissues and distorts their morphology. Temperature is also important. With extremely low temperatures, such as those with liquid helium and liquid nitrogen, a layer of tissue near the surface is frozen hard and acts as an insulator, and the center of the tissue remains unfrozen for sometime. Also, the OCT compound may shatter at the low temperature.

Step 2. Peel off the aluminum, cut off the excess OCT compound, and mount the frozen tissue block on a chuck using fresh OCT compound on the quick freezing stage while making sure that the tissue stays frozen.

Step 3. Cut sections 4–6 μm thick with a cryostat. Brain tissues contain more lipid than others and may require a lower temperature. If the tissue section smears or wrinkles on the edge of a knife during sectioning, the temperature of the cold chamber should be lowered; if the section shatters, the temperature should be raised. Pick up sections by placing a glass microscope slide coated with adhesive materials, which has been kept at room temperature (see Note 1), on the section.

Note 1: We have used three types of coated slides for fresh frozen sections. Known adhesive materials should be used as they often may be a source of nonspecific background and autofluorescence. Methods to prepare the coated slides are described.

Silane-treated slides: Silane-coated slides may be purchased from commercial vendors such as Polysciences (Warrington, PA) and DAKO (Carpenteria, CA), but they can be made easily and inexpensively using the following method. (i) Clean glass microscope slides in 1% HCl–70% ethanol by immersion and agitation for several minutes. (ii) Dip slides in 100% acetone and allow them to air dry. Steps (i) and (ii) may be skipped when precleaned slides are used. (iii) Mix 120 ml of 100% acetone and 2.4 ml of 3-aminopropyltriethoxysilane (Sigma, St. Louis, MO) in a glass container just before use. (iv) Dip slides in the mixture for 10 sec. (v) Wash in 100% acetone (in a separate jar) 1 min for three times. (vi) Air dry. These slides can be kept for several months.

Gelatin-coated glass slides: (i) Dissolve gelatin in 80°C distilled water (1 g/100 ml) while stirring. (ii) Cool the gelatin mixture to 40°C. (iii) Precleaned slides (see above, as for silane-treated slides) are placed in a slide staining basket and dipped into the mixture for 1 min. (iv) Drain off excess solution on a paper towel and place slides in racks to dry. (v) Dry at 60°C for 2–12 hr. The slides can be kept in a dry chamber for several months.

Egg albumin-coated slides: (i) Combine one fresh egg white and 1 ml of concentrated (28%) NH_4OH with 500 ml of distilled water. Stir for 10 min. Filter through a paper towel or six layers of gauze. (ii) Place cleaned slides (see above, as for silane-treated slides) into the freshly made egg albumin solution for 1 min. Drain off excess solution on a paper towel and place slides in racks to dry. (iii) Dry at 60°C for 2–12 hr. About 500 slides can be coated with this egg mixture. The slides can be kept in a dry chamber for several months.

Step 4. Air dry the slides with the tissue sections as quickly as possible. We use a hair dryer without heat for drying. Let the slides dry for a minimum of 30 min at room temperature. The sections may be kept in a dry chamber at deep freezer temperatures (-20 to $-80°C$) for 1 to 2 months. Make sure the slides are brought to room temperature in the dry chamber before proceeding to the next step.

Step 5. Fix the section by immersing the slide in prechilled 100% ethanol or acetone for 10 min at 4°C.

Step 6. Wash in phosphate-buffered saline (PBS: 10 mM phosphate buffer, pH 7.2, with 0.85% NaCl) three times, 5 min each time (see Note 2).

Note 2: When horseradish peroxidase (HRP) is used as a reporting marker, endogenous peroxidase activity should be inactivated by immersing the slide in 1–3% (v/v) H_2O_2 in methanol or in PBS for 10–30 min and washing the slide in PBS two times, 5 min each time.

Step 7. Using a piece of gauze, wipe off the PBS except for the area of the section. Place a drop of blocking solution (for a 10 × 10 mm section about 20 μl) (see Note 3) on the area of the section and carefully mix the solutions with a pipette tip making sure that the tip does not touch the section. Place the slide in a moist chamber (see Note 4) for 10 min or longer at room temperature.

Note 3: Blocking solutions are used to mask charged elements in tissue sections so that specific antibodies will not adhere to sections by nonspecific charge. Several types of solutions are used. The most common is nonimmune serum of the species from which labeled antibody was made. For example, if HRP-labeled sheep anti-rabbit immunoglobulin G (IgG) is used, then use normal sheep serum, and if HRP-rabbit anti-mouse IgG is used, then use normal rabbit serum. Other blocking solutions such as skim milk are also used for the normal serum.

Note 4: To prevent the blocking solutions from drying up, slides should be kept in a moist chamber. We use a disposable plastic dish with a cover such as a Nunc 24.5 × 24.4 cm flat dish (InterMed, Roskilde, Denmark). First place a piece of paper towel soaked with PBS on the bottom of the dish, and then place the slides on the paper towel while making sure that the slides will not touch one another.

Step 8. Wash in PBS for 5 min.

Step 9. Wipe off PBS as in Step 7. Place a drop of antibody solution (for a 10 × 10 mm section about 20 μl) on the area of section and carefully mix the solutions with a pipette tip while making sure that the tip will not touch the section (see Note 5). Place the slide in a moist chamber for 30 min at room temperature.

Note 5: There are several procedural permutations for various immunohistochemical detection antigens. Some representative ones and the steps involved for each of them are listed.

Direct HRP-labeled antibody method (1): This is the simplest method. The antibody applied in Step 9 is labeled with HRP. To date no commerical vendor supplies HRP–anti-NOS and the antibody should be labeled (8). Go to Step 10.

Indirect HRP-labeled antibody method (9): After Step 9 wash in PBS three times, 5 min each time. Wipe off PBS as in Step 9. Place a drop of HRP–anti-IgG solution on the tissue section as in Step 9. Most commercial vendors supply HRP–anti-IgG. Place the slide in a moist chamber for 30 min at room temperature. Go to Step 10.

Peroxidase–antiperoxidase method (PAP) (10): Most commercial vendors supply HRP–anti-HRP complexes. After Step 9, wash in PBS three times, for 5 min each time. Wipe off PBS as in Step 9. Place a drop of anti-IgG solution as in Step 9. If the first antibody used in Step 9 is rabbit, then the antiserum used should be anti-rabbit IgG. If such is a case, the anti-HRP should be made by rabbits. The antibody used must be in excess, otherwise a false-negative may result. Place the slide in a moist chamber for 30 min at room temperature. Wash in PBS three times, 5 min each time. Wipe off PBS as in Step 9. Place a drop of HRP–anti-HRP solution on the section as in Step 9. Place the slide in a moist chamber for 30 min at room temperature. Go to Step 10.

Avidin–biotin–peroxidase complex method (ABC) (11): Most commercial vendors supply avidin–biotin–HRP complexes. After Step 9 wash in PBS three times, 5 min each time. Wipe off PBS as in Step 9. Place a drop of biotin-labeled anti-IgG solution on the section as in Step 9. Place the slide in a moist chamber for 30 min at room temperature. Wash in PBS three times, 5 min each time. Wipe off PBS as in Step 9. Place a drop of avidin–biotin–HRP complex solution on the section as in Step 9. Place the slide in a moist chamber for 30 min at room temperature. Go to Step 10.

Multilayered peroxidase-labeled antibody method (MLP) (12): This is a newly introduced extremely sensitive method. This method utilizes amphireactive HRP-labeled antibodies, that is, the antibody will function as the antibody and as the antigen for another HRP-labeled antibody. For example, if the first antibody is rabbit IgG and the second antibody is HRP–sheep anti-rabbit IgG in the indirect method, then the third antibody will be HRP–rabbit anti-sheep IgG, and the fourth antibody will be the same as the second antibody. The following procedure is given using the above example: Wash in PBS three times, 5 min each time. Wipe off PBS as in Step 9. Place a drop of HRP-sheep anti-rabbit IgG solution on the section as in Step 9. Place the slide in a moist chamber for 30 min at room temperature. Wash in PBS three times, for 5 min each time. Wipe off PBS as in Step 9. Place a drop of HRP-rabbit anti-sheep IgG solution on the section as in Step 9. Place the slide in a moist chamber for 30 min at room temperature. Repeat this procedure two or three times, then go to Step 10.

Indirect fluorescence method (13): With the advent of laser scan microscopes, antibodies labeled with a fluorescent compound are used more frequently. Antibody against the antibody applied in Step 9 is labeled with a fluorescent compound, mainly with fluorescein isothiocynate (FITC) or rhodamine isothiocyanate (TRITC). Most commercial vendors supply FITC–anti-IgG. After Step 8, wash in PBS three times, 5 min each time. Wipe off PBS as in Step 9. Place a drop of FITC–anti-IgG or TRITC–anti-IgG solution on the section as in Step 9. Place the slide in a moist chamber for 30 min at room temperature. Wash in PBS three times, 5 min each time. Apply a coverslip using PBS : glycerol (1 : 3, v/v) containing 0.1% *p*-phenylenediamine as a mounting medium. Examine the slides with a fluorescence microscope.

Step 10. Wash slides with tissue sections in PBS three times, 5 min each time.
Step 11. Develop color at room temperature (Note 6).

Note 6: The sites of HRP are visualized histochemically. Classic 3,3′-diaminobenzidine (DAB) solution (14) deposits brown reaction products, and Ni–Co–DAB (15) deposits blue-black reaction products. The Ni–Co–DAB reagent is more sensitive but produces more background than DAB. Make up the solutions in the following manner.

DAB solution (14): Dissolve 25 mg 3,3′-diaminobenzidine tetrahydrochloride (DAB) in 100 ml of 50 mM Tris-HCl buffer at pH 7.6. If not completely colorless, filter. Adjust to pH 7.6 with 50 mM Tris base or with HCl if necessary. Add 0.1 ml of 5% H_2O_2 to 100 ml of DAB–Tris solution.

Ni–Co–DAB solution (15): Dissolve 50 mg DAB, 2.5 mg $CoCl_2$, and 2.5 mg $NiSO_4(NH_4)_2SO_2$ in 100 ml of sodium phosphate buffer, pH 7.5. Add 0.2 ml of H_2O_2 and filter just before use.

Step 12. Wash in PBS three times, 5 min each time.
Step 13. For counterstaining, see Note 7.

Note 7: If the antigen is mainly in the cytoplasm, as is the case with NOS, counterstain nuclei using veronal acetate buffered 1% methyl green solution, pH 4.0, for about 10 min.

Step 14. Wash slides with sections in distilled water briefly, then dehydrate in increasing concentrations of ethanol (70, 80, 90, 95, 100, and 100%, v/v), 100% ethanol and 100% xylene (1 : 1), and then twice in 100% xylene, 5 min each time. Mount a clean coverslip over the section using Paramount (Fisher Scientific, Fair Lawn, NJ). Examine the slide with an ordinal microscope.

Method II: Sections of Fixed (Unfrozen) Tissues

This method is often used by neuroanatomists because thicker tissue sections allow observation in three dimensions and stained sections can be processed for electron microscopic observation. Major drawbacks to this method are that reactants such as antibodies diffuse through tissue sections sluggishly so that all areas of section may not gain access to the reactants evenly and serial sections are difficult to obtain so that the orientation of tissues may be lost.

Step 1. For fixation of tissues, prepare fixative solution (see Note 8).

Note 8: The following aldehyde-type fixatives are recommended: 4% paraformaldehyde in 0.1 M phosphate buffer, pH 7.3; 2% paraformaldehyde with 0.2% picric acid in 0.1 M sodium phosphate buffer, pH 7.0 [Zamboni fixative (16)]; and 2% paraformaldehyde–lysine–periodate [PLP (17)].

Step 1a. For fixation by immersion in fixative, cut tissues into pieces about 10 mm (width) × 10 mm (width) × 2 mm (thickness) or smaller and fix them in one of the above fixatives (Note 8) for about 4 hr.

Step 1b. For fixation by perfusion, first anesthetize the animals, flash the blood with Ca^{2+}-free PBS, and follow with one of the above fixatives (Note 8) for several minutes. Quickly remove desired organs or tissues and further fix them in the fixative for an additional 2 hr.

Step 2. Wash tissues in PBS for at least 12 hr. Tissues left in fixative for 1 day or longer may need to be washed for 2 or more days.

Step 3. Cut sections 10–50 μm thick using a Vibritome (Oxford Instrument, Oxford, UK). Pick up sections with a small paintbrush and place them in a test tube with PBS.

Step 4. Wash sections in PBS three times for 10 min each time.

Step 5. When HRP is used as a reporting marker, inactivate endogenous peroxidase activity by immersing sections in 1–3% H_2O_2 in PBS for 10–30 min and wash twice in PBS, for 5 min each time; if HRP is not used, skip this step.

Step 6. Decant PBS and add antibody solution. Under gentle agitation, let the antibody permeate for at least 12 hr at 4°C (see Note 9).

Note 9: The size of the antibody molecules greatly influences the rate of penetration of the antibody through the section. Ideally, either Fab or Fab′ fragments of IgG should be used. In any case IgM should be avoided. If the direct HRP-labeled antibody method is used, use of HRP–Fab of IgG or HRP–Fab′ of IgG is recommended. Some investigators add detergent such as 1–3% (v/v) Triton X-100 (18) and 0.05–0.1% saponin (19) to increase the rate of permeation, although there will be some deterioration of morphology. All five methods listed in Note 5 may be used, however PAP and MLP should be avoided because PAP is extremely large and MLP takes too much time.

Direct HRP-labeled antibody method (see Note 5): HRP–IgG is required. Because no commercial vendor supplies HRP–anti-NOS, the antibody should be labeled (8). Go to Step 7.

Indirect HRP-labeled antibody method (20): After Step 6, decant antibody solution and wash with PBS three times, 30 min each time, at 4°C under gentle agitation. Decant PBS, add HRP–Fab solution, and incubate for at least 12 hr at 4°C under gentle agitation. Go to Step 7.

ABC method (11): After Step 6, decant the antibody solution and wash with PBS three times, for 30 min each time, at 4°C under gentle agitation. Decant the PBS, add biotin-labeled anti-IgG solution, and incubate for at least 3 hr at 4°C under gentle agitation. Decant the antibody solution and wash with PBS three times, 30 min each time, at 4°C under gentle agitation. Decant the PBS, add the avidin–biotin–HRP complex solution, and incubate for at least 3 hr at 4°C under gentle agitation. Go to Step 7.

Step 7. Decant the antibody solution and wash with PBS three times, for 30 min each time.

Step 8. Divide tissue sections into two tubes, one for light microscopy and another for electron microscopic observations. For light microscopy go to Step 9 and for electron microscopy go to Step 8a.

Step 8a. Postfix the sections in 2% glutaraldehyde in 0.1 M sodium phosphate buffer, pH 7.2, for 30 min at 4°C under gentle agitation.

Step 8b. Wash in PBS three times, for 30 min each time, at 4°C under gentle agitation.

Step 8c. Develop color in DAB solution (see Note 10).

Note 10: As H_2O_2 diffuses through tissue sections faster than DAB, the DAB solution minus H_2O_2 (see Note 6) is applied first for about 30 min and then followed by the complete DAB solution for 2–3 min at room temperature. The Ni–Co–DAB reagent solution is not recommended for this method.

Step 8d. Wash in distilled water three times, for 10 min each time, at 4°C under gentle agitation.

Step 8e. Incubate in 2% OsO_4 in 0.1 M sodium phosphate buffer, pH 7.2, for 60 min at 4°C under gentle agitation.

Step 8f. Wash in distilled water three times, for 5 min each time, at 4°C under gentle agitation.

Step 8g. Dehydrate in increasing concentrations of ethanol (70, 80, 90, 95, 100, and 100%), ethanol and propylene oxide (1 : 1), 100% propylene oxide, and 1 : 1 propylene oxide and Epon, for 5–10 min each time. Fish out the stained sections and flat embed them using a silicone mold in Epon according to routine electron microscopic procedures.

Step 9. Develop color in DAB solution (see Note 6) for 10 min at room temperature.

Step 10. Wash in PBS three times, 5 min each time.

Step 11. Briefly wash in distilled water and then dehydrate in increasing concentrations of ethanol (70, 80, 90, 95, 100, and 100%), 100% ethanol and 100% xylene (1 : 1), and twice in 100% xylene, for 5 min each time. Fish out the stained sections, place them on a clean glass slide, and mount a clean coverslip over the section using Paramount. Examine the slide with an ordinal microscope.

Method III: Sections of Fixed Frozen Tissues

The majority of immunohistochemical investigations on NOS are carried out using fixed frozen tissues. This method has the advantages of allowing tissues to be oriented, and serially cut sections are not difficult to obtain. These materials are also applicable for localization of NADPH diaphorase and mRNAs.

Step 1. Prepare fixed tissues according to Method II, Step 1.

Step 2. Wash tissues in 50 mM phosphate buffer, pH 7.2 (PB), containing 10% (w/v) sucrose (PB + S) for 4 hr at 4°C under gentle agitation.

Step 3. Wash tissues in PB containing 15% sucrose for 4 hr at 4°C under gentle agitation.

Step 4. Wash tissues in PB containing 20% sucrose for 4 hr at 4°C under gentle agitation.

Step 5. Wash tissues in PB containing 20% sucrose and 10% glycerol for 1 hr at 4°C under gentle agitation (see Note 11).

Note 11: Tissues that have gone through Step 5 will be difficult to section using a cryostat, but an adequate sucrose–glycerol content is required for good ultrastructural preservation. If there is no plan for electron microscopic observation of the tissues, then skip Step 5 and replace PB + S in subsequent steps with PBS.

Step 6. Freeze tissues as in Method I, Step 1.

Step 7. Mount the frozen tissue block on a chuck as in Method I, Step 2.

Step 8. Section tissues using a cryostat. As sucrose-impregnated tissues are somewhat difficult to section, the cold chamber temperature may have to be lowered to $-40°C$ when tissues are washed in 20% sucrose and 10% glycerol in Step 5. To ensure sections adhere to the slides, cut sections 4–6 μm in thickness and pick up sections with a slide coated with gelatin or egg albumin (see Note 4); then go to Step 9. For sections to be suspended in solution, cut sections 10–20 μm in thickness, pick up sections with a small paintbrush, and place them in a test tube with PBS; then go to Method II, Step 4.

Step 9. Air dry tissue sections as in Method I, Step 4. If the sections are to be used later, place a drop of 50% glycerol in PB + S over the sections and keep them in a refrigerator (4°C).

Step 10. Wash with PB + S three times, for 10 min each time.

Step 11. When HRP is used as a reporting marker, inactivate endogenous peroxidase activity by immersing the slides in 1–3% H_2O_2 in PB + S for 10–30 min and wash the slides twice in PB + S for 5 min each time; if HRP is not used, skip this step.

Step 12. Wipe off the PB + S as in Method I, Step 7. Place a drop of blocking material (see Note 3) in PB + S as in Method I, Step 7. Place the slides in a moist chamber (see Note 4) for 10 min or longer at room temperature.

Step 13. Wash with PB + S briefly.

Step 14. Wipe off PB + S as in Method I, Step 7. Place a drop of antibody in PB + S on sections as in Method I, Step 9 (see also Note 9). Place the slide in a moist chamber for 60 min at room temperature for light microscopy and at least 3 hr (preferably overnight) at 4°C for electron microscopy. For light microscopy go to Method I, Step 8, except adjust the durations of antibody applications and PBS washes to two to three times longer. For electron microscopy, go to Step 15 (see Notes 9 and 12).

Note 12: We have no experience using the ABC method for electron microscopy; hence, indirect and direct HRP-labeled antibody methods are described.

Step 15. Wash in PB + S three times, for 10 min each time.

Step 16. Wipe off PB + S as in Method I, Step 7, and apply HRP-labeled Fab of IgG or HRP-labeled Fab' of IgG as in Method I, Step 9. Place the slide in a moist chamber for 3–4 hr at room temperature.

Step 17. Wash in PB + S three times, for 10 min each time.

Step 18. Immerse slides in 2% glutaraldehyde in 0.1 M sodium phosphate buffer pH 7.2 for 30 min at 4°C.

Step 19. Wash in PB + S three times, for 10 min each time.

Step 20. Immerse slides in DAB solution minus H_2O_2 (see Notes 6 and 10) for 30 min at room temperature.

Step 21. Immerse slides in DAB solution with H_2O_2 (see Note 6) for 2 min at room temperature.

Step 22. Wash in PB + S three times, for 10 min each time.

Step 23. Apply a drop of 2% OsO_4 in 0.1 M sodium phosphate buffer, pH 7.2, for 60 min in a moist chamber at room temperature.

Step 24. Wash in PB + S three times, for 10 min each time.

Step 25. Briefly wash in distilled water, then dehydrate in increasing concentrations of ethanol (70, 80, 90, 95, 100, and 100%).

Step 26. Embed the sections by inverting a gelatin capsule filled with fresh Epon over each section.

Step 27. Polymerize the Epon by incubating at 37°C overnight and at 60°C for 24–48 hr. The gelatin capsule and its tissue section can then be removed from the slide by very briefly heating the slide over a Bunsen burner flame (21).

Step 28. Obtain ultrathin sections from the surface of the Epon block and examine them with an electron microscope without counterstaining.

Method IV: Sections of Paraffin-Embedded Fixed Tissues

Some reports (7, 22) have appeared in which sections of fixed tissues embedded in paraffin for localization of NOS are used. Prerequisites for successful localization of NOS in these materials include: (1) tissues must be fixed under controlled conditions (see Note 8) and (2) antigenicities are retrieved (see Note 14) and antibodies known to react with paraffin-embedded materials are used.

Step 1. Prepare fixed tissues according to Method II, Step 1.

Step 2. Wash tissues in PBS.

Step 3. Wash tissues in distilled water briefly, then dehydrate in increasing concentrations of ethanol, clear in 100% xylene, and embed in paraffin (normally in Paraplast) by routine histological methods.

Step 4. Section tissues 4–6 μm thick and adhere on coated slides (see Note 1).

Step 5. Dry tissue sections at room temperature. Often they are dried in an oven in routine histopathology laboratories, but high temperature may denature some of the NOS antigenicity.

Step 6. Deparaffinize the slides first in 100% toluene at 60°C (see Note 13) or at room temperature, then hydrate in a series of decreasing concentrations of ethanol as done in routine histopathology laboratories.

Note 13: Most tissues are embedded in Paraplast. The Paraplast, however, is not completely dissolved in xylene and will often interfere with the reaction between antibody and antigen (23).

Step 7. Wash in PBS three times, for 5 min each time.
Step 8. Perform retrieval of antigenicity (see Note 14). Go to Method I, Step 7.

Note 14: There are several ways to retrieve antigenicity. The two most frequently used methods are digestion with protein hydrolase and heating. For hydrolysis, immerse slides in 0.1% (w/v) trypsin and 0.1% (w/v) $CaCl_2$ in 50 mM Tris-HCl buffer, pH 7.6, 0.05% (w/v) pronase in 50 mM Tris-HCl buffer, pH 7.6, and 0.4% (w/v) pepsin in 0.01 N HCl at 37°C for a predetermined optimal duration. For heating, immerse slides in water and heat the water to boiling either in a microwave oven or on a hot plate. Some investigators autoclave the slide immersed in PBS, but we have no experience with this method for NOS antigen retrieval.

Comments

With the available technology for immunohistochemistry, all three isozymes of NOS may be localized with relative ease at the light microscopic level. Localization at the electron microscopic level has been achieved for some isozymes (24, 25); however, the results are difficult to interpret for soluble antigens such as bNOS, and it should be limited to investigations where the ultrastructural resolution is essential and with a full understanding of the pitfalls (i.e., where antigens are found may not be where those antigens occur *in vivo*).

In performing immunohistochemistry, various controls are mandatory. Negative controls include use of nonimmune sera, unrelated antibodies, etc., in the place of specific sera. For the positive controls we use sections of intestine, which contain all types of NOS: bNOS in Meissner's plexus and Auerbacks's plexus, eNOS on the wall of blood vessels, and iNOS in macrophages in the lamina propria near the tips of villi.

For the usual investigations at the light microscopic level we recommend the following sequence of steps: Method II, Step 1b, with 4% paraformaldehyde → Method III, Steps 2, 3, 4, 6, 7 → Method III, Step 8, section at 4–6 μm thick → Method III, Step 9 → Method III, Step 10 using PBS, Step 11 using PBS, Step 12 using PBS, Step 13 using PBS, Method III, Step 14 using polyclonal antibody against peptides and PBS → Method I, Steps 8, 9, 6, 9, using HRP–anti-IgG (or Method I, Steps 8, 9, 6, 9, using biotin-labeled anti-IgG) → Method I, Step 10 → Method I, Step 11 using DAB + H_2O_2 → Method I, Steps 12, 13, 14.

Acknowledgments

The author thanks Dr. S. Izumi for verifying the adequacy of some of these methods. This work was supported in part by a grant from Japanese Ministry of Education.

References

1. P. K. Nakane, *Ann. N.Y. Acad. Sci.* **254,** 203 (1975).
2. C. J. Lowenstein, C. S. Glatt, D. S. Bredt, and S. H. Snyder, *Proc. Natl. Acad. Sci. U.S.A.* **89,** 6711 (1992).
3. S. P. Janssens, A. Shimouchi, T. Quertermous, D. B. Bloch, and K. D. Bloch, *J. Biol. Chem.* **267,** 14519 (1992).
4. Q. Xie, H. J. Cho, J. Calaycay, R. A. Mumford, K. M. Swiderek, and T. D. Lee, *Science* **256,** 225 (1992).
5. A. Rengasamy, C. Xue, and R. A. Johns, *Am. J. Physiol.* **267,** 1704 (1994).
6. C. Nathan, *FASEB J.* **6,** 3051 (1992).
7. R. V. Lloyd, L. Jin, X. Qian, S. Zhang, and B. W. Scheihauer, *Am. J. Pathol.* **146,** 86 (1995).
8. P. K. Nakane and A. Kawaoi, *J. Histochem. Cytochem.* **22,** 1084 (1974).
9. P. K. Nakane and G. B. Pierce, *J. Histochem. Cytochem.* **14,** 789 (1966).
10. L. A. Sternberger, P. H. Hardy, Jr., and J. J. Cuculis, *J. Histochem. Cytochem.* **18,** 315 (1968).
11. S. M. Hsu, L. Raine, and H. Fanger, *J. Histochem. Cytochem.* **29,** 577 (1981).
12. M. Shin, S. Izumi, and P. K. Nakane, *J. Clin. Lab. Anal.* **9,** 424 (1995).
13. A. H. Coons, *in* "General Cytochemical Methods" (J. F. Danielli, ed.), p. 399. Academic Press, New York, 1958.
14. R. C. Graham and M. J. Karnovsky, *J. Histochem. Cytochem.* **14,** 291 (1996).
15. J. C. Adams, *J. Histochem. Cytochem.* **29,** 775 (1981).
16. M. Stafanini, C. DeMartino, and L. Zamboni, *Nature* (*London*) **216,** 173 (1967).
17. I. W. McLean and P. K. Nakane, *J. Histochem. Cytochem.* **22,** 1077 (1974).
18. A. L. Burnett, S. Saito, M. P. Maguire, H. Yamaguchi, T. S. K. Chang, and D. F. Henley, *J. Urol.* **153,** 212 (1995).
19. A. Tojo, N. J. Guzman, L. C. Garg, C. C. Tisher, and K. M. Madsen, *Am. J. Physiol.* **267,** F509 (1994).

20. P. K. Nakane and G. B. Pierce, *J. Cell Biol.* **33,** 307 (1967).

21. S. Sell, D. S. Linthicum, D. Bass, R. Bahu, B. Wilson, and P. K. Nakane, *in* "Cancer Biology IV, Differentiation and Carcinogenesis" (C. Borek, C. M. Fenoglio, and D. W. King, eds.), p. 294. Stratton Intercontinental Medical Book, New York, 1977.

22. A. L. Jansen, T. Cook, G. M. Taylor, P. Largen, V. Riveros-Moreno, S. Moncada, and V. Catell, *Kidney Int.* **45,** 1215 (1994).

23. A. Delellis, L. A. Sternberger, R. B. Mann, R. M. Banks, and P. K. Nakane, *Am. J. Clin. Pathol.* **71,** 483 (1979).

24. W. Kummer and B. Mayer, *Histochemistry* **99,** 175 (1993).

25. K. Dirkranian, M. Trosheva, S. Nikolov, and P. Bodin, *Acta Histochem.* **96,** 145 (1994).

[6] NADPH Diaphorase Staining

Ted M. Dawson and Valina L. Dawson

Introduction

The histochemical stain NADPH diaphorase (NDP) was originally described by Thomas and Pearse (1, 2). In this method enzymes that possess diaphorase activity reduce tetrazolium dyes in the presence of NADPH, but not NADH, to a dark blue formazan precipitant. Staining with NDP is quite robust and gives Golgi-like staining in neurons. Neurons that stain for NDP are a unique population that resist the toxic effects of excitatory amino acids and hypoxia and survive the degenerative processes of Huntington and Alzheimer's disease in select areas (3). Nitric oxide synthase (NOS; EC 1.14.13.39) catalytic activity accounts for NDP staining (4, 5). Nitric oxide is a prominent vascular and neuronal messenger molecule first identified as endothelium-derived relaxing factor (EDRF) activity (6–8). Nitric oxide is also formed in macrophages and other peripheral blood cells and mediates immune responses (9, 10). Nitric oxide is formed from arginine by NOS, which oxidizes the guanidino nitrogen of arginine, releasing NO and citrulline (11). Several distinct NOS enzymes have been purified and molecularly cloned. In this chapter, we describe the methodology for NDP staining and show its striking colocalization with NOS.

NADPH Diaphorase Staining

Insight into the action of a variety of proteins has come from information about their localization. In a similar manner, the function of NDP in biological systems has been greatly clarified by its localization.

Under appropriate conditions of paraformaldehyde fixation NDP staining can be used to identify all NOS isoforms. For unknown reasons, NOS is resistant to paraformaldehyde fixation, whereas all other NADPH-dependent oxidative enzymes are inactivated by fixatives. The NDP staining is performed by incubating paraformaldehyde-fixed tissue or cultures with 1 mM NADPH, 0.2 mM nitro blue tetrazolium in 0.1 M Tris-HCl buffer (pH 7.2) containing 0.2% Triton X-100 and 0.015% NaN_3 (Table I).

Methods in Neurosciences, Volume 31

TABLE I   NADPH Diaphorase Stain

NDP stain solution
  Buffer:
    0.1 *M* Tris-HCl (pH 7.2)
    0.2% Triton X-100 (v/v)
    7.5 mg Sodium azide/100 ml buffer
  Staining solution (final volume 10 ml):
    1 m*M* NADPH reduced form (Sigma, St. Louis, MO), 8.33 mg/5 ml
    0.2 m*M* Nitroblue tetrazolium (NBT) (Sigma), 1.64 mg/5 ml
  Solubilize NADPH and NBT in separate test tubes to prevent precipitation.
    NBT may need to be sonicated to solubilize. Mix soluble NADPH and
    NBT to form stain mix.
FIX: 4% paraformaldehyde/0.1 *M* PB
  8% Paraformaldehyde (8 g/100 ml Milli-Q water)
  Heat to 80°C for 30 min (do not boil).
  Clear with 1–2 drops of 10 *N* NaOH and replace volume of evaporated
    water.
  Add an equal volume of 0.2 *M* PB (100 ml 0.2 *M* NaH$_2$PO$_4$: 400 ml 0.2 *M*
    Na$_2$HPO$_4$, pH 7.4) and pass through a 0.45-μm filter.

Tissue Preparation

Rats or mice are anesthetized with pentobarbital (100 mg/kg intraperitoneal injection) and perfused with 200 to 300 ml of 50 m*M* phosphate-buffered saline (PBS), pH 7.4 at 4°C, followed by 200 to 300 ml of 4% (w/v) freshly depolymerized paraformaldehyde and 0.1% (v/v) glutaraldehyde in 0.1 *M* phosphate buffer (PB), pH 7.4. The brains are removed and postfixed for 2 hr in 4% paraformaldehyde in PB before cryoprotection in 20% (v/v) glycerol in PB. Alternatively, the brains can be cryoprotected in 30% (v/v) sucrose in PB by sequentially bathing the tissue in 10, 20, and 30% sucrose. Slide-mounted or free floating tissue sections are then permeabilized in 50 m*M* Tris-HCl-buffered saline (1.5% NaCl), pH 7.4 (TBS), containing 0.4% Triton X-100 for 30 min at room temperature. This is followed by incubating the tissues in the NDP stain and rinsing the tissue section three times for 10 min in TBS. The sections are then mounted on chrome/alum-coated glass slides. After dehydration, the sections are coverslipped with Permount (Fisher Scientific, Pittsburgh, PA).

Localization of NADPH Diaphorase

In the brain, the highest density of NDP is evident in the cerebellum and the olfactory bulb (12, 13). The accessory olfactory bulb has even more prominent staining. Other areas of high density staining include the peduncu-

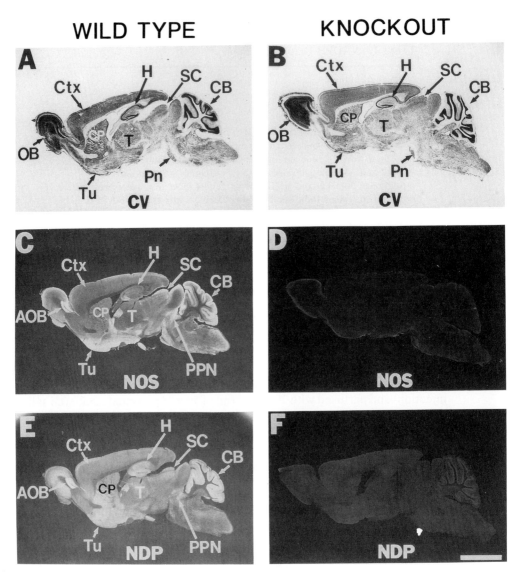

WILD TYPE KNOCKOUT

FIG. 1   Neuronal NOS null mice (KNOCKOUT) are devoid of NDP staining and nNOS immunostaining and have normal cytoarchitecture. Wild-type mice are shown in (A), (C), and (E), and nNOS knockout mice are shown in (B), (D), and (E). The absence of NDP in the nNOS null mice confirms that NOS catalytic activity accounts for NDP staining. CV, Cresyl violet; NOS, neuronal nitric oxide synthase; NDP, NADPH diaphorase staining; AOB, accessory olfactory bulb; CP, caudate-putamen; Ctx, cortex; CB, cerebellum; H, hippocampus; OB, olfactory bulb; Pn, pontine nuclei; T, thalamus; Tu, olfactory tubercle; PPN, pedunculopontine tegmental nucleus; SC, superior colliculus. Bar: 2.5 mm. Reprinted with permission from T. M. Dawson and V. L. Dawson, *Neuroscientist* **1**, 7.

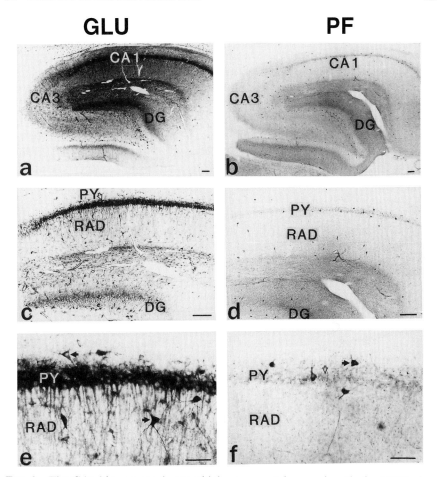

FIG. 2   The CA1 hippocampal pyramidal neurons stain prominently for NDP when tissue sections are fixed in high concentrations of glutaraldehyde (2%) as shown in (a), (b), and (c), but when fixed in the presence of only paraformaldehyde CA1 pyramidal cells do not stain for NDP as shown in (b), (d), and (f) (see Ref. 17). CA1, CA1 field of ammons horn; CA3, CA3 field of ammons horn; DG, dentate gyrus; PY, pyramidal cell layer of CA1; RAD, stratum radiatum; Glu, glutaraldehyde; PF, paraformaldehyde. Bar, 100 μm in a–d and 50 μm in e and f. Reprinted with permission from Dinerman *et al., Proc. Natl. Acad. Sci.* **91,** 4214 (1994).

tion, the Alzheimer's Association, and the Muscular Dystrophy Association. Ted M. Dawson is supported by grants from the USPHS (NS01578, NS26643, NS33277), the American Health Assistance Foundation, the Paul Beeson Physician Scholars in Aging Research Program, and the International Life Sciences Institute.

lopontine tegmental nucleus, the superior and inferior colliculi, the supraoptic nucleus, the islands of calleja, the caudate-putamen, and the dentate gyrus of the hippocampus. In the cerebellum, NDP occurs in glutaminergic granule cell as well as γ-aminobutyric acid (GABA) containing basket cells. In the cerebral cortex, NDP staining neurons are colocalized with somatostatin-, neuropeptide Y (NPY), and GABA-containing cells. In the corpus striatum, NDP neurons also stain for somatostatin and NPY. In the pedunculopontine tegmental nucleus of the brain stem, NDP neurons stain for choline acetyltransferase but do not stain for somatostatin and NPY. Even though there does not seem to be a single neurotransmitter that colocalizes with NDP, all NDP neurons identified colocalize with NOS (4) [Fig. 1 (14)]. The NOS catalytic activity accounts for diaphorase staining because transfection of cultured human kidney 293 cells with neuronal NOS (nNOS) cDNA produces cells that stain for both nNOS and NADPH diaphorase (4).

Throughout the gastrointestinal tract, NDP-staining neurons are present in the myenteric plexus. Staining with NDP also occurs in discrete ganglia cells and fibers in the adrenal medulla where it may regulate blood flow; NDP is also prominent within the posterior pituitary gland in fibers and terminals. In the spinal cord NDP is localized in the substantia gelatinosa and the intermediolateral cell column and neurons around the central canal (15). It has also been localized to the macula densa of the kidney (16).

Staining with NDP has been localized to the endothelium of blood vessels in the periphery and the nervous system. In addition, NDP is concentrated in the hippocampus (17) and is evident in pyramidal cells of the CA1 region and in granule cells of the dentate gyrus (Fig. 2). Endothelial NOS (eNOS) or a closely related isoform accounts for NDP staining in this structure. Using high concentrations of glutaraldehyde fixatives, we found that NDP provided robust staining of pyramidal cells of the CA1 region. Finally, NDP has also been identified in neutrophils, macrophages, microglia, and astrocytes.

Conclusion

Under appropriate conditions, NDP is a simple and sensitive stain which can be used to visualize nNOS, eNOS, and inducible NOS (iNOS). It has facilitated the characterization and study of the function of these proteins.

Acknowledgments

Valina L. Dawson is supported by grants from the U.S. Public Health Service (NS22643, NS33142), the American Foundation for AIDS Research, the National Alliance for Research on Schizophrenia and Depression, the American Heart Associa-

References

1. E. Thomas and A. G. E. Pearse, *Histochemistry* **2**, 266 (1961).
2. E. Thomas and A. G. E. Pearse, *Acta Neuropathol.* **3**, 238 (1964).
3. T. M. Dawson, V. L. Dawson, and S. H. Synder, *Ann. Neurol.* **32**, 297 (1992).
4. T. M. Dawson, D. S. Bredt, D. M. Fotuhi, P. M. Hwang, and S. H. Synder, *Proc. Natl. Acad. Sci. U.S.A.* **88**, 7797 (1991).
5. B. T. Hope, C. J. Michael, K. M. Knigge, and S. R. Vincent, *Proc. Natl. Acad. Sci. U.S.A.* **88**, 2811 (1991).
6. S. Moncada and A. Higgs *N. Engl. J. Med.* **330**, 613 (1993).
7. T. M. Dawson and S. H. Synder, *J. Neurosci.* **14**, 5147 (1994).
8. J. Garthwaite and C. L. Boulton, *Annu. Rev. Physiol.* **57**, 683 (1995).
9. C. Nathan, *FASEB J.* **6**, 3051 (1992).
10. M. A. Marletta, *J. Biol. Chem.* **268**, 12231 (1993).
11. D. S. Bredt and S. H. Snyder, *Annu. Rev. Biochem.* **63**, 175 (1994).
12. D. S. Bredt, C. E. Glatt, P. M. Hwang, M. Fotuhi, T. M. Dawson, and S. H. Synder, *Neuron* **7**, 615 (1991).
13. S. R. Vincent and H. Kimura, *Neuroscience (Oxford)* **46**, 755 (1992).
14. T. M. Dawson and V. L. Dawson, *Neuroscientist* **1**, 7 (1995).
15. S. Saito, G. J. Kidd, B. D. Trapp, T. M. Dawson, D. S. Bredt, D. A. Wilson, R. J. Traystman, S. H. Snyder, and D. F. Hanley, *Neuroscience (Oxford)* **59**, 447 (1994).
16. P. L. Huang, T. M. Dawson, D. S. Bredt, S. H. Snyder, and M. C. Fishman, *Cell (Cambridge, Mass.)* **75**, 1273 (1993).
17. J. L. Dinerman, T. M. Dawson, M. J. Schell, A. Snowman, and S. H. Snyder, *Proc. Natl. Acad. Sci. U.S.A.* **91**, 4214 (1994).

[7] Immunohistochemical Visualization of Cyclic Nucleotides

Jan de Vente, Heather M. Young, and Harry W. M. Steinbusch

Introduction

In 1972 antibodies against cAMP and cGMP became available (1), and they were immediately used in immunocytochemistry (ICC) (2). These early studies showed for the first time a discrete localization of cyclic nucleotides in specific cell somata and processes [see Steiner *et al.* (5) for a review]. However, it became apparent that the technique suffered from several drawbacks. In the case of cAMP it was necessary to use unfixed tissue material (2). Subsequently it was shown that during the ICC procedure for cAMP more than 99% of this cyclic nucleotide was lost from the tissue (3, 4). In addition, it was difficult to visualize the active pool of cyclic nucleotides on pharmacological stimulation (5, 6). In a review on cyclic nucleotide ICC, Steiner *et al.* (5) suggested the possibility that these antibodies recognized cAMP and cGMP bound to the respective protein kinases.

We approached the problem of cyclic nucleotide ICC by taking into account the necessity of fixing the cyclic nucleotides to the tissue proteins, in order to obtain a reliable visualization of these second messengers. We also realized that this approach meant that a new class of antisera to cyclic nucleotides had to be raised.

Fixation of Cyclic Nucleotides

Cyclic nucleotides are small, water-soluble, nonimmunogenic molecules. They therefore need to be conjugated to a carrier protein, and then antibodies can be raised to these haptens. The chemical properties of cyclic nucleotides are not promising in this respect when tissue preservation and cellular morphology have to be taken into account. In the ideal situation, the chemical reaction which fixes the hapten to the immunogenic carrier protein should be the same as when fixing the antigen to the tissue protein in preparing the tissue for ICC. Fixation may alter the structure of the antigen in an important way, as shown for serotonin (6, 7). Antisera used in radioimmunoassays

Methods in Neurosciences, Volume 31

for cAMP and cGMP are raised against conjugates of the 2'-succinylated derivatives (1). Succinylation is not a procedure which can be used in combination with ICC. Rall and Lehne (8) showed that formaldehyde fixation of cAMP to tissue components is possible, although this reaction is not irreversible (9). In addition, tissue fixation prior to cGMP–ICC had no adverse effects and was reported to even improve immunostaining (10). These data were promising enough to follow up, and the kinetics of the reaction of cAMP and cGMP with formaldehyde were studied in more detail (9). It was found that the formaldehyde–protein conjugate of cAMP dissociates rapidly, and the conjugate was not successful in raising antibodies against cAMP. Better results were obtained using acrolein as a fixative for cAMP (11). The cAMP–acrolein–protein conjugate is stable, and antibodies which were specific for acrolein-fixed cAMP were obtained. In addition, cAMP accumulation after dopaminergic stimulation of adenylate cyclase was demonstrated in neurons of the rat frontal cortex (11). However, study of the suitability of this procedure has not been taken further than the initial characterization of these antisera.

Compared to cAMP, the formaldehyde conjugate of cGMP with a protein is more stable (9). After 24 h at 4°C we found negligible dissociation of the conjugate. This conjugate proved successful in raising antibodies to formaldehyde-fixed cGMP.

Fixation efficiency of cyclic nucleotides was first assessed in a gelatin model system. Following the protocol for the processing of tissue for ICC it was found that retention of cGMP in gelatin pieces after formaldehyde fixation was 29%, whereas the retention of cAMP under similar conditions was 0.5% (9). The acrolein fixation efficiency of cAMP in this model was about 3.5%. We established that loss of cGMP from cerebellar slices during fixation was negligible, whereas the loss of cGMP during washing of the tissue sections (three times, 5 min each, at room temperature) before application of the primary antibody was $62 \pm 6\%$. Compared to the rate of dissociation of the cGMP–formaldehyde–protein complex at room temperature (9), and given the fact that the reaction of formaldehyde, cGMP, and protein does not run to completion (8, 9), it can be assumed that loss of the nucleotide from the tissue during the washing steps represents unfixed cGMP.

Specificity of Cyclic Nucleotide Antisera

Specificity testing of antisera is of the utmost importance (12, 13). The cAMP and cGMP antisera fulfilled all the criteria necessary for a specific recognition of cAMP and cGMP, respectively. Application of preimmune sera or preabsorption of the antisera with the respective conjugates resulted in no immunostaining (9, 13). As the cGMP antibodies have a higher affinity for the formaldehyde-fixed cGMP than for free cGMP (13), preabsorption of the cGMP antibody with free cGMP needs to be done with a high concentration of the nucleotide (10 mM) (13). Both the cAMP and cGMP antisera were tested for specificity in a gelatin model system (11, 14). In the case of the cGMP antibodies this was done independently in another laboratory (14). In addition, the cGMP antisera were tested in a nitrocellulose model (13). No cross-reactivity was observed of the antisera with other cyclic nucleotides or ATP, ADP, AMP, adenosine, GTP, GDP, GMP, guanosine, or NADP (9, 13, 14). A final test in the application of antisera is to apply them in biological model systems. A parallel must be found between ICC visualization of cGMP and biochemical measurements of the content of cGMP in a number of tissues. Thus increases in cGMP were evaluated in the aortic smooth muscle (13), the kidney (14, 15), the superior cervical ganglion (16), the cerebellum (17, 18), the hippocampus (19, 20), and the carotid body (21).

Guanylate Cyclase

Cyclic GMP is synthesized by the enzyme guanylate cyclase (GNC). Two families of isoenzymes can be distinguished: the particulate, plasma membrane-bound GNC (pGNC) and the cytosolic, soluble GNC (sGNC) (22). In mammals pGNC is activated by peptide hormones like atrial natriuretic peptide (ANP) and congeners (22). It is still a matter of debate which agent is the principal activator of sGNC in biological tissues, however, evidence is accumulating that nitric oxide (NO) is the primary factor regulating sGNC activity (23, 24).

Biochemical (25), ICC (26), and *in situ* hybridization studies of mRNA (27, 28) revealed an abundant presence of sGNC in the central nervous system (CNS). Soluble GNC is found in virtually all parts of the CNS, although there are important regional differences in the amount of enzyme being present. Immunochemistry and mRNA studies reveal the localization of sGNC in the CNS in some detail (26–28). These methods do not localize the active enzyme. Biochemical assay of sGNC yields precise data on NO-sensitive GNC activity but lacks details on localization. Thus cGMP–ICC demonstrates the localization of NO-sensitive GNC in great morphological detail, even at the electron microscopical level (Fig. 2C) (29, 30).

Localization of Cyclic GMP in Biological Tissues

Central Nervous System

For the purpose of studying localization of NO-mediated cGMP production in rat brain, the *in vitro* incubated brain slice became the method of choice (9, 20). In a detailed study on the localization of ANP-responsive pGNC in the rat brain, it was found that ANP-responsive, cGMP-producing cells were present in discrete regions in the CNS (Fig. 1E) (31). A large majority of these cells were characterized as astrocytes, and only in a few instances was a response to ANP found in neuronal cells (31). In the moth *Manduca sexta* cGMP production was visualized in identified CNS neurons following exposure to the insect peptide eclosion hormone (32).

Nitric oxide-mediated cGMP production was found to be present in varicose fibers throughout the CNS, with only a few responsive cell somata (Fig. 1C,D) (26, 28). The exceptions to this statement are the olfactory bulb (Fig. 1D) and the cerebellum (Fig. 1F). In addition there are only a few brain regions in which only a small number of varicose fibers was observed, for example, the substantia nigra pars reticularis and the pallidum. It has been demonstrated that NO functions as an intercellular messenger in the rat brain (18, 20), as originally proposed by Garthwaite *et al.* (33, 34).

In the immature rat brain, the NO–cGMP signal transduction pathway seems to be expressed differently, as cGMP is found in high levels in cell somata and dendrites in all brain regions (Fig. 1A,B). After postnatal day 10 this pattern in the localization of cGMP slowly changes into the situation found in the adult (17, 20). The underlying molecular mechanism of this change in the functioning of the NO–cGMP signal transduction pathway is not yet clear.

Enteric Nervous System

The gastrointestinal tract was one of the first locations in which it was demonstrated that NO acts as a neurotransmitter (35). Several biochemical studies have shown that NO donors and stimulation of enteric inhibitory nerves both cause increases in cGMP levels in the gastrointestinal tract (36). Cyclic GMP–ICC has revealed the cell types within the intestinal wall responsible for the increase in cGMP levels after tissue exposure to NO, NO donors, or electrical stimulation (29, 30). Following electrical stimulation or exposure to NO, vascular smooth muscle cells and pericytes in the intestinal wall show strong cGMP immunoreactivity (cGMP-IR; Fig. 2A). Cyclic GMP is also induced in subpopulations of neurons in the myenteric (Fig. 2B,H) and submucous plexus (Fig. 2D), and in interstitial cells of Cajal (Fig.

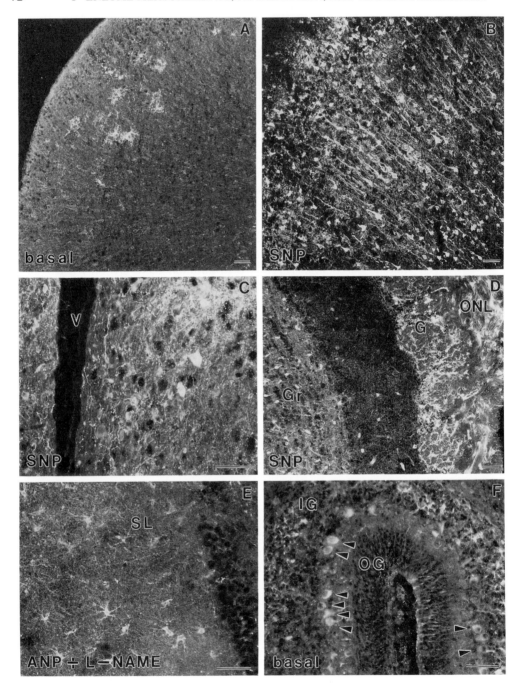

2B). In contrast to many parts of the CNS, no cGMP-IR is observed in enteric glial cells. In the submucous plexus, double-label studies revealed that cGMP-IR is localized exclusively in neurons containing vasoactive intestinal peptide (VIP) (Fig. 2D,E). This finding represents the first demonstration of cGMP being localized to a functionally defined class of neuron. The VIP neurons are surrounded by many NOS-containing nerve terminals (Fig. 2F, G). Ultrastructural studies have revealed that NOS-containing neurons receive many synaptic inputs from NOS-containing nerve terminals (37, 38), suggesting that some enteric neurons could be both sources and targets of NO. However, in tissue processed to reveal both cGMP-IR and NADPH diaphorase (NOS) activity, no double-labeled neurons are observed (Fig. 2H,I). This indicates either that NO released from many nerve terminals in myenteric ganglia might not play a role in transmission to adjacent neurons, or that NO acts through a cGMP-independent mechanism at many myenteric neuronal–neuronal connections. Consistent induction of cGMP in intestinal smooth muscle cells is only observed following pretreatment with both of the phosphodiesterase inhibitors, isobutylmethylxanthine (IBMX) and M&B 22948 (2-o-propoxyphenyl-δ-azapurin-6-one, Sigma) (30).

Materials and Methods

Coupling of Cyclic GMP to Carrier Proteins

The cGMP conjugate for immunization is prepared in a reaction mixture consisting of 0.1 M phosphate buffer (pH 6.0) that contains 1 mg/ml carrier protein, for example, bovine thyroglobulin, 5 mM cGMP, and 4% (w/v) freshly depolymerized paraformaldehyde. The reaction is performed over-

---

FIG. 1 Cyclic GMP immunoreactivity (cGMP-IR) in areas of the rat CNS after incubation *in vitro* followed by formaldehyde fixation and cryostat sectioning. (A, B) Parietal cortex of the rat at postnatal day 8. Slices were incubated in the absence of drugs (A) or in the presence of 1 mM isobutylmethylxanthine (IBMX) and 0.1 mM sodium nitroprusside (SNP) (B). (C) Area of the paraventricular nucleus. Slice was incubated in the presence of 1 mM IBMX and 0.1 mM SNP. V, Third ventricle. (D) Slice from the olfactory bulb incubated in the presence of 1 mM IBMX and 0.1 mM SNP. G, Glomerular layer; Gr, granule cell layer; ONL, olfactory nerve layer. (E) Hippocampal slice incubated in the presence of 1 mM IBMX, 1 μM ANP, and 0.1 mM N^G-nitro-L-arginine as an NOS inhibitor. SL, Stratum lacunosum. (F) Slice from the cerebellar cortex at postnatal day 8 incubated in the presence of 1 mM IBMX. IG, Internal granule cell layer; OG, outer granule cell layer. Darts point to Purkinje cells. Bars: 20 μm.

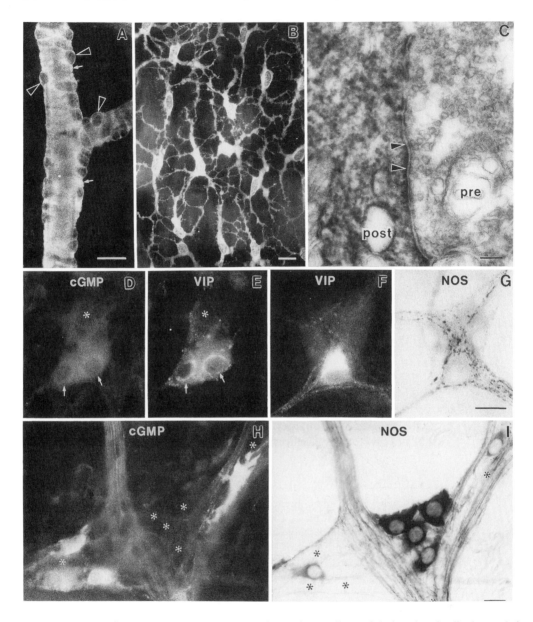

FIG. 2   (A) Whole-mount preparations of a small arteriole in a longitudinal muscle/ myenteric plexus preparation from guinea pig colon following exposure to SNP and processing for cGMP-immunohistochemistry. Smooth muscle cells (white arrows) and pericytes (arrowheads) show cGMP-IR. Bar: 20 μm. (B) Whole-mount prepara-

night at room temperature. The conjugate is purified by dialysis or repeated ultrafiltration at 4°C in 10 mM phosphate buffer (pH 7.0). The mole/mole ratio of cyclic GMP/thyroglobulin is approximately 75 (mean of 12 preparations).

Immunization

Before immunization, preimmune serum of each animal is collected and stored. Rabbits are immunized with 0.5 or 1.0 mg of conjugate emulsified in 0.5 ml of saline and 0.5 ml of Freund's complete adjuvent. The antigen preparation is injected intradermally in eight sites on the back of the animal. Four weeks after priming, the first booster injection (same antigen concentration emulsified in Freund's incomplete adjuvant) is given intramuscularly. Booster injections are repeated every 3 weeks. Sera are tested for cGMP antibodies 10 days after the booster injections. Rabbits are bled when the antisera showed a sufficient response in the gelatin model system. Blood is obtained in tubes without heparin or citrate and left at room temperature for 30 min and thereafter at 4°C overnight. Subsequently the blood is centrifuged for 10 min at 2000 g. The serum is pipetted off and incubated with 3 mg/ml formaldehyde-treated thyroglobulin for 30 min at 37°C and left overnight at 4°C. Thereafter the serum is centrifuged for 10 min at 10,000 g, and the antiserum is decanted and incubated for 30 min at 56°C. The antiserum is stored at −70°C.

---

tion of longitudinal muscle from the guinea pig ileum showing cGMP-IR subserosal interstitial cells of Cajal. Bar: 20 μm. (C) Electron micrograph of synapse (arrowheads) on a cGMP-IR neuron from the myenteric plexus of the guinea pig ileum. The DAB reaction product is not obviously associated with any particular organelle. post, Postsynaptic, cGMP-IR side of synapse; pre, presynaptic side. Bar: 200 nm. (D, E) Paired fluorescent micrographs of a whole mount of a submucous ganglion in guinea pig ileum following exposure to SNP and processing for cGMP (D) or VIP (E) immunohistochemistry. Neurons showing cGMP-IR (arrows) are also VIP-IR. Neurons that do not show cGMP-IR (asterisk) also are not VIP-IR. (F, G) Paired micrographs of a whole mount of a submucous ganglion in the guinea-pig ileum following processing for VIP immunohistochemistry (F) and NADPH diaphorase histochemistry (G) to reveal NOS. A VIP-immunoreactive neuron is surrounded by many NOS-containing nerve terminals. Bar (D–G): 25 μm. (H, I) Paired micrographs of a myenteric ganglion from the guinea pig colon. The tissue was processed for cGMP-IR, photographed, and then subsequently processed for NADPH diaphorase activity. None of the neurons that showed cGMP-IR contained NOS. The locations of the NOS-containing neurons are indicated in H by white asterisks, and the locations of the cGMP-IR neurons in I are indicated by black asterisks. Bar: 20 μm.

Serum Specificity Studies Using Gelatin Model System

The gelatin model system used was described by Schipper and Tilders (6). A solution of 10% (w/v) gelatin in distilled water is prepared. Solutions of cAMP, ATP, ADP, AMP, adenosine, cGMP, GTP, GDP, GMP, guanosine, and NADP (0.2 μM to 10 mM) are made up in 0.2 M phosphate buffer (pH 7.4) and 2 M sucrose, and then mixed with equal volumes of the warm gelatin solution. Aliquots of these solutions are pipetted into Tygon tubing. Tubes are placed at 4°C for 1 hr; thereafter, the gels are fixed for 16 h at room temperature in a solution of 4% (w/v) freshly depolymerized paraformaldehyde in 0.1 M phosphate buffer (pH 7.4) containing 1 M sucrose. The gels are removed from the tubing, embedded in Tissue Tek OCT compound (Miles, Elkhart, IN) and frozen in isopentane prechilled with liquid nitrogen. Using a cryostat, 10 μm thick sections are cut and thawed onto glass slides coated with chrome–alum gelatin. Slides are dried at room temperature and are washed three times, 5 min each time, in Tris-buffered saline (TBS), pH 7.4. Routinely each slide is incubated for 18 hr at 4°C with antiserum diluted 1 : 300 in TBS containing 0.3% Triton X-100 (TBS-T). Thereafter slides are washed at least three times, 15 min each time, with TBS and incubated for 45 min at room temperature with fluorescein isothiocyanate- or CY3-conjugated sheep anti-rabbit immunoglobulins. Finally, slides are washed again as described and mounted in TBS–glycerin (1 : 3, v/v).

In addition, specificity is assessed by applying preimmune sera to the gelatin sections at the same dilutions as used for the antisera. No immunofluorescence is observed with the preimmune sera. Further, cGMP antisera are appropriately diluted in TBS-T and incubated in the presence of 10 mM cGMP or the cGMP–formaldehyde–thyroglobulin conjugate (1 mg/ml). These mixtures are incubated for 1 hr at 37°C, followed by 18 hr incubation at 4°C. Possible precipitates are removed by centrifugation at 10,000 g for 15 min at 4°C. The preabsorbed antisera are applied to the gelatin section as described above.

As a second model system the spot–blot model system in combination with vapor fixation of cGMP can be used. For this, 2 mg/ml of thyroglobulin in distilled water is mixed 1 : 1 (v/v) with each concentration of a serial dilution of the nucleotides described above, resulting in a solution containing 1 mg/ml thyroglobulin which is 1 μM to 10 mM in the respective nucleotides. Aliquots of 3 μl are spotted onto strips of nitrocellulose paper and air dried. Strips are vapor fixed with formaldehyde for 4 hr at 75°C. Subsequently the strips are soaked for 1 hr in a solution of 1% bovine serum albumin (BSA) to block antibody binding to the nitrocellulose. After washing three times, 5 min each, with TBS, the strips are incubated overnight at 4°C or for 1 hr at 37°C with the respective antibodies diluted in TBS. All strips are washed three times, 15 min each, in TBS, TBS-T, and TBS, respectively, and then incubated 30 min with the secondary antibody. After being subjected to the same washing procedure a second time, the strips are incubated with rabbit peroxidase-antiperoxidase complex for 30 min. After washing immunoreactivity is made visible by means of a diaminobenzidine reaction.

In Vitro Incubation of Biological Tissue and Visualization of cGMP Immunoreactivity

Depending on the tissue used, slight modifications may be necessary for the following *in vitro* procedure, which was devised for brain tissue slices. After decapitation, brains are removed immediately and transferred into ice-cold incubation buffer of the following composition (mM): NaCl, 121.1; KCl, 1.87; KH_2PO_4, 1.17; $MgSO_4$, 1.15; $NaHCO_3$, 24.9; $CaCl_2$, 1.2; glucose, 11.0; aerated with 5% CO_2/95% O_2 (v/v) (pH 7.4). Incorporation of a phosphodiesterase inhibitor is a matter of choice; good results may be obtained by using 1 mM IBMX (Fig. 1A,B,F). A combination of phosphodiesterase inhibitors may be necessary to obtain a good response (30). Brain tissue is placed on damp filter paper on a precooled (4°C) stainless steel table, and transverse slices (300 μm thick) are chopped using a McIlwain tissue chopper. With the aid of a microscope, separate slices from each other, while keeping the slices submerged in ice-cold Krebs–Ringer bicarbonate buffer (KRB), aerated with 5% CO_2/95% O_2 (v/v). Slices are stored in a multiwell tissue culture chamber and are kept on ice during this procedure. Thereafter slices are brought slowly to room temperature and subsequently to the incubation temperature of 37°C. Routinely, brain slices may be incubated for 30 min in KRB at 37°C before adding drugs. 3′,5′-Cyclic nucleotide phosphodiesterase inhibitors are usually present from the start of the incubation. Similarly, inhibitors of NOS like N^G-nitro-L-arginine, N^G-methyl-L-arginine, and 7-nitroindazole may be included in the KRB. The incubation time in the presence of NO donor compounds and drugs stimulating or inhibiting NOS will be dependent on the compound used and need to be established. An incubation time of 10 min in the presence of 0.1 mM sodium nitroprusside as an NO donor is sufficient to elicite a strong cGMP response.

Incubations are terminated by adding the ice-cold fixative solution to the incubation medium (final concentration 4% formaldehyde, 5% sucrose, and 0.1 M phosphate, pH 7.4). The formaldehyde needs to be freshly prepared by depolymerizing paraformaldehyde. Fixation time is 2 hr, followed by 1.5 hr wash in cold TBS containing 5–15% sucrose (pH 7.4). Immediately thereafter 10 μm thick sections are cut on a cryostat, and then thawed onto chrome–alum gelatin-coated slides. The slides are stored at −20°C. For immunohistochemistry the slides are processed as described above.

Electron Microscopy

After fixing the tissue as described above, the tissue is washed in 50% (v/v) ethanol (four times, 10 min each), left overnight in 0.1 *M* phosphate buffer, and then placed in 0.1% sodium cyanoborohydride in 0.1 *M* phosphate buffer for 30 min. Subsequently the tissue is washed with several changes of phosphate-buffered saline (PBS) and processed for cGMP immunocytochemistry using the PAP/DAB method to visualize the cGMP. The tissue is postfixed in 1% glutaraldehyde in 0.1 *M* phosphate buffer, pH 7.3, for 1 hr, washed in PBS, exposed to 1% OsO_4 in phosphate buffer for 1 hr, washed in distilled water, stained with 2% uranyl acetate, dehydrated through graded ethanol solutions, and embedded flat in Araldite. Ultrathin sections are cut and stained with lead citrate.

Acknowledgment

The research described in this contribution was partly supported by the National Health and Medical Research Council of Australia (HMY).

References

1. A. L. Steiner, C. W. Parker, and D. M. Kipnis, *J. Biol. Chem.* **247**, 1106 (1972).
2. H. J. Wedner, B. J. Hoffer, E. Battenberg, A. L. Steiner, C. W. Parker, and F. E. Bloom, *J. Histochem. Cytochem.* **20**, 293 (1972).
3. R. A. Ortez, R. W. Sikes, and H. G. Sperling, *J. Histochem. Cytochem.* **28**, 263 (1980).
4. R. Cumming, S. Dickison, and G. Arbuthnott, *J. Histochem. Cytochem.* **28**, 54 (1980).
5. A. L. Steiner, S. H. Ong, and H. J. Wedner, *Adv. Cyclic Nucleotide Res.* **7**, 115 (1976).
6. J. Schipper and F. J. H. Tilders, *J. Histochem. Cytochem.* **31**, 12 (1983).
7. C. Milstein, B. Wright, and C. Cuello, *Mol. Immunol.* **20**, 113 (1983).
8. T. W. Rall and R. A. Lehne, *J. Cyclic Nucleotide Res.* **8**, 243 (1982).
9. J. de Vente, H. W. M. Steinbusch, and J. Schipper, *Neuroscience (Oxford)* **22**, 361 (1987).
10. V. Chan-Palay and S. L. Palay, *Proc. Natl. Acad. Sci. U.S.A.* **76**, 1485 (1979).
11. J. de Vente, J. Schipper, and H. W. M. Steinbusch, *Histochemistry* **99**, 457 (1993).
12. F. Berkenbosch and F. J. H. Tilders, *Neuroscience (Oxford)* **23**, 823 (1987).
13. J. de Vente, J. Schipper, and H. W. M. Steinbusch, *Histochemistry* **91**, 401 (1989).
14. R. L. Chevalier, R. J. Fern, M. J. Garmey, S. S. El-Dahr, R. A. Gomez, and J. de Vente, *Am. J. Physiol.* **262**, F417 (1992).
15. H. S. Berkelmans, G. A. Burton, J. Schipper, and J. de Vente, *Histochemistry* **96**, 143 (1991).

16. J. de Vente, J. Garssen, F. J. H. Tilders, H. W. M. Steinbusch, and J. Schipper, *Brain Res.* **411,** 120 (1987).

17. J. de Vente, J. G. J. M. Bol, H. S. Berkelmans, J. Schipper, and H. W. M. Steinbusch, *Eur. J. Neurosci.* **2,** 845 (1990).

18. E. Southam, R. Morris, and J. Garthwaite, *Neurosci. Lett.* **137,** 241 (1992).

19. J. de Vente, J. J. Wietsma, J. G. J. M. Bol, J. Schipper, S. K. Malhotra, and H. W. M. Steinbusch, *Neurochem. Res. Commun.* **4,** 25 (1989).

20. J. de Vente and H. W. M. Steinbusch, *Acta Histochem.* **92,** 13 (1992).

21. Z. Z. Wang, L. J. Stensaas, J. de Vente, B. Dinger, and S. J. Fidone, *Histochemistry* **96,** 523 (1991).

22. D. L. Garbers, *J. Biol. Chem.* **264,** 9103 (1989).

23. R. G. Knowles and S. Moncada, *Biochem. J.* **298,** 249 (1994).

24. E. Southam and J. Garthwaite, *Neuropharmacology* **32,** 1267 (1993).

25. J. A. Ferrendelli, *Adv. Cyclic Nucleotide Res.* **9,** 453 (1978).

26. J. Zwiller, M. S. Ghandour, M. O. Revel, and P. Basset, *Neurosci. Lett.* **23,** 31 (1981).

27. I. Matsuoka, G. Giuili, M. Poyard, D. Stengel, J. Parma, G. Guellaen, and J. Hanoune, *J. Neurosci.* **12,** 3350 (1992).

28. T. Furuyama, S. Inagaki, and H. Takagi, *Mol. Brain Res.* **20,** 335 (1993).

29. H. M. Young, K. McConalogue, J. B. Furness, and J. de Vente, *Neuroscience (Oxford)* **55,** 583 (1993).

30. C. W. Shuttleworth, C. Xue, S. M. Ward, J. de Vente, and K. M. Sanders, *Neuroscience (Oxford)* **56,** 513 (1993).

31. J. de Vente, J. G. J. M. Bol, and H. W. M. Steinbusch, *Eur. J. Neurosci.* **1,** 436 (1989).

32. J. Ewer, J. de Vente, and J. Truman, *J. Neurosci.* **14,** 7704 (1994).

33. J. Garthwaite, *Trends Neurosci.* **14,** 60 (1991).

34. J. Garthwaite, S. L. Charles, and R. Chess-Williams, *Nature (London)* **336,** 385 (1988).

35. H. Bult, G. E. Boeckxstaens, P. A. Pelckmans, F. H. Jordaens, Y. M. Van Maercke, and A. G. Herman, *Nature (London)* **345,** 346 (1990).

36. M. Barnette, T. J. Torphy, M. Grous, C. Fine, and H. S. I. Ormsbee, *J. Pharmacol. Exp. Ther.* **249,** 524 (1989).

37. I. J. Llewellyn-Smith, Z. Song, M. Costa, D. S. Bredt, and S. H. Snyder, *Brain Res.* **577,** 337 (1992).

38. H. M. Young, J. B. Furness, and J. M. Povey, *J. Neurocytol.* **24,** 257 (1995).

[8] Detection of Downstream Components of Cyclic GMP Cascade by *in Situ* Hybridization Analysis

Michael F. Goy, Zhiping Li, Ellen M. Zimmermann, and David R. Repaske

Introduction

It has been known for many years that cyclic GMP levels can be regulated by depolarizing agents and neurotransmitters in the mammalian brain (1–5). It is now apparent that the enzyme nitric oxide synthase (NOS; SEC 1.14.13.39) is one of the key molecular mediators of this process. A constitutive form of NOS is widely distributed in the brain, and its diffusible reaction product (nitric oxide) is a potent stimulator of the cytoplasmic form of guanylate cyclase. Allosteric activation of NOS by calcium permits tight coupling of depolarization- or transmitter-triggered changes in the level of cytoplasmic calcium to downstream changes in the level of cyclic GMP.

Detailed histochemical and immunocytochemical mapping studies have provided a complex picture of the central nervous system distributions of NOS and the cytoplasmic guanylate cyclase (6–8). However, in considering possible functions of cyclic GMP in specific populations of neurons, it is equally important to map the distributions of the members of a growing list of cyclic GMP-regulated effector molecules. These include two isoforms of the cyclic GMP-dependent protein kinase (PKG) (9, 10), two isoforms of the cyclic GMP-inhibited phosphodiesterase (cGI-PDE) (11), at least one form of the cyclic GMP-stimulated phosphodiesterase (cGS-PDE) (12), a family of cyclic GMP-activated channels (13), and at least one cyclic GMP-inhibited channel (14). This chapter describes the use of *in situ* hybridization histochemistry as a tool for evaluating the distributions of such molecules by localizing expression of the mRNAs encoding them.

The technique of *in situ* hybridization for the detection of specific mRNA species in thin-sectioned tissue was first introduced in 1981 (15). Since then, the method has been used widely and successfully to map the expression of genes in a variety of tissues obtained from vertebrate and invertebrate species (16–20). Indeed, because of powerful advances in gene cloning methodologies it is frequently easier and faster to develop a nucleotide probe than it is to develop an antibody. We have entered an era where expression of the product

Methods in Neurosciences, Volume 31

of an interesting gene is often characterized at the mRNA level before it is characterized at the protein level. This idea is well exemplified by the cyclic GMP effector molecules: only the type I isoform of PKG has been character- ized immunocytochemically in nervous tissue (21), whereas the mRNA ex- pression patterns of the type II isoform of PKG (22, 23), the cGS-PDE (24), two cGI-PDE isoforms (11), and at least two different cyclic GMP-activated channels (25, 26) have all been investigated in some detail.

There are many parallels between *in situ* hybridization and the more famil- iar techniques of immunocytochemistry. Both methods are particularly useful for identifying anatomical sites of expression, and even for revealing expres- sion at the level of single cells. Both methods rely on probes that are capable of specifically recognizing a target molecule of interest. For immunocyto- chemistry, the probe is a monoclonal or polyclonal antiserum that interacts with epitopes in a target protein, whereas for *in situ* hybridization the probe is a polynucleotide that is complementary in sequence to a target mRNA. For both methods, the target molecule must be immobilized in a tissue section using fixation techniques that preserve both the morphology of the tissue and the ability of the target molecule to interact with the probe. Finally, users of both methods must pay particular attention to the specificity of probe–target interactions. Potential problems include nonspecific binding of the probe to "sticky" regions of the tissue and specific but incidental interactions of the probe with cross-reacting molecules that share regions of common structure with the target of interest. Specificity controls are critically important in studies of this type, a topic that is considered in more detail below.

The *in situ* hybridization technique must also take into account two issues that do not pertain to immunocytochemical studies. First, gene expression does not necessarily imply protein expression. Thus, without independent corroboration from Western blots, immunocytochemical studies, or func- tional assays (e.g., for enzyme activity), a positive hybridization signal can- not by itself be taken as evidence that a cell expresses a particular protein. This sets limits on the ability to use *in situ* hybridization data to infer the functional status of a cell. Second, even if an mRNA species is known to be translated into protein, the subcellular localization of that protein cannot be predicted from the distribution of the mRNA that encodes it. This is an important issue for any type of cell, but it is particularly acute for cells with highly polarized morphologies, like neurons, because proteins and peptides can be selectively targeted to parts of the cell that are located very far away from the cell bodies where the mRNA hybridization signal is actually de- tected.

Despite these caveats, *in situ* hybridization has proved quite useful in investigating tissue- and cell-specific expression of a variety of genes. This

chapter provides experimental details for an ^{35}S-riboprobe-based hybridization technique that has been used productively in our laboratories, and taught
successfully to many colleagues. The method offers both high sensitivity
and resolution adequate to localize hybridization signals to specific brain
regions (and in some cases to specific cells). However, the reader is urged
to consult reviews of alternate methods that can be employed in situations
calling for enhanced sensitivity [as required for rare transcripts (27, 28)] or
enhanced resolution [as required for definitive single-cell localization
(28–32)].

Unfortunately, many of the basic techniques required for microscopy and
recombinant DNA technology are beyond the scope of this review. Compre-
hensive volumes, such as *Histopathologic Methods* (33), *Light Microscopy
in Biology* (34), *Current Protocols in Molecular Biology* (35), and *Molecular
Cloning* (36), provide helpful background in these areas. In addition, the
Appendix provides recipes and detailed procedures required for several of
the methods described in this chapter. These appear in the order that they
are referred to in the text.

A Note Concerning RNA

In situ hybridization involves the use of a nucleotide probe to detect mRNA
in a tissue section. RNases are highly active and highly stable enzymes that
degrade RNA and are present in most biological tissues. It is therefore critical
to freeze tissues used for *in situ* hybridization procedures immediately on
harvesting, and to store them continuously at −80°C. Otherwise, endogenous
RNases may quickly destroy the signal you are trying to detect.

In addition, it is essential to guard against inadvertent introduction of
RNases from external sources into reagents, glassware, and tissue sections.
RNases present on human skin are readily transferred through physical
contact. RNases are also frequently employed in molecular biology protocols
and thus can easily be present and enzymatically active throughout the
laboratory as invisible spots of contamination. (Many laboratories restrict
the use of RNases to specific areas and pieces of glassware.) Several basic
precautions are mandatory. Gloves should be worn at all times and changed
frequently. All glassware should be autoclaved or pretreated with commercial
irreversible RNase inhibitors (such as RNasezap, available from Ambion,
Austin, TX), and all water should be treated before use with 0.1% diethyl
pyrocarbonate (DEPC), which irreversibly denatures RNases, followed by
autoclaving to denature the DEPC (see Appendix). It is helpful to cover
working surfaces with clean bench coverings. Use individually wrapped
pipettes to make solutions. We have found that gloves, bench coverings,

and plasticware that have not been touched by bare hands arrive from the manufacturer essentially RNase-free.

Choice of Probe for *in Situ* Hybridization Reactions

The main types of probes that have been used for *in situ* hybridization are enzymatically synthesized cRNA and cDNA probes (the former are often referred to as riboprobes) and manually synthesized oligonucleotides. This review focuses on riboprobes, but in this section we also briefly consider some of the salient properties of the other types of probes.

Riboprobes are highly sensitive and selective reagents for the detection of specific mRNA species. Because riboprobes are generated from plasmid templates they can be produced in essentially unlimited quantities. The optimal length for these probes is 200–500 bases: this provides a high degree of specificity while still permitting adequate tissue penetration. Riboprobes are most commonly synthesized from radioactive ribonucleotides using double-stranded vectors that contain strand-specific bacteriophage promoters and a cDNA insert corresponding in sequence to the target mRNA (Fig. 1); this allows probe sequences corresponding to the antisense (noncoding) and sense (coding) strands to be produced independently (16). The antisense probe is then used to determine the distribution of the immobilized target RNA in the tissue, while the sense probe is used to assess nonspecific probe–tissue interactions.

In contrast, when cDNA probes are synthesized, both the antisense and sense strands are produced simultaneously, and, because these strands tend to anneal to each other at least as readily as to the immobilized target RNA, the sensitivity of the hybridization reaction is greatly decreased. Furthermore, RNA–DNA complexes are less stable than RNA–RNA complexes, so hybridizations performed with cDNA probes must be washed at lower temperatures than hybridizations performed with riboprobes, leading to higher levels of nonspecific binding. For these reasons, riboprobes are heavily favored over cDNA probes for most *in situ* applications.

Manually synthesized oligonucleotides are also usually considered less desirable than riboprobes. The cost of a synthetic oligonucleotide increases linearly with length, but yield decreases geometrically. Taking this into consideration, it is usually most practical to synthesize oligonucleotides in the range of 20–50 bases. Because the strength of the probe–target interaction is proportional to length, short oligonucleotide probes must be hybridized and washed under less stringent conditions than those used with longer riboprobes, resulting in higher backgrounds and lower sensitivity. In addition, use of a short probe increases the likelihood of cross-reactivity with

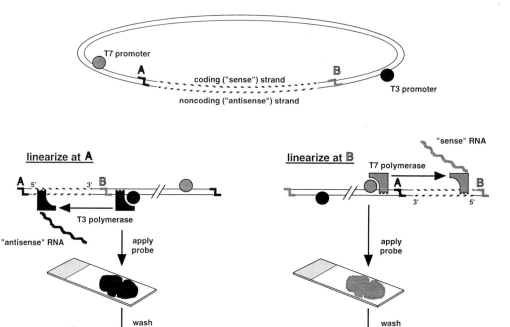

FIG. 1   Production of sense and antisense riboprobes. (Top) Schematic diagram of a double-stranded plasmid containing a cDNA insert that encodes the target sequence of interest (dashed lines). The plasmid is propagated in and isolated from *Escherichia coli*. To make a riboprobe, either strand of the cDNA insert can be transcribed into RNA by an RNA polymerase enzyme moving along that strand in the 3' to 5' direction. Because the plasmid contains strand-specific RNA polymerase promoter sites (gray and black balls), each strand can be transcribed independently, simply by adding the appropriate polymerase enzyme. Before RNA transcription, the plasmid is linearized by restriction endonuclease digestion at the distal end of the insert, to ensure that the RNA product contains primarily an insert sequence. Many different kinds of plasmids are available for riboprobe production; note that each type will have its own characteristic set of promoter and endonuclease restriction sites. In this example, as shown at left, T3 RNA polymerase produces an antisense RNA beginning at the T3 RNA polymerase promoter, after restriction endonuclease cleavage of the plasmid at site A. Similarly, as shown at right, a sense riboprobe is produced using T7 RNA polymerase and the T7 RNA polymerase promoter, after restriction endonuclease cleavage of the plasmid at site B. (Bottom right) Sense riboprobe is applied to tissue sections but fails to hybridize to target mRNA because its sequence is identical (not complementary) to the sequence of the target mRNA. After washing,

irrelevant RNA species that happen to share a small region of sequence homology with the target RNA. Despite these disadvantages, a synthetic oligonucleotide is often the probe of choice in a laboratory that is not accustomed to handling bacteria and plasmids. Synthetic oligonucleotide sequences are completely user-defined, and a probe can be generated without the need to obtain, propagate, purify, and manipulate a vector containing a cloned nucleotide sequence. Oligonucleotide probes are conveniently labeled by using T4 polynucleotide kinase to transfer a single radioactive phosphate from γ-labeled ATP to the 5' terminus or by using terminal deoxynucleotidyltransferase to catalyze the addition of radioactive deoxynucleotides to the 3' terminus (35, 36).

Along with choosing the type of probe, it is also crucial to determine the optimal sequence of the probe. Multigene families, alternately spliced mRNAs, and conserved functional domains are so common that the issue of cross-reactivity cannot be ignored. The first prerequisite is to subject the sequence of the target RNA to a homology search on a computerized database. Optimally, the probe sequence should be chosen from a region of the target that has little or no homology to other known genes. If splice variants have been characterized, then specific sequences should be selected to distinguish among them. As additional considerations, it is useful to avoid sequences that contain internal palindromes (because they will self-anneal) and sequences with GC contents above 60% (because they tend to bind to RNA relatively nonspecifically).

Once a suitable probe sequence has been chosen, its cross-reactivity can be evaluated by applying the probe to a Northern blot of RNA extracted from the tissue of interest; if the probe is absolutely specific, it should only hybridize to a single species of RNA of the appropriate size. In practice, many probes detect several alternately spliced forms of the target mRNA. If this is revealed by the Northern blot, it must be kept in mind when interpreting *in situ* results. If possible, the probe should be tested on Northern blots in parallel with probes for related mRNAs to ensure that each probe is specific for a single target mRNA (Fig. 2). A Northern blot will also provide

---

any probe that remains associated with the section (depicted in gray) is likely to be nonspecifically bound. (Bottom left) Antisense riboprobe is applied to tissue sections. In addition to nonspecific interactions, the probe also remains hybridized to complementary target mRNA sequences during the wash; that is, the probe associated with the section (depicted in black) represents the sum of nonspecific and specific binding. Thus, to recognize an authentic hybridization signal, the first important control involves side-by-side comparison of sense- and antisense-hybridized sections.

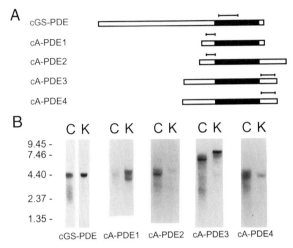

FIG. 2   Evaluating probe specificity when the target mRNA encodes a protein that belongs to a family of related proteins. (A) Diagram of the domain structure of the rat cGS-PDE and four cAMP-specific PDE isozymes (cA-PDE1, cA-PDE2, cA-PDE3, and cA-PDE4). Each shares a highly conserved 270 amino acid catalytic domain, indicated by the solid portion of the bar. The cGS-PDE riboprobe was derived from the nucleotides encoding a portion of this catalytic domain, as indicated by the line above the bar. The cAMP-specific PDE probes were designed from nucleotide sequences outside of the catalytic domain, unique for each isozyme. (B) Antisense riboprobes specific for these five PDE isoforms were hybridized to Northern blots of rat cerebral cortex (C) and kidney (K). The pattern of hybridization to these tissues is unique for each probe, demonstrating that each probe does not detect the mRNA for the other related enzymes under the conditions of this experiment. Molecular sizes (in kilobases) are given at left.

information about the abundance of the target mRNA in the tissue of interest, which can be helpful for troubleshooting.

Choice of Isotope to Use for Labeling *in Situ* Hybridization Probes

The isotopes most commonly used for labeling riboprobes for *in situ* hybridizations are ^{35}S, ^{33}P, and [3]H. Each has advantages and disadvantages. Sulfur-35 is used most frequently, and it is the isotope that we routinely employ in our own work. It has a relatively long half-life (87 days), and can therefore be used to synthesize probes for several months after purchase.

It emits a moderately energetic β particle (0.17 MeV) that provides a strong signal on both X-ray film and emulsion-coated slides. The energy of the released particle is absorbed by the emulsion within a few hundred micrometers, thereby providing reasonably good spatial resolution. However, the signal from an ^{35}S-labeled probe does spread somewhat away from the source, making it difficult to evaluate mRNA expression at the single-cell level. Phosphorus-33 has an emission intensity comparable to that of ^{35}S (0.25 MeV), and therefore provides similar spatial resolution and signal strength. An advantage of ^{33}P is that it often generates a probe with lower background binding than does ^{35}S, though the short half-life (25 days) and high cost of ^{33}P make it somewhat less economical than ^{35}S. For very high resolution, in order to determine with precision which cell or which portion of a cell contains the target mRNA, ^{3}H can be employed. However, the low energy (0.02 MeV) of the ^{3}H β particle requires very long exposure times, especially for transcripts expressed at low levels, and detection on X-ray film is not possible.

Labeling of Riboprobes with ^{35}S

In our laboratories, the protocol described below yields primarily full-length ^{35}S-labeled transcripts of high specific activity. Note that it is also possible to purchase commercial labeling kits from a variety of suppliers that produce high quality riboprobes using very similar techniques. Before starting this procedure, a plasmid must be obtained and appropriately linearized. General principles of linearization are described in the legend to Fig. 1, and more complete details can be found elsewhere (35, 36).

Nucleotide cocktail

> 1.0 μl 10 mM GTP
> 1.0 μl 10 mM ATP
> 1.0 μl 10 mM CTP
> 0.2 μl 10 mM UTP (optional; including a small amount of cold UTP improves the production of full-length transcripts)
> 3.3 μl 100 mM dithiothreitol (DTT)
> 2.3 μl H$_2$O (DEPC-treated)
> Add at the last minute 1.7 μl RNasin (40 units/μl)

Reaction mix

> 100 μCi [^{35}S] UTP ($>$1000 Ci/mM) before proceeding, dry without heating in a Speed-Vac

3.0 μl nucleotide cocktail

1.0 μl linearized plasmid (1 μg)

1.0 μl 5× transcription buffer (usually provided by the supplier of the polymerase enzyme)

Add at the last minute 1.0 μl RNA polymerase [use SP6, T7, or T3, as appropriate (Fig. 1); 15 units/μl]

1. Nucleotide and dithiothreitol stock solutions should be prepared in advance (or purchased commercially) and stored at −20°C. Use them to make up fresh nucleotide cocktail on the day that the labeling reaction will be performed. Then use the cocktail to prepare the reaction mix. Follow safe radiation handling procedures.

2. Incubate the reaction mix at 37°C for 45–60 min.

3. Place the reaction tube at room temperature and add 45 μl of DEPC-treated water. Separate labeled riboprobe from unincorporated label on a Quick-Spin column (Boehringer Mannheim, Indianapolis, IN), following the manufacturer's instructions. The probe will be eluted in a final volume of approximately 50 μl.

4. Add 0.5 μl of 5 M DTT (final concentration 50 mM) and freeze aliquots of probe at −80°C until needed. We prefer to use labeled probe within 2 days, but have performed successful hybridizations with probes that have been stored for up to 7 days.

5. To evaluate the success of the labeling reaction, mix 1 μl of labeled probe with scintillation fluid and determine the counts per minute (cpm) in a scintillation counter. Convert counts per minute to disintegrations per minute [dpm = cpm ÷ (% counter efficiency × 0.01)]. For a typical 300-base pair probe, if labeling and probe purification have been successful, recovery will be on the order of 1 to 6 × 10^6 dpm/μl.

6. To check probe integrity, mix 1 μl of labeled probe with 10 μl of sample buffer and electrophorese at constant voltage (100 V) on a 5% acrylamide 8 M urea gel until the bromphenol blue tracking dye has migrated two-thirds of the way to the bottom of the gel (for a detailed description of urea gel electrophoresis, consult Ref. 36; the solutions that we use for preparing gels are given in the Appendix). Turn off the power, separate the gel plates, transfer the gel to Whatman (Clifton, NJ) 3MM filter paper, dry under vacuum (optional), wrap in plastic wrap, and expose to X-ray film at room temperature. A high quality labeling reaction (Fig. 3A) will produce a sharply defined radioactive band of the appropriate size. A smear or no predominant band usually indicates probe degradation. The presence of bands that are larger than the expected size may indicate incomplete linearization of the plasmid template. The presence of bands that are smaller than the expected size indicates incomplete transcript

FIG. 3   Analysis of riboprobe by denaturing polyacrylamide gel electrophoresis. (A) cGS-PDE riboprobe that is homogeneous and full-length. (B) cGS-PDE riboprobe with a predominance of partial-length product and some full-length product. See text for full discussion.

formation (Fig. 3B). These smaller transcripts are generally not a problem as long as the predominant transcript is of the expected size. If the predominant transcript is smaller than the expected size, you may have accidentally chosen to linearize the plasmid with a restriction enzyme that has cut at a site within the probe sequence. Alternatively, the plasmid may not contain the insert that you think it does. This should be resolved before proceeding further.

Tissue Preparation

Fixation and Sectioning

Aldehyde fixatives provide good retention of cellular RNA and inactivate or immobilize endogenous RNases that might otherwise degrade the target RNA. We usually apply fixative after the tissue has been sectioned and mounted (a technique called postfixation). Until an unfixed section goes into fixative it must be kept frozen, to avoid degradation of RNA.

Frozen tissues are routinely sectioned on a cryostat. Such fresh frozen sections work quite well for *in situ* hybridization studies; however, they cannot be used very successfully when *in situ* hybridization and immunocytochemical techniques are applied in tandem for colocalization studies. For such studies, tissue should be fixed before it is removed from the animal (37). In addition, frozen sections lack the degree of histological detail that can be obtained with paraffin-embedded tissue. If paraffin sections are required, then tissue must also be obtained from an animal that has been perfused with fixative. Although not described here, established procedures for performing *in situ* hybridization on paraffin sections are available (18).

To Obtain Fresh-Frozen Tissue Sections

1. Freeze freshly dissected brain tissue by placing it directly onto a bed of dry ice. Other tissues, if small, can be embedded in O.C.T. Compound (Miles Inc., Elkhart, IN) in a glass shell vial (Fisher Scientific, Pittsburgh, PA) and frozen in isopentane cooled to between −40° and −50°C. Store frozen tissues at −80°C until they are to be sectioned. Carefully break the shell vial to retrieve the tissue.
2. Use a razor blade to trim the tissue into blocks while it is still frozen. Pay attention to the orientation of the tissue, and trim it so as to expose the region intended for study. Attach each tissue block to a cryostat chuck with OCT. Cut the tissue into 10-μm sections in a cryostat at −20°C and transfer the sections to prepared slides (use Fisher Superfrost Plus slides) or ordinary glass slides that have been subbed (see Appendix). Store the mounted tissue sections at −80°C.
3. Before hybridization, remove slides from −80°C and place them immediately into 4% formaldehyde (see Appendix) for 45–60 min at room temperature.
4. Wash extensively in phosphate-buffered saline (PBS) (at least three times).

Deproteination

Deproteination increases access of the probe to target mRNAs in the tissue. We deproteinate by treating the section with proteinase K. This is followed by exposure to acetic anhydride, which acetylates amino groups in the tissue and reduces electrostatic binding of the probe (38).

To Deproteinate Tissue Sections

1. A proteinase K stock solution (10 mg/ml in distilled, deionized H_2O) should be prepared in advance and stored at $-20°C$. This stock solution can be thawed and refrozen without loss of activity. For each set of slides to be processed, prepare a fresh working solution (1 μg/ml proteinase K in 100 mM Tris-HCl, pH 8.0, 50 mM EDTA) and put it in a standard glass slide-staining dish (Fisher Scientific, Pittsburgh, PA). Two hundred ml is sufficient to completely cover a rack of immersed slides. Prewarm the dish containing the protease solution to 37°C in a covered water bath.

2. Introduce a rack of slide-mounted sections into the dish containing the proteinase K and replace the cover on the water bath. Incubate for exactly 10 min. Longer incubation times (or higher protease concentrations) can cause significant tissue damage. If damage does occur at this step, cut back on the concentration of proteinase K (we have achieved success with levels as low as 0.01 μg/ml).

3. Wash the sections for 1 min with DEPC-treated water.

4. Transfer the sections to a glass slide-staining dish that has been placed on a magnetic stirrer. The dish should contain a stir bar and 200 ml of 0.1 M triethanolamine, pH 8.0. Add 0.5 ml of acetic anhdride (0.25%, v/v) with rapid stirring. Stir for 10 min at room temperature.

5. Wash for 5 min in 2$\times$ SSC (see recipe in Appendix).

6. Dehydrate the sections by passing them through the following ethanol solutions, in sequence: 50, 70, 80, 95, 95, 100, and 100%. Leave the slides in each solution for 1 min. Allow the slides to air dry after removing them from the final 100% solution. (Note: The 100% ethanol solutions need to be water free and should be replaced after use each day; the other ethanol solutions can be reused for a period of several days.) Do not allow sections to sit once they have air dried; move quickly to the hybridization step.

Hybridization Reaction

Hybridization

Radioactive riboprobes are diluted with formamide-containing hybridization buffer to give the desired number of dpm/slide. This varies between 1×10^5 and 3×10^6 dpm/slide, depending on the abundance of the target RNA and the strength of the probe–target interaction. A good starting point for most probes is 2×10^6 dpm/slide. Fifty microliters should be applied to each slide if using 24 × 40 mm coverslips; 30 μl if using 22 × 30 mm coverslips.

In the Appendix are two different recipes for hybridization buffer that differ only in the percentage of formamide. The addition of formamide to the buffer allows hybridization to be performed at lower temperatures while still maintaining high stringency. Generally, 50% formamide buffer is used for short oligonucleotide probes, and 75% formamide buffer is used for riboprobes. However, the appropriate one to use actually depends on the strength of the probe–target interaction, which can be determined numerically, using the formula given in Eq. (1) for calculating the melting temperature (T_m). The T_m varies with salt and formamide concentration, and also varies from probe to probe, depending on probe length and nucleotide composition. For a new probe, choose the hybridization buffer that gives a calculated T_m in the range of 80°–92°C. Then perform the hybridization and wash steps at a temperature that is 30°C below the T_m. For riboprobes, the hybridization and wash temperature should be kept in the range of 50°–62°C. Temperatures below 50°C lead to unacceptably high backgrounds while temperatures above 62°C are likely to destroy the tissue.

To Perform Hybridization Reaction

1. We perform hybridizations in a humidified box, constructed as shown in Fig. 4B. Prepare the box by placing strips of water-soaked Whatman 3MM filter paper into the troughs in the box so that the wet strips will lie on the bottom of the box, underneath the slides.

2. Hybridization buffer is very viscous and requires careful handling. Thaw the buffer, and mix it well by inverting the tube several times. Enlarge the fluid-delivery opening of a disposable 1 ml capacity (blue) plastic pipette tip by cutting a few millimeters off of the end with a clean ethanol-washed razor blade, and slowly draw the calculated amount of hybridization buffer into the tip. Discharge the buffer into a snap-cap microcentrifuge tube. Allow a minute for buffer clinging to the sides of the tip to drain, and add it to the solution in the tube. Add the calculated amount of riboprobe to the tube, vortex well to mix, and centrifuge briefly to bring all the solution to the bottom of the tube.

3. Lay coverslips out in widely spaced columns on clean paper towels. You should include a few more coverslips than the number of sections that you plan to incubate, to allow for errors. Enlarge the fluid-delivery opening of a disposable 200 μl capacity (yellow) plastic pipette tip by cutting a few millimeters off of the end with a clean ethanol-washed razor blade. Use the tip to apply 50 μl of the probe–hybridization mix to the center of each coverslip (again being careful each time to ensure that the viscous

A   B

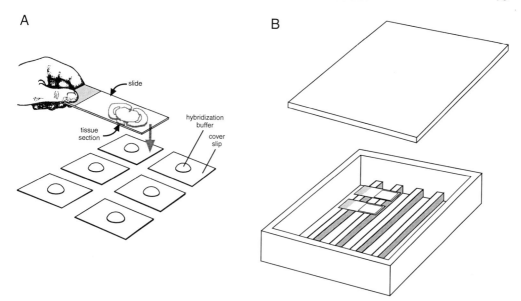

FIG. 4   Hybridization technique. (A) Hybridization buffer, containing sense or anti-sense riboprobe, is placed on coverslips. The slide is held with the fixed tissue section face down and is carefully lowered until the section contacts the hybridization buffer. The coverslip adheres to the slide and allows the hybridization buffer to cover the tissue section in a thin layer. (B) After carefully checking for absence of air bubbles over the tissue section, each slide is placed in a humidified chamber (tissue section facing up). Once the slides have been placed in the chamber, it is sealed to prevent evaporation of the hybridization solution. The entire sealed chamber is incubated at the selected hybridization temperature in an incubator or oven.

material that drains off of the sides of the tip is included). Try to avoid creating bubbles.

4. Grasp a slide at the frosted glass end and hold it with the mounted section facing downward. Lower the slide so that the section itself touches the drop of liquid on a coverslip (Fig. 4A). The coverslip will adhere to the slide, and the solution will quickly spread out to fill the space between the slide and the coverslip. Then turn the slide over and check for air bubbles. If bubbles are present, but do not overlap the tissue section, they can be ignored; if bubbles do overlap the tissue section, they will need to be removed by gently working them toward the edge of the coverslip, or by removing the coverslip and replacing it with a new one. Once the slide has been coverslipped, place it in the humidified box (Fig. 4B). Repeat this for all the slides in the set, and seal the box with electrical tape.

5. Incubate the sections at a selected hybridization temperature, usually in the range of 50°–62°C. A standard air incubator, such as is commonly used for bacterial plates, is adequate for this purpose. Once initial results have been obtained, the conditions can be modified as required to improve the signal. A hybridization temperature of 55°C for 18–24 hr is a good starting point for most riboprobes. Oligonucleotide probes should be hybridized at a lower temperature, usually around 42°C. Although hybridizing at 42°C leads to higher backgrounds than hybridizing at 55°C, the short length of most oligonucleotide probes requires low hybridization temperatures to avoid disruption of probe–target complexes.

Posthybridization

After hybridization, sections are washed in a series of decreasing salt concentrations to reduce background and remove unhybridized probe. For riboprobes, we also routinely treat with RNase A at this point; this digests single-stranded (unhybridized) RNA, making radioactivity from unhybridized probe much easier to wash out of the tissue. The RNase treatment usually leads to a dramatic reduction of nonspecific background labeling.

To Process Tissue Sections after Hybridization

1. At the end of the hybridization period, remove the sealed, humidified box from the incubator and open it carefully. Place two slides back to back and hold them by the frosted glass end. Immerse them in a beaker filled with 2× SSC so that the coverslips fall off. If a coverslip adheres tightly to a slide, use a pair of forceps to gently slide it off laterally (if you try to pry it off, you will probably destroy the tissue section). Swish gently to rinse the sections, then place the slides into a glass slide rack. Repeat for the remaining slides in the set. This rinse contains the bulk of your radioactive probe, so dispose of it appropriately.
2. Wash all slides in two additional changes of 2× SSC for at least 5 min each time.
3. Incubate the sections in 10 mM Tris-HCl, pH 8.0, containing 0.5 M NaCl and 200 μg/ml RNase A for 30 to 60 min at 37°C. The RNase solution can be stored at 4°C, and used many times. Please note: RNase should be used with caution as it quickly digests RNA from any source. This includes tissue RNA on slides prior to hybridization and RNA extracted from tissue for use in Northern analysis. RNase is resistant to denaturation. Therefore, use RNase in a limited laboratory area in glassware dedicated for its use. This glassware should not go through general labora-

tory washing procedures and should not be taken into other parts of the laboratory.

4. The conditions for further washing steps must be optimized for each probe. Perform a trial run at an initial arbitrary stringency. A good starting point for most riboprobes consists of washing in 2× SSC at room temperature for 5 min, 1× SSC at room temperature for 5 min, 0.5× SSC at room temperature for 5 min, four times in 0.2× SSC at 55°C (hybridization temperature) for 15 min each, and finally 0.2× SSC at room temperature for 5 min. Sections are then dehydrated in graded ethanol solutions (as described above in the Deproteination section) and air dried. If this washing procedure leads to unacceptably high nonspecific background, then the stringency of the washes should be increased by elevating the wash temperature and/or decreasing the salt concentration. If it leads to an unacceptably weak specific hybridization signal, then the stringency of the washes should be decreased.

Detection

For quick assessment of the distribution of labeled probe within the tissue, sections can be placed directly on X-ray film (with the section facing the film) and exposed at room temperature in a light-tight cassette. Exposure time should vary from a few hours to several days, depending on the abundance of the message and the specific activity of the probe. If discrete populations of cells are consistently well-labeled, they can be seen as darkened regions on the film (Fig. 5A). For detection of radioactivity associated with individual cells, however, sections must be emulsion-dipped; a cluster of silver grains in the emulsion above an individual cell is indicative of high levels of probe binding (Fig. 5B,C).

Procedures for Dipping, Exposing, and Developing Slides

1. Thaw an aliquot of emulsion [Kodak NTB2 (Eastman Kodak, Rochester, NY) prepared as described in the Appendix] in a darkroom in a 42°C water bath. Transfer approximately 20 ml to a standard, 5-slide plastic mailer (Baxter) clamped and suspended in the 42°C water bath. A darkroom safelight (Kodak number 2 filter) can be used so long as care is taken to expose the emulsion and the dipped slides to a minimum of direct illumination.

2. Test-dip a few blank slides by inserting them into the emulsion with a single, smooth down-and-up movement. Hold the test slides up to the safelight to check that the emulsion has been applied evenly and to ensure that bubbles are not adhering to the slides. Discard the test slides.

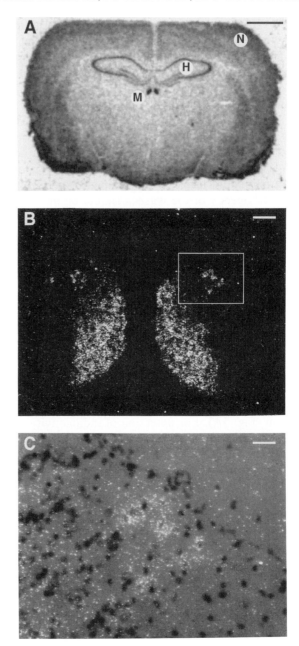

FIG. 5   Hybridized probe can be detected both by X-ray film and by photographic emulsion. (A) An X-ray film image of a rat brain coronal section hybridized with

3. Dip sections into the emulsion (as above) and set them aside to dry. An empty Kimble 20-ml scintillation vial box makes a convenient slide holder; after dipping, each slide can be placed at an angle (tissue section facing forward) into one of the small compartments formed by the cardboard dividers. (Do not hold slides with sections on them up to the safelight after coating with emulsion, or you risk increasing the background! Instead, periodically test-dip and examine a blank slide, as above.) Once all the sections have been dipped, they should be allowed to dry in complete darkness at room temperature for 2 hr, then transferred to slide boxes (Fisher). For each box of slides, wrap a tablespoon of Drierite in a Kimwipe and place it at one end of the box (leave a couple of vacant slots to provide space). Put the top on each box, wrap with several layers of aluminum foil to ensure that it is light-tight, and store at 4°C for the desired exposure time. We usually plan experiments so that we have enough sections to examine each brain region after a short (1–5 day) and long (7–28 day) exposure interval. The duration of the long interval can be adjusted on the basis of the results obtained from the short interval. With experience, the initial X-ray film exposure can be used to estimate the duration of the short interval.

---

antisense cGS-PDE probe (Bar: 2 mm). Film was placed directly over the tissue section (prior to emulsion application) and exposed for 2 days. Regions of dense riboprobe hybridization can be detected by eye in the neocortex (N), hippocampus (H), and medial habenular nucleus (M). (B) The same slide used to generate the X-ray film image at the top was dipped in photographic emulsion and exposed for 7 days, then lightly counterstained, coverslipped, and examined by dark-field microscopy. This photomicrograph shows the medial habenular nucleus at high magnification (Bar: 100 μm). Silver grains (white dots) are scattered densely over the ventral two-thirds of the nucleus. The dorsal third is much more sparsely labeled but shows isolated regions where silver grains are clustered at high density (centered in the white box). (C) Higher magnification of the boxed region from (B) (Bar: 25 μm). The image was obtained under combined bright-field/dark-field illumination, which allows simultaneous viewing of silver grains (white dots) and cell nuclei (dark spots). It is apparent that cell density is reasonably uniform in this region of the habenula, indicating that the cGS-PDE transcript is expressed by some cells but not others. The size of the silver grain clusters ($\sim$20–30 μm) suggests that each may represent an individual cGS-PDE-expressing cell.

Thus the X-ray film provides an overview of the riboprobe signal and allows a relatively rapid assessment of the success of the experiment. It also provides a guide to the necessary length of exposure of the emulsion, as discussed in the text. The photographic emulsion provides much finer detail and allows visualization of small areas of hybridization that may not be detectable by the eye on the X-ray film.

4. Unwrap the exposed, emulsion-dipped slides in the dark (safelight only) and put them into glass slide racks. Immerse the racked slides sequentially in standard slide staining dishes containing Kodak D-19 (2 min), water (20 sec), and Kodak Rapid Fix (3 min), all at room temperature. The sections should be gently agitated by hand at all times. After fixation, the lights can be turned on. Wash the sections in running water for 20 min, counterstain with hematoxylin if desired (see Appendix), dehydrate through graded alcohols (as described above), clear in xylene (2 changes of 3 min each), and cover with dust-free coverslips mounted with DPX (Gallard Schlesinger, Carle Place, NY). Before viewing under the microscope, scrape the emulsion off the back of the slide with a sharp razor blade (be extremely careful, as it is quite easy to cut yourself at this stage) and clean the back of the slide with liquid glass cleaner (Windex) and a Kimwipe.

5. When viewing sections, it is common to switch back and forth between dark-field illumination (where silver grains show up as intense white dots on a dark background, Fig. 5B) and bright-field illumination (where the anatomical features of the tissue section are easily visualized but silver grains show up only as faint black dots superimposed on the stained structures). We have also found it convenient to generate a single bright-field–dark-field image (Fig. 5C) by combining a low level of conventional bright-field illumination with high-intensity light delivered perpendicular to the axis of viewing (Darklite Illuminator, Micro Video Instruments, Avon, MA).

Interpretation and Controls

Like any other well-designed experiment, each *in situ* hybridization reaction should include positive and negative controls. The best positive control is the use of a probe–tissue combination that has worked well in previous studies, preferably employing a probe that detects an abundant transcript that can be readily visualized overnight on X-ray film. Examples of combinations that provide reliable positive controls include the cGS-PDE in brain (24), the H^+-/K^+-ATPase in stomach (39), the IGF-1 propeptide in liver (40), and the guanylin propeptide in colon (41). If these controls fail, then there is a fundamental problem with the sections, the probe, the reagents, or some other aspect of the technique.

Negative controls should also be incorporated into each experiment. These include parallel hybridization reactions performed on adjacent tissue sections with alternate probes and conditions. For example, if the pattern of tissue labeling displayed by the antisense probe is authentic, then the labeling should be abolished when the sense probe is substituted. Similarly, the pattern of labeling should be abolished when the antisense probe is applied to a section that has been treated with RNase prior to hybridization. To accommodate these basic controls, we typically design an experiment to include three pairs of consecutive tissue sections: one pair is hybridized with the antisense probe, one pair is hybridized with the sense probe, and one pair is pretreated with RNase A and then hybridized with the antisense probe. This provides two sections of each type, one to be developed after a short exposure and one after a long exposure. As an additional negative control, it is important, if the target RNA is a member of a multigene family, to show that antisense probes that selectively recognize other members of the family show a pattern of hybridization that is distinct from the one observed with the riboprobe of interest. These controls, and others, are described in more detail below.

Sense Probe

Hybridize a tissue section with a sense riboprobe (the sequence of a sense riboprobe is complementary to that of the antisense riboprobe and identical to that of the coding sequence of the target mRNA, see Fig. 1). The sense probe will have the same GC content (one important determinant of "stickiness") as the antisense probe, and thus it provides a particularly appropriate control for nonspecific binding. If a signal is observed with the sense probe, then any portion of the antisense signal that overlaps the sense signal is likely to be artifactual. Note that the converse result (failure to observe labeling with the sense probe) is necessary, but not sufficient, to verify that your antisense signal is real. Additional controls should be run to help confirm this, as described below. An example of a successful sense probe control for the cGS-PDE is shown in Fig. 6: the clear signal that is apparent with the antisense probe (Fig. 6A) is absent when an adjacent section is hybridized with the sense probe (Fig. 6B).

RNase Pretreatment of Tissue

Fixed tissue can be incubated in RNase A (200 μg/ml) for 60 min at 37°C before proceeding with deproteination and hybridization. This treatment should remove all mRNA in the tissue. The subsequent proteinase K step will then inactivate the RNase (note that without proteinase treatment the control will not be valid, because residual RNase will simply digest the probe during the hybridization step). If the antisense riboprobe still shows localized patterns of binding to the section after RNase treatment, then either nonspecific probe–tissue interactions or incomplete predigestion of RNA has occurred. An example of a successful RNase pretreatment control for the cGS-PDE is shown in Fig. 6C: compared to the hybridization pattern observed with the antisense probe on a normal tissue section (Fig. 6A), the pattern of

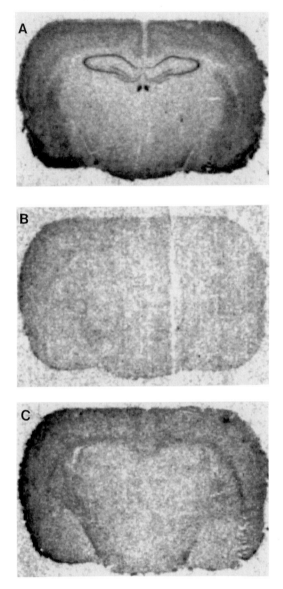

FIG. 6 Negative controls for *in situ* hybridization. Rat brain coronal sections were hybridized with antisense (A) or sense (B) probes for the cGS-PDE, or pretreated with RNAse (C) before hybridizing with the antisense probe. The autoradiograph in (A) is the same as that presented in Fig. 5A, and the autoradiographs in (B) and (C) are from adjacent brain sections. The strong antisense probe signal observed in (A) appears to be specific, as it is not observed after hybridization with the sense probe

antisense probe binding to the RNase-treated section is markedly attenuated (though a faint signal remains that may be due to incomplete penetration of RNase into the tissue).

Use of Northern Blots

Northern blots can be used with caution to help evaluate *in situ* hybridization results. A tissue that does not give a signal on a Northern blot would not be expected to provide an abundant signal when tested with *in situ* techniques. Conversely, an abundant signal on a Northern blot should be expected to correlate with at least some areas of abundant signal on a tissue section. Furthermore, Northern blots can provide evidence in support of the specificity of an *in situ* signal, as described above and illustrated in Fig. 2. However, it is important to keep in mind that hybridization and wash conditions for blots and for tissue sections, even when designed to be similar, are not directly comparable.

Multigene Families

If the target RNA encodes a protein that shares significant homology with other proteins, then it is important to investigate whether the probe is detecting only the expected transcript. An initial way to evaluate this possibility is to compare two adjacent tissue sections, one hybridized with the probe of interest and a second hybridized with a probe for a related transcript. For example, as mentioned in the legend to Fig. 2, all currently cloned PDE isozymes share a region of homology that comprises a catalytic (cyclic nucleotide-hydrolyzing) domain. Probes that have been used to investigate expression of the cGS-PDE and a 63-kDa calmodulin-regulated PDE (CaM-PDE) are both derived from this conserved domain (Fig. 7A), and thus could potentially cross-react in an *in situ* hybridization reaction. In fact, when the cGS-PDE and the CaM-PDE antisense probes are compared on adjacent sections of brain tissue, they do, indeed, detect mRNA in partially overlapping structures. The cGS-PDE signal is present in neocortex, piriform cortex, striatum, and throughout the hippocampus (Fig. 7B); the CaM-PDE signal is also present in piriform cortex and striatum but is restricted to the dentate

---

and is markedly attenuated on RNase pretreatment. The RNase-treated section does show some residual signal, especially associated with fissures in the tissue and at the edges of the section. The residual signal could be due to nonspecific "stickiness" of these regions of the tissue, though the absence of signal in (B) suggests that it might also be a consequence of inadequate penetration by the RNase. (Scale equivalent to that shown in Fig. 5.)

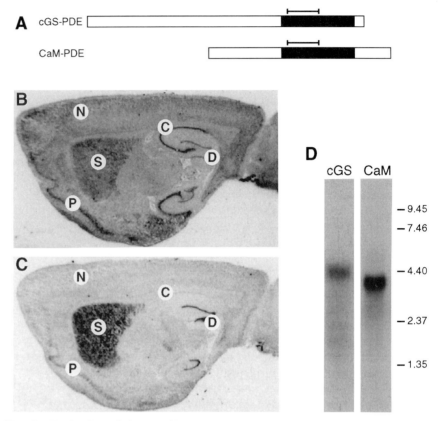

FIG. 7 Evaluation of the specificity of antisense hybridization, using probes for related mRNAs. (A) Schematic diagram of the cGS-PDE and CaM-PDE transcripts, with the conserved region that encodes the catalytic domain indicated by the solid bars. Probes for cGS-PDE and CaM-PDE were made from the conserved region, as indicated by the lines above the bars. (B) *In situ* hybridization of the cGS-PDE antisense probe to a sagittal section of rat brain (scale approximately equivalent to that shown in Fig. 5). The probe hybridizes to neocortex (N), piriform cortex (P), striatum (S), and hippocampus [dentate gyrus (D) and areas CA1–CA3 (C)]. The sense and RNase pretreatment controls described in the text reveal that these signals are specific (data not shown). (C) *In situ* hybridization of CaM-PDE antisense probe to an adjacent sagittal section. The probe hybridizes to piriform cortex (P), striatum (S), and hippocampus [dentate gyrus (D), but not areas CA1–CA3 (C)]. The implications of the overlapping distributions of these hybridization signals are discussed in the text. (D) Northern blot of rat mRNA from the forebrain probed with cGS-PDE (left) and CaM-PDE (right) antisense probe. The striatum is the only positively hybridizing brain region represented in this mRNA. The cGS-PDE and the CaM-PDE probes each detect a single band on the blot, but in each case the size of the mRNA is unique. These results support the conclusion that the *in situ* cGS-PDE and CaM-PDE signals are specific, as discussed in the text.

gyrus of the hippocampus and does not have a major signal in neocortex (Fig. 7C). Because the CaM-PDE signal is absent from several areas where the cGS-PDE signal is strongly represented, it appears that the CaM-PDE probe is not cross-reacting with the cGS-PDE transcript. However, the converse cannot be concluded from these data. It is possible that some or all of the cGS-PDE signal in piriform cortex, striatum, and dentate gyrus is due to cross-reactivity with the CaM-PDE transcript. In this case, the abundant signal obtained with both probes in the striatum provides a good opportunity to test this idea with a Northern blot. Figure 7D shows a blot of mRNA purified from an isolated region of forebrain that includes striatum but no other group of cells that hybridizes strongly with either the cGS-PDE or the CaM-PDE probes. On the blot, each probe detects an mRNA species of a unique size, with no evidence of cross-detection. This demonstrates that the probes do not cross-react under the conditions of the Northern blot, and also that transcripts encoding both the cGS-PDE and the CaM-PDE are present in the striatum. Therefore, the Northern blot supports the conclusion that these probes are each specifically detecting their own mRNA in the *in situ* hybridization experiments of Fig. 7B,C.

Confirmation of Melting Temperature

The melting temperature (T_m) is the point at which half of the duplex molecules formed during hybridization have dissociated into single strands. If a signal is real it will melt off in a predictable fashion (showing the steep sigmoidal temperature dependence typical of a T_m curve) and the T_m obtained from the melting curve will be close to the theoretical value obtained from Eq. (1) below. If the probe is nonspecifically bound, it will melt off in a linear fashion, and the experimentally determined T_m is likely to be very different from the calculated T_m. Thus it can be useful, though quite labor-intensive, to compare the calculated T_m with the experimentally determined temperature at which the signal is lost from the tissue. The calculation of T_m is also useful for determining the hybridization and wash temperatures that are theoretically optimal for a new probe (though this calculation should only be used as a first approximation, and refined empirically). For RNA–RNA hybrids greater than 50 nucleotides in length, T_m is given by Eq. (1) [see *Molecular Cloning* (36)]:

$$T_m \, (^\circ C) = 79.8 + [58.4 \times (G + C)] + [11.8 \times (\%G + C)^2] \\ + [18.5 \log(M)] - 820/L - 0.35 \, (\% \, F) - \% \, m \tag{1}$$

where (G + C) is the G–C content of the probe (mole fraction); M, monovalent cation concentration (for the 75% formamide hybridization buffer given in the appendix, M is 0.674); L, length of duplexes formed (number of base pairs); $\% F$, $\%$ formamide; and $\% m$, $\%$ mismatch (as, for example, would occur when using a probe from one species on tissues from another).

As an example for rat cGS-PDE riboprobe applied to rat tissues, where $(G + C) = 0.5$, $M = 0.674$, $L = 351$ bases, $\% F = 75$, and $\% m = 0$, the calculation is

$$T_m = 79.8 + 58.4(0.5) + 11.8(0.5)^2 + 18.5(\log 0.674)$$
$$- 820/351 - 0.35(75) - 0$$
$$T_m = 80.2°C$$

Theoretically, the optimal hybridization temperature is $T_m - 30°C$, or $50.2°C$ in this example.

Common Sense

Common sense is the most important control to ensure proper interpretation of *in situ* hybridization results. These questions should be considered when examining putative signals: Is the signal over cells, as opposed to areas where there are holes in the tissue (or no tissue)? Background is considered to be the density of exposed grains over the slide where no tissue is present. Is the signal exclusively along the edges of tissue sections and at the edges of holes in the tissue sections? Frequently such signals are artifactual, arising from probe that has become trapped between the section and the slide. Is the signal present over certain cells and not others? A signal that is evenly distributed over a complex tissue (like the brain) is suspect. For example, suppose a section hybridized with your antisense probe is uniformly labeled and dark, while a section hybridized with your sense probe is uniformly labeled and light. A trivial explanation for this difference could be that the specific activity of the antisense probe was simply higher than the specific activity of the sense probe. Does the distribution of the transcript that you are detecting make physiological sense? Is it consistent with what is known about protein distribution from enzyme assays, immunocytochemical measurements, or receptor binding studies? Although there are several possible reasons why the localization of a transcript may not match that of the translated protein (e.g., the protein has been transported away from the site of synthesis or secreted into extracellular space), a mismatch should be reason to investigate further.

Summary

In situ hybridization is a powerful technique for localizing specific mRNA species in tissue sections. As we have tried to point out above, there are potential pitfalls in performing the technique and interpreting the results.

However, with careful attention to detail and appropriate controls, the technique can be used with confidence and precision.

Appendix: Miscellaneous Procedures and Solutions for *in Situ* Hybridization

DEPC-treated water

1. Add 0.5 ml of diethyl pyrocarbonate (DEPC) to 500 ml of distilled, deionized water.
2. Heat to 37°C for 8 hr, shaking once or twice.
3. Autoclave for 30 min.

5× TBE

54 g Tris base
27.5 g boric acid
20 ml 0.5 M NaEDTA (pH 8.0)
Bring to 1 liter with distilled, deionized H_2O

Dilute the stock solution 1:5 in distilled, deionized H_2O to make 1× TBE.

Solutions for 5% Acrylamide 8 M Urea Gels

This recipe provides sufficient gel solution for a single minigel cast in a Hoefer minigel apparatus (8 × 11 cm, using 1.5-mm spacers). Run the gel at constant voltage (50–150 V). At 100 V the starting current will be about 10 mA and the gel will take approximately 150 min to run. In 5% acrylamide, the migration rate of the bromphenol blue tracking dye (blue band) is comparable to that of a 50-base pair polynucleotide, while the migration rate of the xylene cyanol tracking dye (purple band) is comparable to that of a 150-base pair polynucleotide.

15 ml of Gel Solution

1.88 ml 40% (w/v) acrylamide:bisacrylamide (19:1) (National Diagnostics, Atlanta, GA)
2.5 g ultrapure urea
3 ml 5× TBE
Add water (DEPC-treated) to bring the volume to 15 ml
120 μl 10% ammonium persulfate
Add at the last minute, 16 μl TEMED

Electrode Buffer
    $1\times$ TBE

Sample Buffer
    96% formamide
    20 mM NaEDTA (pH 8.0)
    0.05% bromphenol blue
    0.05% xylene cyanol

Subbed Slides

1. Arrange the slides in glass slide racks, so that both sides of the slides are exposed.

2. Dissolve a handful of Alconox in a tub of warm tap water. Place the racks containing the slides into the soapy solution (be sure all slides are submerged). Agitate the tub continuously for 1 hr at slow speed on an orbital shaker.

3. Discard the soapy water and rinse the slides extensively with tap water.

4. Discard the last rinse and cover slides with 80% ethanol for 1 hr.

5. Wash the slides in distilled, deionized H_2O. Dip the slides in polylysine (see below for recipe). Dry the slides overnight in a 60°–70°C oven and store at $-20°C$.

Polylysine

    1.38 g Tris base
    50 mg polylysine (Sigma, St. Louis, MO, or equivalent)
    900 ml distilled, deionized H_2O
    Adjust pH to 8.0, then bring to 1 liter with distilled, deionized H_2O

4% Formaldehyde

1. Dissolve 5 g $NaH_2PO_4 \cdot H_2O$ and 22.7 g Na_2HPO_4 in 1 liter of distilled, deionized H_2O.

2. Heat a second liter of distilled, deionized H_2O to 55°–60°C. Add 80 g of paraformaldehyde (powder). While stirring, add drops of 10 N NaOH until the solution clears (about 25 drops). Be sure to keep the temperature below 60°C (at higher temperatures there is significant polymerization of the formaldehyde monomers, resulting in less efficient tissue penetration and less effective fixation).

3. Mix the two solutions. Stir until cool.

4. Filter through Whatman No. 2 filter paper.

5. Measure the pH and adjust to the range of pH 7.0 to 7.4.

6. Store in a light-tight container at 4°C.

1 M KPO₄ Buffer

    143.2 g K_2HPO_4
    24.2 g KH_2PO_4
    Bring to 1 liter with DEPC-treated H_2O; heat to dissolve

Phosphate-Buffered Saline (PBS)

    18.7 g NaCl
    40 ml 1 *M* KPO_4 buffer
    1.8 liters of DEPC-treated H_2O
    Adjust pH to the range of pH 7.0 to 7.4, then bring to 2 liters with
      DEPC-treated H_2O

Proteinase K Buffer

    13.8 g Tris base
    19 g Na_4EDTA
    800 ml distilled, deionized H_2O
    Adjust to pH 8.2, then bring to 1 liter with distilled, deionized H_2O

10× Triethanolamine, pH 8.0

    133 ml triethanolamine (7.53 *M*)
    800 ml DEPC-treated H_2O
    Adjust to pH 8.0 with HCl, then bring to 1 liter with DEPC-treated H_2O

20× SSC

    701.2 g NaCl
    353 g trisodium citrate hydrate
    3.5 liters DEPC-treated H_2O
    Adjust to pH 7.0–7.4, then bring to 4 liters with DEPC-treated H_2O

Dilute the stock solution appropriately in DEPC-treated H_2O to make SSC solutions of different strengths.

Formamide-Containing Hybridization Buffer

Determine which of these buffers is appropriate for a given probe by the method described in the text (see Hybridization Reaction). Each recipe makes 10 ml of solution. Hybridization buffers are also available commercially from many sources.

75% Formamide

    7.5 ml formamide
    1.5 ml 20× SSC
    200 μl 50× Denhardt's solution (see recipe below)

200 μl yeast tRNA (10 mg/ml)
500 μl 1 M sodium phosphate, pH 7.4
100 μl 1 M dithiothreitol (for ^{35}S-labeled probes only)
1 g dextran sulfate
Rotate and heat to 37°C to dissolve solids; store at −20°C (this solution
    is stable to repeated freezing and thawing)

50% Formamide

5 ml formamide
2.5 ml DEPC-treated H$_2$O
1.5 ml 20× SSC
200 μl 50× Denhardt's solution (see recipe below)
200 μl yeast tRNA (10 mg/ml)
500 μl 1 M sodium phosphate pH 7.4
100 μl 1 M dithiothreitol (for ^{35}S-labeled probes only)
1 g dextran sulfate
Rotate and heat to 37°C to dissolve solids; store at −20°C (this solution
    is stable to repeated freezing and thawing)

50× Denhardt's Solution (also available premixed from Sigma)

5 g Ficoll (Type 400)
5 g polyvinylpyrrolidone
5 g bovine serum albumin (Fraction V)
Bring to 500 ml with DEPC-treated H$_2$O

RNase Buffer

0.345 g Tris base
7.25 g NaCl
200 ml distilled, deionized H$_2$O
Adjust to pH 8.0, then bring to 250 ml with distilled, deionized H$_2$O

RNase A Stock

250 ml RNase buffer
50 mg RNase A (final concentration 200 μg/ml)

Emulsion: Kodak NTB2

The emulsion is highly light-sensitive. We dilute and divide the emulsion
into aliquots in total darkness, because accidental exposure to light will ruin
it. Although the manipulations are a bit tricky, this allows us to avoid possible
deleterious effects of darkroom safelights (whose emission spectrum is not
always known to darkroom users). If a communal stock of emulsion is used,

coat a blank slide and develop it before proceeding with valuable tissue sections; examining the density of silver grains on this slide will determine whether someone has accidentally exposed the emulsion to light.

Because the emulsion is semisolid, even at room temperature, we keep a 42°C water bath in the darkroom to melt the emulsion before dividing it, and also to ensure that each aliquot is completely liquefied before dipping slides into it. Detergent (mild baby soap such as Dreft) is added as a wetting agent, to help prevent bubbles and to thin the emulsion for more even coating of slides.

In the Light

Make up a 1 mg/ml solution of mild baby soap (obtained from a local supermarket). Dissolve the soap with stirring and mild heat, then filter the solution. Measure 118 ml and put it into a 600-ml beaker.

In the Dark

Warm 118 ml of emulsion (the standard volume supplied by Kodak) in a 42°C water bath. When it has completely melted, add it to the beaker containing the soap solution. Stir well and divide the mixture into eight plastic centrifuge tubes (50 ml capacity). Fill each tube approximately halfway (about 30 ml/ tube). Wrap each aliquot with three layers of aluminum foil and store them at 4°C.

Hematoxylin Staining

1 Liter of Hematoxylin Stain

    50 g aluminum ammonium sulfate
    1 liter distilled deionized H_2O
    1 g hematoxylin crystals
    0.2 g sodium iodate
    1 g citric acid (anhydrous)
    50 g chloral hydrate

Dissolve aluminum ammonium sulfate in water, without heat. Add and dissolve the hematoxylin. Add, in order, sodium iodate, citric acid, and chloral hydrate; stir until all components are completely in solution. Store in a dark brown bottle at room temperature. Filter before each use through Whatman No. 1 filter paper.

Procedure

1. Pass sections through a distilled water rinse (5 min).
2. Stain sections in hematoxylin (0–5 min, depending on desired degree of tissue staining). Note that the best images of silver grains will be obtained

from unstained or very lightly stained sections; silver grains are easily over-whelmed by light scattered from heavily stained structures within the section.

3. Wash sections in cool, running tap water for 20 min.

4. Dehydrate in 95 and 100% ethanol (two changes for 3 min each).

5. Clear in xylene (two changes for 3 min each).

6. Mount coverslips using DPX.

Acknowledgments

This chapter is dedicated to Julius Adler on the occasion of his sixty-fifth year. We gratefully acknowledge the support of the UNC Center for Gastrointestinal Biology and Disease (NIH Center Grant P30 DK34987) and the many contributions made by Stanley J. Watson in the course of developing the methods reported here. Thanks are also due to Bonnie Taylor-Blake for reading the manuscript from a histologist's point of view.

References

1. J. A. Ferrendelli, *Adv. Biochem. Psychopharmacol.* **15,** 303 (1976).

2. N. D. Goldberg and M. K. Haddox, *Annu. Rev. Biochem.* **46,** 823 (1977).

3. G. I. Drummond, *Adv. Cyclic Nucleotide Res.* **15,** 373 (1983).

4. J. De Vente, J. G. J. M. Bol, and H. W. M. Steinbusch, *Eur. J. Neurosci.* **1,** 436 (1989).

5. S. R. Vincent and B. T. Hope, *Trends Neurosci.* **15,** 108 (1992).

6. P. K. NaKane, Chapter [5], this volume.

7. T. M. Dawson and V. L. Dawson, Chapter [6], this volume.

8. G. Giuili, A. Luzi, M. Poyard, and G. Guellaen, *Dev. Brain Res.* **81,** 269 (1994).

9. J. D. Scott, *Pharmacol. Ther.* **50,** 123 (1991).

10. M. D. Uhler, *J. Biol. Chem.* **268,** 13586 (1993).

11. R. R. Reinhardt, E. Chin, J. Zhou, M. Taira, T. Murata, V. C. Manganiello, and C. A. Bondy, *J. Clin. Invest.* **95,** 1528 (1995).

12. W. K. Sonnenburg, P. J. Mullaney, and J. A. Beavo, *J. Biol. Chem.* **266,** 17655 (1991).

13. U. B. Kaupp, *Trends Neurosci.* **14,** 150 (1991).

14. S. S. Kolesnikov and R. F. Margolskee, *Nature (London)* **376,** 85 (1995).

15. P. Hudson, J. Penschow, J. Shine, G. Ryan, H. Niall, and J. Coghlan, *Endocrinology (Baltimore)* **108,** 353 (1981).

16. K. H. Cox, D. V. DeLeon, L. M. Angerer, and R. C. Angerer, *Dev. Biol.* **101,** 485 (1984).

17. A. F. M. Moorman, P. A. J. De Boer, J. L. M. Vermuelen, and W. H. Lamers, *Histochem. J.* **25,** 251 (1993).

18. J. N. Wilcox, *J. Histochem. Cytochem.* **41,** 1725 (1993).

19. G. D. Pratt and M. Kokaia, *Trends Pharmacol. Sci.* **15,** 131 (1994).
20. R. Lehmann and D. Tautz, *Methods Cell Biol.* **44,** 575 (1994).
21. S. M. Lohmann, U. Walter, P. E. Miller, P. Greengard, and P. DeCamilli, *Proc. Natl. Acad. Sci. U.S.A.* **78,** 653 (1981).
22. T. Jarchau, C. Hausler, T. Markert, D. Pohler, J. Vandekerckhove, H. R. De-Jonge, S. M. Lohmann, and U. Walter, *Proc. Natl. Acad. Sci. U.S.A.* **91,** 9426 (1994).
23. A. E.-D. El-Husseini, C. Bladen, and S. R. Vincent, *J. Neurochem.* **64,** 2814 (1995).
24. D. R. Repaske, J. G. Corbin, M. Conti, and M. F. Goy, *Neuroscience (Oxford)* **56,** 673 (1993).
25. I. Ahmad, T. Leinders-Zufall, J. D. Kocsis, G. M. Shepherd, F. Zufall, and C. J. Barnstable, *Neuron* **12,** 155 (1994).
26. A. E.-D. El-Husseini, C. Bladen, and S. R. Vincent, *NeuroReport* **6,** 1331 (1995).
27. P. Komminoth, P. U. Heitz, and A. A. Long, *Verh. Dtsch. Ges. Pathol.* **78,** 146 (1994).
28. D. G. Wilkinson, *Curr. Opin. Biotechnol.* **6,** 20 (1995).
29. J. B. Lawrence and R. H. Singer, *Cell (Cambridge, Mass.)* **45,** 407 (1986).
30. B. Bloch, *J. Histochem. Cytochem.* **41,** 1751 (1993).
31. J. G. Szakacs and S. K. Livingston, *Ann. Clin. Lab. Sci.* **24,** 324 (1994).
32. B. Rosen and R. S. P. Beddington, *Trends Genet.* **9,** 162 (1993).
33. L. G. Luna, "Histopathologic Methods and Color Atlas of Special Stains and Tissue Artifacts." American Histolabs, Gaithersburg, Maryland, 1992.
34. A. J. Lacey, "Light Microscopy in Biology: A Practical Approach." IRL Press, Oxford, 1989.
35. F. M. Ausubel, R. Brent, R. E. Kingston, D. D. Moore, J. G. Seidman, J. A. Smith, and K. Struhl, "Current Protocols in Molecular Biology." Wiley, New York, 1995.
36. J. Sambrook, E. F. Fritsch, and T. Maniatis, "Molecular Cloning: A Laboratory Manual." Cold Spring Harbor Laboratory, Cold Spring Harbor, New York, 1989.
37. K. Sugaya and M. McKinney, *Brain Res. Mol. Brain Res.* **23,** 111 (1994).
38. S. Hayashi, I. C. Gillam, A. D. Delaney, and G. M. Tener, *J. Histochem. Cytochem.* **26,** 677 (1978).
39. K. Bamberg, S. Nylander, K. G. Helander, L. G. Lundberg, G. Sachs, and H. F. Helander, *Biochim. Biophys. Acta* **1190,** 355 (1994).
40. E. M. Zimmermann, R. B. Sartor, R. D. McCall, M. Pardo, D. Bender, and P. K. Lund, *Gastroenterology* **105,** 399 (1993).
41. Z. Li and M. F. Goy, *Am. J. Physiol. Gastrointest. Liver Physiol.* **265,** G394 (1993).

[9] Detection of Heme Oxygenase-1 and -2 Transcripts by Northern Blot and *in Situ* Hybridization Analysis

James F. Ewing

Introduction

Production of the gaseous heme ligand nitric oxide (NO) and its activation of guanylate cyclase has been associated with various activities in the systemic organs and the brain (1–4). Likewise, it has been shown that carbon monoxide (CO), another gaseous ligand, regulates cGMP production in the brain (5–7) and in the heart (8), and hence, may function as an intracellular messenger (9, 10).

Synthesis of CO, a putative neurotransmitter (11), is catalyzed by the heme oxygenase (HO) enzyme system (12). The α-meso carbon of heme tetrapyrrole (Fe-protoporphyrin IX) is the source of CO. This system, which represents the only means for the formation of CO found in mammalian species, consists of two isozymes, HO-1 and HO-2 (13, 14), that are the products of different genes (15, 16). The isozymes differ in numerous ways, including their regulation, tissue distribution, and antigenic epitopes. Difference in the latter property has made possible the production of polyclonal antibody preparations specific to the HO-1 or HO-2 protein (14, 17, 18).

The protein HO-1, also known as hsp32, is a stress protein that is normally only present at low levels in the brain (19–21). However, under conditions of environmental (20–23) and chemical stress (6, 24) robust expression of hsp32 is observed among discrete cell populations. In contrast, HO-2 is far more prevalent in the mammal brain, and thus constitutes the primary component of the impressive CO-generating capability of the central nervous system (CNS) (18). While this isozyme is refractory to many stressors and chemical stimuli (6, 13, 17, 20, 24), it is responsive to glucocorticoids (23, 25). The HO-2 protein is generally associated with neurons (5) and displays a developmentally regulated pattern of expression (19).

In most brain regions, HO-2 colocalizes with guanylate cyclase (26), including the hippocampal complex, the Purkinje cells, and neurons in molecular and granule layers of the cerebellum, where a robust expression of the protein and its transcripts is detectable (5). Indeed, colocalization, in addition to the undetectable levels of nitric oxide synthase (NOS) in Purkinje cells of

Methods in Neurosciences, Volume 31

cerebellum as well as CA1 pyramidal cells of hippocampus (27, 28), has been a strong argument for the suspected role of CO as a regulator of cyclic GMP. Nitric oxide synthase generates NO, the recognized activator of the heme-dependent isozyme of guanylate cyclase (1, 2, 4, 29, 30).

Studies colocalizing both the CO- and NO-generating systems have demonstrated distinct but overlapping patterns of expression in the normal brain (31). This section details some of the methodology currently being employed to assess the expression and regulation of the heme oxygenase system in the brain.

RNA Hybridization

Cloning and characterization of genes for rat HO-1 (32) and HO-2 (33) has allowed for derivation of specific cDNA probes. In turn, Northern blot analysis using these cDNA probes has allowed characterization of transcripts coding for heme oxygenase isozymes in the brain (19). This technique of transferring electrophoretically separated RNA to nitrocellulose filters and subsequent analyzing specific messenger RNA (mRNA) using radiolabeled cDNA probes (34, 35) has been described in detail elsewhere (36). This chapter seeks only to summarize the procedure as it specifically pertains to the analysis of HO mRNAs in brain tissue.

Tissue Preparation and Isolation of RNA from Brain

Prepare brain tissue by rapidly removing the organ and snap freezing in a dry ice–methanol bath. Tissue may be stored frozen at −80°C prior to use. Isolate total RNA using a modification of the guanidine isothiocyanate method (37) as follows: homogenize the crushed, frozen brain tissue (0.5–0.8 g) on ice in 5 ml of 50 mM Tris, pH 7.5, containing 4 M guanidine isothiocyanate, 10 mM EDTA and 5% (v/v) 2-mercaptoethanol using three bursts of a Polytron homogenizer for 15 sec each. The homogenate should be pelleted at 2600 g for 10 min (4°C) and the supernatant retained. To the clarified homogenate, add 1 ml of 10% (w/v) N-laurylsarcosyl and heat the sample for 2 min at 65°C in a water bath. Add cesium chloride (0.6 g solid CsCl) to give a final concentration of 1.2 M, and then layer the homogenate onto 5 ml of a 5.7 M cesium chloride solution containing 0.1 mM EDTA, pH 7.0, in a siliconized polyallomer tube. Fractionate the sample in an ultracentrifuge at 25,000 rpm (maintained at 18°C) for 18 hr in a Beckman SW41Ti swinging bucket rotor (Beckman Instruments, Columbia, MD). Redissolve the RNA pellet overnight at 4°C in 2 ml of 5 mM EDTA, pH 7.0, containing 0.5% (w/v) N-laurylsarcosyl and 5% (v/v) 2-mercaptoethanol. Extract the RNA with an equal volume of freshly prepared 25:24:1 (v/v/v) Tris-saturated phenol/chloroform/isoamyl alcohol. Now centrifuge the sample at 10,000 g for 5 min. Reextract the top aqueous phase with an equal volume of freshly prepared 24:1 (v/v) chloroform/isoamyl alcohol. Centrifuge the sample as before, and again collect the top aqueous phase. The aqueous RNA solution should now be brought up to 0.3 M sodium acetate concentration with a 3 M, pH 5.2, stock solution of the salt, and then precipitated with 2.5 volumes of 100% ethanol with subsequent incubation of the sample in a dry ice–methanol bath for at least 30 min. Collect the precipitate by centrifugation (15,000 g for 30 min), wash the pellet with 80% ice-cold ethanol, and then recentrifuge (10,000 g for 10 min). The precipitated total RNA should finally be dissolved in 50–100 μl RNase-free water.

Poly(A)$^+$ Selection

Selection of the poly(A)$^+$ fraction by oligo(dT)-cellulose affinity chromatography has been described in detail elsewhere (38). Briefly, heat up to 1 mg total RNA from brain in a 70°C water bath for 10 min and then rapidly chill on ice. Bring the sample up to 0.5 M lithium chloride concentration with a 10 M solution of the salt. Bind the poly(A)$^+$ RNA to a 0.5-ml column of oligo(dT)-cellulose equilibrated in 10 mM Tris, pH 7.5, containing 0.5 M lithium chloride, 1 mM EDTA, and 0.1% (w/v) sodium dodecyl sulfate with three sequential passages of the crude RNA extract through the column at modest flow. The bound fraction should be washed once with 1 ml of 10 mM Tris, pH 7.5, containing 0.15 M lithium chloride, 1 mM EDTA, and 0.1% (w/v) SDS. Elute poly(A)$^+$ RNA from the column with 1 ml of 2 mM EDTA containing 0.1% (w/v) SDS. Bring the poly(A)$^+$ RNA solution up to 0.3 M sodium acetate concentration with a 3 M, pH 5.2, stock solution of the salt, precipitate with 2–5 volumes of 100% ethanol, and subsequently incubate the sample in a dry ice–methanol bath for at least 30 min. Collect the precipitate by centrifuging (27,000 g, for 45 min), wash the pellet with 80% (v/v) ice-cold ethanol, and then centrifuge (27,000 g, for 20 min). Finally, dissolve the precipitated poly(A)$^+$ RNA in 20 μl of 10 mM Tris, pH 8.0, containing 1 mM EDTA and quantify by standard spectrophotometric technique using UV absorbance (260/280 nm ratio). Samples may be stored at −80°C for months prior to use.

Electrophoresis of Poly(A)$^+$ RNA

Electrophoresis of RNA [4 μg poly(A)$^+$RNA] should be carried out in horizontal slab gels of 1.2% agarose in 20 mM 3-(N-morpholino) propanesulfonic acid (MOPS) that contain 48 mM sodium acetate and 1.11% (v/v) formaldehyde. Transfer the fractionated RNA onto a Nytran membrane (Schleicher and Scheull, Keene, NH) by capillary transfer (36). Fix RNA to the membrane by heating the filter for 1.5 hr at 80°C *in vacuo*. The gel may be stained with ethidium bromide after the blotting procedure to confirm that the transfer of RNA is complete.

Hybridization of ^{32}P-Labeled HO cDNA Probe

Hybridization of filter-bound RNA should be performed using DNA restriction fragments specific for HO-1 or HO-2, and labeled by random primer labeling technique with ^{32}P-labeled deoxycytidine to a specific activity greater than 5×10^8 cpm/μg. It is notable that the best results are obtained with ^{32}P-labeled HO-1/HO-2 cDNA probe prepared fresh and purified by spin column chromatography using Sephadex G-25 (36).

The HO-1 hybridization probe we have utilized is a 568-bp HO-1 fragment corresponding to nucleotides +86 to +654 (32) generated via polymerase chain reaction (PCR) and subsequently cloned into a PBS$^+$ vector as has been previously described (19). The HO-2 hybridization probe was the full-length (1300 bp) HO-2 cDNA that was purified from a rat testis DNA library (33). The filter should be prehybridized at 42°C for at least 4 hr in 25 mM potassium phosphate buffer, pH 7.4, containing 50% (v/v) deionized formamide, 5× SSC (0.75 M NaCl, 75 mM sodium citrate, pH 7.0), 5× Denhardt's solution [0.1% (w/v) Ficoll and 0.1% (w/v) polyvinylpyrrolidone], and 1 mg/ml heat-denatured fish sperm DNA (prehybridization solution). Hybridize the filters overnight on a platform rocker at 42°C with 1×10^6 cpm/ml of the appropriate heat-denatured ^{32}P-labeled HO cDNA in prehybridization solution containing 10% (w/v) dextran sulfate. Wash the filters at 48°C in 1× SSC (0.15 M NaCl, 15 mM sodium citrate, pH 7.0) containing 0.1% (w/v) SDS followed by 0.25× SSC (37.5 mM NaCl, 3.75 mM sodium citrate, pH 7.0) containing 0.1% (w/v) SDS in a shaking water bath. Damp filters should be placed between two pieces of plastic wrap and evaluated for positive hybridization by autoradiography. Quantification of autoradiographic signals may be obtained by densitometric analysis.

An assessment using RNA standards from GIBCO-BRL (Grand Island, NY) should reveal a single HO-1 homologous transcript of about 1.8 kb in the adult rat brain (19). The adult rat brain expresses two homologous HO-2 transcripts about 1.3 and 1.9 kb in size (19). It has been established that 1.3- and 1.9-kb transcripts differ from one another primarily due to the presence of a nearly 600-nucleotide 3' extension of monocoding sequence on the larger mRNA (16). It is notable that HO-2 homologous transcripts in the brain are subject to age-dependent regulation; the level of HO-2 mRNA expressed in the neonatal brain is much lower than that observed in the adult brain (19).

Localization of HO-1 and HO-2 mRNA in Brain

In situ hybridization has been successfully applied to reveal the cellular localization of HO gene expression in the brain (5). A nonisotopic approach using digoxigenin labeling of cDNA affords a highly sensitive technique without the hazards inherent with radioisotopes. Furthermore, immuno-chemical detection of the HO-1/HO-2 cDNA–mRNA hybrids can be completed in 1 day, whereas techniques utilizing autoradiographic detection of radiolabeled probes can take weeks to obtain proper exposure. Using this method, specific mRNA can be detected in a heterogeneous cell population where only a subpopulation may be transcribing the HO-1/HO-2 gene.

Preparation of the cDNA Probe Reagents

The cDNA probes are prepared by adaption of the PCR utilizing digoxigenin dUTP as previously described (39) with modification.

Reagents

The following reagents are needed: 100 mM Tris-HCl, pH 8.0, containing 500 mM KCl, and 15 mM $MgCl_2$ (10× PCR buffer); 1 mM digoxigenin-11–dUTP (Boehringer Mannheim, Mannheim, Germany); 100 μM oligonucleotide primers (sense and antisense); deoxynucleoside triphosphates (dNTPs: dATP, dGTP, dCTP, dTTP); plasmid template (33) [10–20 μg/ml in 10 mM Tris-HCl, pH 8.0, containing 1 mM EDTA (TE)]; Tris–Acetate–EDTA (TAE) buffer; 5 mg/ml ethidium bromide; 5 units/μl *Taq* polymerase (U.S. Biochemical, Cleveland, OH); 3 M sodium acetate, pH 5.2; absolute ethanol; mineral oil; Gene Clean kit (Bio101, La Jolla, CA).

Probe Design HO-1

The digoxigenin-labeled HO-1 cDNA most successfully used for *in situ* histo-chemical analysis was an approximately 765-bp fragment corresponding to HO-1 nucleotides +71 to +833 as described by Shibahara and co-workers

(32). The two oligonucleotide primers used were 5′ TGCACATCCGTGCA-
GAGAAT 3′, homologous to HO-1 cDNA nucleotides +71 to +90, and 5′
AGGAAACTGAGTGTGAGGAC 3′, complementary to HO-1 cDNA nucleotides +814 to +833. This HO-1 cDNA probe was selected such that it would
not hybridize to HO-2 mRNA.

Probe Design HO-2

The digoxigenin-labeled HO-2 cDNA most successfully used for *in situ* histo-
chemical analysis was a 536-bp fragment corresponding to HO-2 nucleotides
−32 to +504 as described by Rotenberg and Maines (33). The two oligonucleotide primers used were 5′ GAAGTGAGGGCAGCACAAAC3′, homologous
to HO-2 cDNA nucleotides −32 to −13, and 5′ CTTCTTCAGCACCTGG-
CCT 3′, complementary to HO-2 cDNA nucleotides +486 to +504. This
HO-2 cDNA probe was selected such that it would not hybridize to HO-1
mRNA. The HO-2 cDNA probe was, however, complementary to regions
common to both 1.3- and 1.9-kb homologous transcripts. Thus, HO-2 *in
situ* hybridization utilizing this probe represents examination of total HO-2
mRNA expression.

Amplification and Purification of HO-1/HO-2 cDNA Fragments

Into a 0.5-ml Eppendorf tube, add 76.5 μl distilled water, 10 μl of 10× PCR
buffer [1× PCR buffer is 10 mM Tris-HCl, pH 8.3, 50 mM KCl, 1.5 mM
MgCl$_2$, and 0.01% (w/v) gelatin], 10 μl of 2 mM dNTPs (dATP, dGTP, dCTP,
dTTP), and 1 μl (100 pmol) each of sense and antisense primers. To this,
add 0.5 μl *Taq* polymerase (5 units/μl, U.S. Biochemical) and 1 μl (~20 ng)
of template plasmid. Mix and overlay with 70 μl mineral oil. Amplification is
carried out in a Programmable Thermal Controller (MJ Research, Watertown,
MA) as follows: Denature at 94°C for 2 min; then amplify via 30 cycles each
of 94°C for 1 min, 57°C for 1.5 min, 70°C for 4 min; hold 70°C for 10 min;
and cool to 4°C. Add 100 μl chloroform, vortex, and centrifuge for 15 sec.
Transfer the top aqueous phase to a clean tube and then remove 5 μl for
analysis on a 1% (w/v) agarose gel in 1× TAE containing 0.5 μg/ml ethidium
bromide. If the HO-1/HO-2 cDNA fragment has been suitably amplified,
then to the remainder of the aqueous phase, add 10 μl of 3 M sodium acetate
and 300 μl absolute ethanol to precipitate the amplified HO-1 or
HO-2 fragment. Vortex the mixture and then place it on dry ice for 10–20
min prior to centrifuging (10,000 g for 15 min). Decant the supernatant and
rinse the pellet containing the cDNA with 70% ethanol. Recentrifuge the

sample for 5 min, air dry the pellet, and then resuspend in 20 μl TE. Run in two lanes of 1% (w/v) agarose gel in 1× TAE. Excise the band corresponding to about 765 and 536 bp, respectively, from the agarose gel and purify the HO-1/HO-2 fragment using Gene Clean (Bio101) as per the manufacturer's instructions. Quantitate DNA concentration by comparing with standards in a 1% (w/v) agarose gel, and store sample at −20°C.

Digoxigenin Labeling of Single-Stranded Probes

Single-stranded sense and antisense probes are prepared and purified as described above for preparation of the double-stranded template with the following modifications:

1. Only one of the two primers (HO-1, +71 to +90; HO-2, −32 to −13 for sense; and the complement of +814 to −833 and +486 to +504, respectively, for antisense) is used in the reaction.
2. The template is 5 μl (2 ng/μl; ~10 ng) of the 765- and 536-bp fragment, respectively, described above.
3. Digoxigenin–dNTP labeling mix (1 mM dATP, 1 mM dCTP, 1 mM dGTP, 0.65 mM dTTP, 0.35 mM digoxigenin-11–dUTP) replaces the standard 10 mM dNTP stock.

Perfusion, Fixation, and Pretreatment of Brain Tissue

Transcardial Perfusion and Fixation of Brain

Reagents

Prepare the following reagents: phosphate-buffered saline (PBS): into 700 ml of distilled water, dissolve 8 g NaCl, 0.2 g KCl, 1.1 g dibasic sodium phosphate, and 0.2 g monobasic potassium phosphate, and then bring to a 1 liter final volume with distilled water; standard fixative: 4% (w/v) paraformaldehyde in 0.1 M phosphate buffer containing 1.5% (w/v) sucrose; graded ethanols: 100–50% (v/v); 0.2 N HCl; xylene; 2× SSC (30 mM citric trisodium containing 0.3 M sodium chloride, pH 7.0); proteinase K solution: 10 μg/ml proteinase K (Sigma, St. Louis, MO) in PBS (this fresh solution is warmed to 37°C for 30 min before use). Note: Care should be taken to ensure that all glassware and reagents are free of RNase, which will destroy target mRNAs.

Procedure

For animals more than 14 days old, transcardial perfusion may be performed using a peristaltic pump. Depending on the individual perfusion setup, it may be necessary to adjust the tubing diameter, pump type, and intravenous catheter to give proper flow of solutions through the animals. For neonatal animals less than 14 days old, the transcardial perfusion may be performed manually using a 30 cm^3 syringe fitted with a 25-gauge needle.

Anesthetize animals by interperitoneal injection of pentobarbital (50 mg/kg), and begin the surgery when no toe-pinch reflex is observed. Open the abdominal cavity with a midline incision to the sternum. Access the heart by diagonal incisions through the rib cage to either side of the sternum and extending to each side of the neck. Clamp back the sternum, and occlude the descending aorta with hemostatic forceps to restrict the flow of fluids to the upper body. Now make a small incision in the right atria. Insert an intravenous cathether (18-gauge for adult rats; 22-gauge for neonates) into the left ventricle so that the tip is positioned in the aortic arch; begin the perfusion process by clearing the adult tissue of blood with 250 ml of 0.9% saline delivered over 10–15 min. Following clearing, fixation of the adult brain is accomplished by perfusion with 500 ml of standard fixative delivered over 20–30 min. Commonly, turgor and twitching of the upper extremities is observed on initial contact of the tissue and fixative. The neonatal brain is perfused with 100–200 ml of fixative. The brain should be removed from the skull and postfixed for 1 hr at room temperature. Care must be taken that excessive fixation does not occur, since this will render tissue useless for *in situ* hybridization. Place tissue in cassettes, dehydrate, and infiltrate with paraffin. Cut tissue to 5- to 10-μm thickness onto Superfrost Plus slides (Fisher Scientific, Pittsburgh, PA) and store at room temperature prior to use.

Pretreatment of brain sections should be performed to increase diffusion of the HO-1/HO-2 cDNA probe into the tissue as follows. Clear the brain tissue of paraffin by incubating twice for 10 min each in xylene. Then rehydrate the tissue by sequential incubations of 3 min each in graded ethanol solutions from 100 to 50% (v/v) at increments of 10% (v/v). Equilibrate the tissue in PBS with three washes for 3 min each. Next, denature tissue in 0.2 N of HCl for 20 min, and then incubate in 2× SSC for 15 min. Equilibrate the tissue in PBS twice for 3 min each, and then enzymatically deproteinate it by incubation (30 min, 37°C) with proteinase K (10 μg/ml) in PBS. Deproteinated tissue should be rinsed of excess enzyme in PBS, then fixed in 4% (w/v) paraformaldehyde for 5 min at 4°C. Finally, rinse the tissue briefly in PBS prior to dehydration by sequential incubations of 3 min each in graded ethanol solutions from 50 to 100% (v/v) and then allow it to air dry.

Hybridization and Posthybridization Treatment of Brain

Reagents

Assemble the following reagents: 20× SSC: 3 M NaCl, 0.3 M citric trisodium, pH 7.0, from this stock solution make 4×, 2×, 1×, 0.5×, 0.25×, and 0.05×

SSC by appropriate dilution with RNase-free water); yeast tRNA: 10 mg/ml divided into aliquots and stored frozen ($-20°C$); fish sperm DNA preparation: fish sperm DNA (Sigma) is dissolved in water to 10 mg/ml and sheared by trituration through a tuberculin syringe sequentially fitted with needles from 18- to 27-gauge, and then briefly sonicated (aliquots of fish sperm DNA are stored frozen at $-20°C$ until use); deionized formamide: prepare by stirring formamide with 1% (w/v) Bio-Rad AG501-X8 resin (Bio-Rad, Richmond, CA) for 1 hr at room temperature, then separate formamide from the resin by filtration through general usage filter paper, and store at $-20°C$ prior to use.

Procedure

Pipette a buffer containing 50% (v/v) formamide, 4× SSC, 1× (v/v) Denhardt's solution, and 0.25 mg/ml heat-denatured fish sperm DNA onto the dry tissue section and incubate the slides in a humidified chamber for 1 hr at room temperature. Rinse tissue for 15 min in 2× SSC and then dehydrate with graded ethanol solutions (50 to 100%, v/v) as described above. Pipette hybridization solution containing the digoxigenin-labeled HO-1/HO-2 cDNA probe diluted to 2 ng/μl in prehybridization solution onto the dry tissue, apply paraffin coverslips, and incubate slides for 16 hr at 37°C in humidified chambers. Posthybridization treatment of brain tissue is performed by successive incubation of tissue in 2× SSC (40 min, 25°C), 1× SSC (40 min, 25°C), 0.5× SSC (40 min, 37°C), 0.25× SSC (40 min, 25°C) and 0.05× SSC (40 min, 25°C), respectively, prior to immunochemical detection of digoxigenin-labeled HO-1/HO-2 cDNA–mRNA hybrids.

Immunochemical Detection of Digoxigenin-Labeled HO-1/HO-2 cDNA–mRNA Hybrids

Reagents

The following reagents are needed: buffer I: 100 mM Tris, pH 7.5, containing 150 mM NaCl; buffer II: 100 mM Tris, pH 9.5, containing 100 mM NaCl and 50 mM MgCl$_2$; buffer I containing 2% (v/v) normal sheep serum (NSS; GIBCO-BRL) and 0.5% (w/v) bovine serum albumin (BSA); antidigoxigenin Fab fragments conjugated to alkaline phosphatase (Boehringer Mannheim); NBT solution: nitro blue tetrazolium chloride in 70% (v/v) dimethyl sulfoxide (DMSO) (Boehringer Mannheim); BCIP solution: 10 mg 5-bromo-4-chloro-3-indolyl phosphate (*p*-toluidine salt; Boehringer Mannheim) is dissolved in 200 μl *N,N*-dimethylformamide just prior to use, and the solution is stored in dark; developer: to 50 ml of buffer II add 250 μl NBT solution, 176.5 μl

BCIP solution, and 12 mg levamisole (Sigma; prepare developer immediately before use and store in dark).

Procedure

Rinse tissue briefly in buffer I and block with the same buffer containing 2% (v/v) NSS and 0.5% (w/v) BSA (30 min, 25°C). The blocking solution should be replaced with 1/500 antidigoxigenin antibody conjugated to alkaline phosphatase diluted in buffer I containing 2% (v/v) NSS and 0.5% (w/v) BSA. Incubate the tissue with the primary antibody at 25°C for 1 hr in a humidified atmosphere. Rinse slides of excess primary antibody twice with buffer I (10 min, 25°C). Thereafter, equilibrate slides for 10 min in buffer II. Next visualize antibody–antigen complexes by incubating slides in buffer II containing 0.61 mM nitro blue tetrazolium, 0.52 mM 5-bromo-4-chloro-3-indolyl phosphate, and 0.024% (w/v) levamisole in the dark. Development time may vary and should be monitored periodically using a microscope. Terminate the reaction by washing the slides in distilled water prior to coverslipping with aqueous mounting medium. Store slides out of direct light.

Controls

In addition to the HO-1/HO-2 antisense probe, the appropriate HO-1/HO-2 sense probe should be run in parallel and at the same concentration. The latter constitutes a negative test, or control, for HO-1/HO-2 mRNA immunostaining. Also, a control in which no probe is added to the tissue should be tested under *in situ* hybridization conditions to confirm the absence of endogenous alkaline phosphatase activity in brain sections.

References

1. J. Garthwaite and G. Garthwaite, *J. Neurochem.* **48,** 29 (1987).
2. M. A. Marletta, *Trends Biochem. Sci.* **148,** 488 (1989).
3. S. Mancada, R. M. J. Palmer, and E. A. Higgs, *Pharmacol. Rev.* **43,** 109 (1991).
4. T. M. Dawson, V. L. Dawson, and S. H. Snyder, *Ann. Neurol.* **32,** 297 (1992).
5. J. F. Ewing and M. D. Maines, *Mol. Cell. Neurosci.* **3,** 559 (1992).
6. M. D. Maines, J. Mark, and J. F. Ewing, *Mol. Cell. Neurosci.* **4,** 398 (1993).
7. A. Verma, D. J. Hirsch, C. E. Glatt, G. V. Ronnett, and S. H. Snyder, *Science* **259,** 381 (1993).
8. J. F. Ewing, V. S. Raju, and M. D. Maines, *J. Pharmacol. Exp. Ther.* **271,** 408 (1994).
9. C. F. Stevens and Y. Wang, *Nature* (*London*) **364,** 147 (1993).
10. M. Zhuo, S. Small, E. A. Kandel, and R. D. Hawkins, *Science* **260,** 1946 (1993).
11. M. D. Maines, *Mol. Cell. Neurosci.* **4,** 389 (1993).

12. M. D. Maines, "Heme Oxygenase: Clinical Applications and Functions." Boca Raton, Florida, 1992.
13. M. D. Maines, G. M. Trakshel, and R. K. Kutty, *J. Biol. Chem.* **261,** 411 (1986).
14. I. Cruse and M. D. Maines, *J. Biol. Chem.* **263,** 3348 (1988).
15. S. Shibahara, R. M. Müller, and T. Yoshida, *Eur. J. Biochem.* **179,** 557 (1989).
16. W. K. McCoubrey, Jr., and M. D. Maines, *Gene* **139,** 155 (1994).
17. G. M. Trakshel, R. K. Kutty, and M. D. Maines, *J. Biol. Chem.* **261,** 11131 (1986).
18. G. M. Trakshel and M. D. Maines, *J. Biol. Chem.* **264,** 1323 (1989).
19. Y. Sun, M. O. Rotenberg, and M. D. Maines, *J. Biol. Chem.* **265,** 8212 (1990).
20. J. F. Ewing and M. D. Maines, *Proc. Natl. Acad. Sci. U.S.A.* **88,** 5364 (1991).
21. J. F. Ewing, S. N. Haber, and M. D. Maines, *J. Neurochem.* **58,** 1140 (1992).
22. J. F. Ewing, C. M. Weber, and M. D. Maines, *J. Neurochem.* **61,** 1015 (1993).
23. M. D. Maines, B. C. Eke, C. M. Weber, and J. F. Ewing, *J. Neurochem.* **64,** 1769 (1995).
24. J. F. Ewing and M. D. Maines, *J. Neurochem.* **60,** 1512 (1993).
25. C. M. Weber, B. C. Eke, and M. D. Maines, *J. Neurochem.* **63,** 953 (1994).
26. J. DeVente, G. J. M. Bol, and H. W. M. Steinbusch, *Brain Res.* **504,** 332 (1989).
27. H. H. Schmidt, C. D. Gacne, M. Nokong, J. S. Pollock, M. F. Miller, and F. Murad, *J. Histochem. Cytochem.* **40,** 1439 (1992).
28. S. R. Vincent and H. Kimura, *Neuroscience (Oxford)* **46,** 755–784 (1992).
29. L. J. Ignarro, B. Ballot, and K. S. Wood, *J. Biol. Chem.* **259,** 6201 (1984).
30. R. M. J. Palmer, A. G. Ferrige, and S. Moncada, *Nature (London)* **327,** 524 (1987).
31. S. R. Vincent, S. Das, and M. D. Maines, *Neuroscience (Oxford)* **63,** 223 (1994).
32. S. Shibahara, R. M. Müller, H. Taguchi, and T. Yoshida, *Proc. Natl. Acad. Sci. U.S.A.* **82,** 7865 (1985).
33. M. O. Rotenberg and M. D. Maines, *J. Biol. Chem.* **265,** 7501 (1990).
34. J. C. Alwine, D. J. Kemp, and G. R. Stark, *Proc. Natl. Acad. Sci. U.S.A.* **74,** 5350 (1977).
35. P. S. Thomas, *Proc. Natl. Acad. Sci. U.S.A.* **77,** 5201 (1980).
36. J. Sambrook, T. Maniatis, and E. F. Fritsch, "Molecular Cloning: A Laboratory Manual." Cold Spring Harbor Laboratory, Cold Spring Harbor, New York, 1982.
37. J. M. Chirgwin, A. E. Przybyla, R. J. MacDonald, and W. J. Rutter, *Biochemistry* **18,** 5294 (1979).
38. R. E. Kingston, *in* "Current Protocols in Molecular Biology" (F. M. Ausubel, R. Brent, D. D. Moore, J. A. Smith, J. G. Seidman, and K. Struhl, eds.), pp. 4.5.1–4.5.3, Wiley, New York, 1987.
39. M. Iwamura, G. Wu, P. Abrahamsson, P. A. DiSant'agnese, A. T. K. Cockett, and L. J. Deftos, *J. Urol.* **43,** 667 (1994).

Section II

Enzyme Expression, Purification, and Gene Regulation

[10] Purification of Endothelial Nitric Oxide Synthase

Jennifer S. Pollock and Masaki Nakane

Introduction

Endothelial nitric oxide (NO) synthase was found to be located mainly (>90%) in the particulate fraction of cultured and native bovine aortic endothelial cells (1, 2). In addition, endothelial NO synthase was found mainly (>85%) in the particulate fraction of cultured human aortic endothelial cells (3), and there were also substantial amounts (65%) found in the particulate fraction of cultured human umbilical vein endothelial cells (4). On the other hand, it was found in equal distribution between the soluble and particulate fractions from human placental tissue (4). It has been determined that endothelial NO synthase is myristoylated at the N terminus, thus accounting for its preponderance to the membrane fraction (5–7).

Purification of Endothelial Nitric Oxide Synthase from Bovine Aortic Endothelial Cells

The endothelial isoform from cultured and native bovine aortic endothelial cells (BAEC) can be purified by a simple two-step purification scheme (8). Purification of the solubilized bovine particulate enzyme preparation by affinity chromatography on adenosine $2',5'$-bisphosphate coupled to Sepharose ($2',5'$-ADP-Sepharose) followed by FPLC (fast protein liquid chromatography) Superose 6 (Pharmacia, Piscataway, NJ) gel filtration chromatography results in a 3419-fold purification from the crude particulate fraction with a 12% recovery. This scheme also results in a single protein band after denaturing polyacrylamide gel electrophoresis that corresponds to 135 kDa.

The BAEC utilized in these preparations are from passages 6–15. No differences are noted in the yield of purified particulate endothelial NO synthase from these passages. However, it is noted that at passages higher than 18 the proportion of soluble and particulate enzymatic activities changes. The percentage of total activity in the soluble fractions increases to approximately 20%, with 80% in the particulate fraction; therefore, the purifications are accomplished on low passage cells.

The BAEC are cultured to confluence in roller bottles, scraped, washed in phosphate-buffered saline (PBS), and homogenized with Dounce homogenizer in ice-cold buffer 1 containing 50 mM Tris-HCl (pH 7.4), 0.1 mM EDTA, 0.1 mM EGTA, and 0.1% 2-mercaptoethanol in the presence of protease inhibitors [1 mM phenylmethylsulfonyl fluoride (PMSF), 2 μM leupeptin, and 1 μM pepstatin A]. All steps are performed at 4°C or on ice, and the purification protocol is accomplished as quickly as possible to obtain the highest activity possible. The crude homogenate is then centrifuged at 100,000 g for 60 min, and the particulate fraction is resuspended with buffer 1 including 10% glycerol, protease inhibitors, and 1 M KCl and then centrifuged at 100,000 g for 30 min. We find that the KCl wash washes away many "sticky" proteins. The KCl-washed particulate fraction is then solubilized with the detergent 3-[(3-cholamidopropyl)dimethylammonio]-1-propane sulfonate (CHAPS, 20 mM) in buffer 1 including 10% glycerol and protease inhibitors for 20 min at 4°C with gentle rotation of the sample and centrifuged at 100,000 g for 30 min. The CHAPS extract can be frozen at this point to await further purification.

The CHAPS extract is further fractionated by 2′,5′-ADP-Sepharose chromatography followed by FPLC Superose 6 gel filtration. In all subsequent steps of the purification 10 mM CHAPS is included in the buffers to facilitate enzyme solubility. The CHAPS extract is loaded onto a 150- to 200-μl column of preswollen 2′,5′-ADP-Sepharose (Pharmacia) and recirculated at least four times. The column is then washed with 1 ml of buffer 1 including 10% glycerol, 10 mM CHAPS, and 0.5 M NaCl and then with 1 ml of buffer 1 including 10% glycerol, 10 mM CHAPS. The endothelial NO synthase activity is eluted with two 500-μl aliquots of buffer 1 including 10% glycerol, 10 mM CHAPS, and 10 mM NADPH. The eluate volume is kept to a minimum to ensure the highest possible protein concentration as well as a minimum volume to inject onto the gel permeation column. Concentration of the NADPH eluate has been attempted; however, the loss in activity is great without exogenous bovine serum albumin to stabilize the activity. The NADPH eluate from the 2′,5′-ADP Sepharose column is immediately injected on the FPLC Superose 6 gel permeation column equilibrated with buffer 2 containing 10 mM Tris-HCl (pH 7.4), 1 mM EDTA, 0.1% 2-mercaptoethanol, 100 mM NaCl, 10% glycerol, and 10 mM CHAPS. Endothelial NO synthase elutes anomalously on the Superose 6 column near the total included volume due to nonspecific interactions with the column material. Superose 6 chromatography yields a pure enzyme preparation that is partially tetrahydrobiopterin-dependent. When Superose 12 is utilized for purification, the enzyme preparation is not completely pure but is substantially dependent on exogenous tetrahydrobiopterin.

The purification of endothelial NO synthase from the soluble fraction of cultured bovine aortic endothelial cells is attempted utilizing the same purification scheme as the particulate enzyme, although the buffers do not include 10 mM CHAPS. Several preparations are pooled, protease inhibitors are added, and the mixture is loaded onto a 400-μl column of 2',5'-ADP-Sepharose and circulated overnight. The column is then washed with 3 ml of buffer 1 containing 0.5 M NaCl and then with 3 ml of buffer 1 alone. The soluble enzymatic activity is eluted with six 500-μl volumes of buffer 1 containing 10 mM NADPH. A sample of the NADPH eluate is applied to a FPLC Superose 6 gel permeation column equilibrated with buffer 2 excluding CHAPS. The soluble endothelial NO synthase activity also elutes near the total included volume of the column at the same retention volume as purified particulate endothelial NO synthase.

Native BAEC are obtained from 50–60 fresh bovine aortas transported from the slaughterhouse in ice-cold PBS. The aortas are cleared of connective tissue, cut longitudinally, and mounted with the luminal surface uppermost. The endothelial cells are removed by scraping and collected into ice-cold PBS. The cells are washed twice in PBS before being resuspended in buffer 1 and homogenized. Homogenization, solubilization, and purification follow the same protocol as for the cultured BAEC.

Protein is determined according to Bradford (9) with bovine serum albumin as a standard. The purity of enzyme preparations is assessed with 7.5% polyacrylamide gels [as described by Laemmli (10)] and visualized with silver staining. Purified enzyme preparations are stored at −70°C in the presence of 10% glycerol and exogenous bovine serum albumin to stabilize the activity for several months. Nitric oxide synthase activity is assessed by determining the conversion of [³H]arginine to [³H]citrulline (11) and by bioassay for the formation of NO using the stimulation of cGMP formation in RFL-6 cells in the presence of arginine, cofactors, and superoxide dismutase (8).

Purification of Endothelial Nitric Oxide Synthase from Human Placental Tissue

The endothelial isoform from human placenta is purified by a three-step purification scheme as outlined by Garvey et al. (4). Purification of the solubilized human enzyme preparation by affinity chromatography on 2',5'-ADP-Sepharose followed by affinity chromatography on calmodulin-agarose and ion-exchange chromatography on a Mono Q HR 5/5 column results in a 3800-fold purification from the crude placental extract with an 11% recovery. This scheme also results in a single protein band after denaturing polyacrylamide gel electrophoresis that corresponds to 135 kDa. The following is a brief summary of the purification of endothelial NO synthase from human placenta (4).

Placenta is cleaned of the amnion and chorion, washed with isotonic saline, and homogenized in a Waring blender in 3 volumes of buffer A containing 20 mM HEPES (pH 7.8), 0.1 mM EDTA, 5 mM dithiothreitol (DTT), 0.2 M sucrose including a protease inhibitor (0.1 mM PMSF). Nitric oxide synthase is extracted with 5 mM CHAPS, mixed for 30 min, and then centrifuged at 27,500 g for 30 min. No separation of the soluble and particulate fractions is accomplished for the bulk preparations. Affinity purification of the NO synthase is accomplished with the addition of 10 ml of 2′,5′-ADP-Sepharose to the CHAPS extract solution, and the slurry is allowed to equilibrate slowly before pouring into a column. The packed 2′,5′-ADP-Sepharose is washed with buffer A, buffer A including 0.5 M NaCl, and finally with buffer A again. The NO synthase activity is eluted with 2 mM NADPH in buffer A. The pooled NADPH eluate is then loaded onto a 0.5-ml column of calmodulin-agarose preequilibrated with buffer A containing 1 mM CaCl$_2$. The calmodulin-agarose column is washed with 20 volumes of buffer A containing 1 mM CaCl$_2$ and 1% Nonidet P-40 (NP-40) then with 10 volumes of buffer A containing 1 mM CaCl$_2$. The NO synthase activity is eluted with buffer A containing 5 mM EDTA and 1 M NaCl. The EDTA eluate is pooled, concentrated, and desalted in buffer A containing 0.1% Tween 20. The partially purified NO synthase preparation is applied to a Mono Q HR 5/5 column equilibrated with buffer A containing 0.1% Tween 20, and the column is then washed with 5 ml of buffer A containing 0.1% Tween 20. An elution gradient is programmed from 0 to 500 mM NaCl in buffer A containing 0.1% Tween 20, with the NO synthase activity eluting at about 300 mM NaCl. The activity fractions are pooled, desalted, and concentrated to 0.5 mg/ml.

Protein concentration is determined by the method of Bradford (9) with bovine serum albumin utilized as the standard. The purity of enzyme preparations is assessed with 7.5% polyacrylamide gels as described by Laemmli (10). Purified enzyme preparations are stored at −70°C. Nitric oxide synthase activity is monitored by the conversion of [^{14}C]arginine to [^{14}C]citrulline as described previously (11).

References

1. U. Förstermann, J. S. Pollock, H. H. H. W. Schmidt, M. Heller, and F. Murad, *Proc. Natl. Acad. Sci. U.S.A.* **88,** 1788 (1991).
2. J. A. Mitchell, U. Förstermann, T. D. Warner, J. S. Pollock, H. H. H. W. Schmidt, M. Heller, and F. Murad, *Biochem. Biophys. Res. Commun.* **176,** 1417 (1991).

3. J. S. Pollock, unpublished results.
4. E. P. Garvey, J. V. Tuttle, K. Covington, B. M. Merrill, E. R. Wood, S. A. Baylis, and I. G. Charles, *Arch. Biochem. Biophys.* **311,** 235 (1994).
5. J. S. Pollock, V. Klinghofer, U. Förstermann, and F. Murad, *FEBS Lett.* **309,** 402 (1992).
6. L. Busconi and T. Michel, *J. Biol. Chem.* **268,** 8410 (1993).
7. W. C. Sessa, C. M. Barber, and K. R. Lynch, *Circ. Res.* **72,** 921 (1993).
8. J. S. Pollock, U. Förstermann, J. A. Mitchell, T. D. Warner, H. H. H. W. Schmidt, M. Nakane, and F. Murad, *Proc. Natl. Acad. Sci. U.S.A.* **88,** 10480 (1991).
9. M. M. Bradford, *Anal. Biochem.* **72,** 248 (1976).
10. U. K. Laemmli, *Nature (London)* **227,** 680 (1970).
11. H. H. H. W. Schmidt, J. S. Pollock, M. Nakane, L. D. Gorsky, U. Förstermann, and F. Murad, *Proc. Natl. Acad. Sci. U.S.A.* **88,** 365 (1991).

[11] Purification of Brain Nitric Oxide Synthase from Baculovirus Overexpression System and Determination of Cofactors

Bernd Mayer, Peter Klatt, Christian Harteneck,
Barbara M. List, Ernst R. Werner, and Kurt Schmidt

Introduction

Oxidation of L-arginine to L-citrulline and nitric oxide (NO) is catalyzed by at least three different NO synthase (NOS) isozymes: the constitutively expressed neuronal (n) and endothelial (e) isoforms, which require Ca^{2+}/calmodulin for activity, and a Ca^{2+}-independent form, which is expressed in most mammalian cells in response to cytokines and endotoxin (1). Albeit regulated in a different manner, the NOS isoforms identified so far are biochemically similar. They exist as homodimers under native conditions, with subunit molecular masses ranging from 130 kDa [endothelial (e) and inducible (i) NOSs] to 160 kDa (neuronal NOS). L-Arginine oxidation and reductive activation of molecular oxygen is catalyzed by a cysteine-ligated P450-like prosthetic heme group localized in the N-terminal half of the protein (2, 3). This part of the enzyme is thus referred to as the oxygenase domain of NOS. The C-terminal half of the enzyme, separated from the oxygenase domain by a binding site for calmodulin, contains the flavins FAD and FMN which serve to shuttle reducing equivalents from the cofactor NADPH to the heme. This reductase domain of NOS shows pronounced sequence similarities to cytochrome P450 reductase (4) and exhibits similar catalytic activities (5). These results suggest that NOSs represent self-sufficient cytochrome P450s with both reductase and oxygenase activities associated with one single protein. In contrast to the classic microsomal cytochrome P450 systems, however, NOSs require the pteridine tetrahydrobiopterin as additional cofactor with as yet unknown function (6–8).

We have worked out a method for overexpression of recombinant rat brain NOS in a baculovirus–insect cell system (9). Here we describe protocols for large-scale purification of the enzyme and for determination of the tightly bound prosthetic groups heme, FAD, FMN, and tetrahydrobiopterin.

Methods in Neurosciences, Volume 31

Infection of Sf9 Cells with Recombinant Baculovirus

The cDNA encoding for rat brain NOS (Genbank accession numbers X59949) (4), a generous gift from Drs. D. S. Bredt and S. H. Snyder, is cloned in the baculovirus expression vector pVL1393 according to standard laboratory protocols as described previously (9). Fall armyworm ovary cells (*Spodoptera frugiperda*; Sf9) are obtained from ATCC (Rockville, MD). Cells are grown in TC-100 medium (Sigma, St. Louis, MO) supplemented with 10% fetal calf serum (SEBAK GmbH, Suben, Austria), amphotericin B (1.25 mg/liter), penicillin (100,000 U/liter), streptomycin (100 mg/liter), and 0.2% Pluronic F-68 (Sigma) under continuous shaking (60–70 rpm) using an orbital shaker. Optimal growth (doubling time of ~24 hr) is achieved at cell densities ranging from 0.5×10^6 to 2×10^6 cells/ml and a temperature of $27° \pm 1°C$. The recombinant virus is generated by cotransfection of Sf9 cells with the expression vector and BaculoGOLD baculovirus DNA (Dianova, Hamburg, Germany) using the lipofection method (10). Positive viral clones are isolated by plaque assay and identified by their ability to direct the expression of NOS as revealed by appearance of a 150- to 160-kDa band in immunoblots of the cell extracts using a polyclonal antibody against brain NOS (ALEXIS, Läufelfingen, Switzerland) for detection. The purified virus is then amplified and titer estimated by plaque assay.

For expression of NOS, the culture medium is supplemented with 20% fetal calf serum, and Sf9 cells (4.5×10^9 cells/3000 ml) are infected with the recombinant baculovirus at a ratio of 5 pfu/cell in the presence of 4 mg/liter hemin (Sigma). Stock solutions (2 mg/ml) of hemin are prepared in 0.4 M NaOH/ethanol (1 : 1, v/v). After 48 hr, cells are harvested by centrifugation for 3 min at 1000 g and washed with 500 ml of serum-free TC-100 medium. Finally, cells are resuspended in buffer A (see solutions and buffers below) to give a final volume of 100 ml, and used for enzyme preparation as described below.

Enzyme Purification

The protocol for purification of rat brain NOS involves two sequential affinity chromatography steps. First, the enzyme is bound to 2′,5′-ADP-Sepharose 4B (Pharmacia, Piscataway, NJ) and eluted with excess 2′-AMP. Subsequently it is chromatographed in the presence of Ca^{2+} over calmodulin-Sepharose 4B (Pharmacia) and eluted by chelating Ca^{2+} with excess EGTA.

The following buffers and solutions should be prepared 1 day before enzyme preparation in double-distilled water and be adjusted to pH 7.4 at ambient temperature (20°–25°C). The buffers are passed through a 0.22-μm filter and kept overnight at 4°–6°C.

Solutions and Buffers

2-Mercaptoethanol should be added to buffers A–D immediately before use.

> EDTA (volume 1.5 ml): 0.5 M EDTA (298 mg Na$_4$EDTA · H$_2$O)
> CaCl$_2$ (volume 10 ml): 200 mM CaCl$_2$ (296 mg CaCl$_2$ · 2H$_2$O)
> Buffer A (volume 500 ml): 50 mM triethanolamine (TEA) hydrochloride (4.64 g) adjusted to pH 7.4 with 5 M NaOH, 0.5 mM EDTA (0.5 ml EDTA stock), 12 mM 2-mercaptoethanol (0.5 ml)
> Buffer B (volume 500 ml): 50 mM TEA hydrochloride (4.64 g), 0.5 M NaCl (14.62 g) adjusted to pH 7.4 with 5 M NaOH, 12 mM 2-mercaptoethanol (0.5 ml)
> Buffer C (volume 500 ml): 20 mM Tris-HCl (1.21 g Tris base), 150 mM NaCl (4.38 g) adjusted to pH 7.4 with 6 M HCl, 2 mM CaCl$_2$ (5 ml CaCl$_2$ stock), 12 mM 2-mercaptoethanol (0.5 ml)
> Buffer D (volume 250 ml): 20 mM Tris-HCl (0.61 g Tris base), 150 mM NaCl (2.19 g), 4 mM EGTA (0.38 g) adjusted to pH 7.4 with 6 M HCl, 12 mM 2-mercaptoethanol (0.25 ml)

Column Chromatography

The geometry of columns is probably not critical, but bed volumes should be adapted to the scale of the enzyme preparations to avoid dilution of the protein or overloading of the columns. We have used two 20 × 2.5 cm Econo-Columns (Bio-Rad, Richmond, CA) equipped with flow adaptors and filled with the affinity resins to give bed volumes of 35 and 50 ml for 2′,5′-ADP-Sepharose and calmodulin-Sepharose, respectively. Flow rates are 2 ml/min throughout the enzyme preparation. Columns are equilibrated with four bed volumes of buffer A (2′,5′-ADP-Sepharose) or C (calmodulin-Sepharose) prior to enzyme purification.

Regeneration of the columns is performed at flow rates of 1 ml/min. The 2′,5′-ADP-Sepharose is washed with 100 ml of alkaline buffer (0.1 M Tris, 0.5 M NaCl, 0.1% (w/v) NaN$_3$, pH 8.5) followed by 100 ml of acidic buffer (0.1 M sodium acetate, 0.5 M NaCl, 0.1% (w/v) NaN$_3$, pH 4.5). This cycle is repeated three times, and finally the column is washed with 100 ml of double-distilled water containing 0.1% (w/v) NaN$_3$ for storage. Calmodulin-Sepharose is regenerated with 300 ml of 50 mM Tris, 2 mM EGTA, 1 M NaCl, and 0.1% NaN$_3$, pH 7.5, and finally equilibrated with 150 ml of 20 mM Tris, 2 mM CaCl$_2$, 150 mM NaCl, and 0.1% NaN$_3$, pH 7.5, for storage.

Purification Procedure

Infected cells suspended in buffer A and kept on ice (see above) are sonicated four times, 10 sec each time, at 150 W, homogenates are centrifuged for 15 min at 30,000 g [Sorvall (Bloomington, DE) SS-34 rotor; 17,000 rpm], and the supernatant is collected. To increase recovery of NOS by about 50%, the pellet is washed once with buffer A (final volume 50 ml). The combined supernatant is loaded onto the equilibrated 2′,5′-ADP-Sepharose column, which is subsequently washed with 100 ml of buffer B. Elution of bound enzyme is performed with 150 ml of 20 mM 2′-AMP (Sigma; containing ~50% of inactive 3′-AMP) in buffer B, and fractions (3 ml) containing NOS (visible as red-brown color) are pooled to give about 100 ml of 2′,5′-ADP Sepharose eluate. The 2′,5′-ADP-Sepharose eluate is adjusted to 2 mM $CaCl_2$ by addition of $CaCl_2$ stock solution (1%; v/v) and passed over the calmodulin-Sepharose column. The column is washed with 150 ml of buffer C, and the enzyme is eluted with 100 ml of buffer D. Fractions (3 ml) containing NOS (visible as red-brown color) are pooled to give about 30 ml of eluate, containing approximately 3 mg of purified NOS per milliliter. Appropriate aliquots of the enzyme should be stored at −70°C.

To monitor enzyme purification, the various fractions should be assayed for conversion of [³H]arginine to [³H]citrulline as described in detail (11). It should be noted that the pure enzyme turned out to be unstable if diluted or assayed in the absence of either serum albumin (2–10 mg/ml) or certain detergents. Thus, we routinely dilute the enzyme in the presence of 1 mM CHAPS, giving final concentrations of 0.2 mM CHAPS in enzyme assays.

Table I shows the results of a representative enzyme purification. Homogenization of 4.5×10^9 cells yielded about 3 g of soluble protein with a specific NOS activity of 0.048 μmol citrulline mg^{-1} min^{-1}. Chromatography over 2′,5′-ADP-Sepharose yielded about 120 mg of protein with a specific activity

TABLE I   Purification of Rat Brain Nitric Oxide Synthase from Recombinant Baculovirus-Infected Sf9 Cells[a]

Fraction	Volume (ml)	Protein (mg)	Total activity (μmol min^{-1})	Specific activity (μmol mg^{-1} min^{-1})	Purification (-fold)	Yield (%)
30,000 g supernatant	85	2940	141	0.048	1	100
2′,5′-ADP-Sepharose	95	122	112	0.92	19	79
Calmodulin-Sepharose	32	92	91	0.99	21	64

[a] The NOS enzyme was purified from 4.5×10^9 Sf9 cells, which had been infected with rat brain NOS–recombinant baculovirus for 48 hr in 3000 ml of culture medium supplemented with 4 mg/I hemin. Enzyme activity was measured at 37°C for 10 min as conversion of [³H]arginine (0.1 mM) to [³H]citrulline.

of 0.9 μmol citrulline mg^{-1} min^{-1}, which is similar to the activities previously described for NOS purified from rat and porcine brain (12, 13). Further purification over calmodulin-Sepharose resulted in about 2-fold concentration of the protein and slightly enhanced specific activities. Loss of enzyme was about 25% in this step, giving an overall yield of 64% of NOS activity over crude supernatants. In the calmodulin-Sepharose eluate, the NOS concentration was 2.9 mg/ml, corresponding to a 18 μM solution of the 160-kDa subunits.

Determination of Enzyme-Bound Cofactors

Determination of NOS-bound cofactors is essential for characterization of the purified enzyme. In the following, we describe high-performance liquid chromatography (HPLC) methods for the quantitative measurement of heme, the flavins FAD and FMN, and tetrahydrobiopterin. We use the HPLC system from Merck (Vienna, Austria) (LiChroGraph L-6200) equipped with a low pressure gradient controller, UV/Vis (L-4250) and fluorescence (F-1050) detectors, and an autosampler (AS-4000). Chromatography is performed on a 250 × 4 mm C_{18} reversed-phase column fitted with a 4 × 4 mm C_{18} guard column (LiChrospher 100 RP-18, 5-μm particle size).

Determination of Heme

Solvent

Solvent A: acetonitrile–water–trifluoroacetic acid, 60:40:0.1 (v/v/v)

Procedure

For the sensitive determination of enzyme-bound heme by HPLC and UV detection, we have modified a previously described method (14) as follows. Protein solutions are thoroughly mixed with 4 volumes of solvent A, and aliquots (30 μl) are injected onto the C_{18} reversed-phase column described above. Elution is performed isocratically with solvent A at a flow rate of 1 ml/min, and heme is detected by its absorbance at 398 nm (retention time 5.4 min). The method is calibrated with freshly prepared solutions of horse heart myoglobin. Heme calibration curves are linear from 0.1 to 10 μM

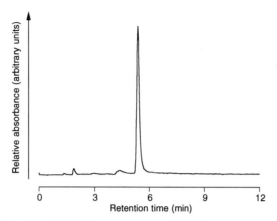

FIG. 1   HPLC chromatogram of NOS-bound heme.

(3–300 pmol per 30-μl injection). Figure 1 shows a representative HPLC chromatogram obtained with rat brain NOS (0.12 mg/ml) purified from Sf9 cells that had been infected with recombinant baculovirus in the presence of 4 mg/ml hemin. Quantitative evaluation of the chromatogram showed that this particular NOS preparation contained 0.94 $\pm$ 0.03 (mean $\pm$ SEM, $n = 3$) equivalent heme per subunit.

Determination of the Flavins FAD and FMN

Solvents

Solvent A: 10 mM K$_2$HPO$_4$ adjusted to pH 6.0 with 85% H$_3$PO$_4$
Solvent B: methanol

Procedure

The noncovalently bound flavins FAD and FMN are released from NOS to the supernatant by heat denaturation of the enzyme (95°C for 5 min in the dark). It is essential to use well-sealed vials for this procedure in order to avoid uncontrolled reduction of sample volume. Subsequently, samples are cooled to 0°–4°C, and to recover water, which may have condensed at the top of the vials, vials are spun down by centrifugation. Samples are kept at 4°C in the dark until HPLC analysis. Then 30 μl of sample is injected onto the reversed-phase column, and a linear gradient from 80% solvent A and 20% solvent B ($t = 0$) to 50% solvent A and 50% solvent B ($t = 10$ min) is generated at a flow rate of 1 ml/min. Prior to the injection of the next sample, the gradient is reversed within 10 min, followed by a further 10 min of column equilibration. FAD and FMN are detected by fluorescence at excitation and emission wavelengths of 450 and 520 nm, respectively. The method is calibrated with solutions of authentic FAD and FMN freshly prepared in the buffer used for dilution of NOS and subjected to heat treatment as described above. The FAD calibration curves are linear from 10 nM–1 μM (0.3–30 pmol per 30-μl injection). Fluorescence of FMN is about 3.5-fold higher than that of FAD, allowing reliable detection of FMN down to 3 nM (0.1 pmol per 30-μl injection).

Figure 2 shows a representative HPLC chromatogram obtained with rat brain NOS (16 μg/ml), purified from Sf9 cells that had been infected with a recombinant baculovirus in the presence of 0.1 mM riboflavin. The flavins FAD and FMN are well separated with retention times of 8.8 and 9.9 min, respectively. Quantitative evaluation of the peak areas showed that the FAD and FMN content of this particular NOS preparation was 0.74 ± 0.03 and 0.80 ± 0.02 equivalent per subunit (mean values ± SEM, $n = 3$), respectively.

Determination of Tetrahydrobiopterin

The reduced pteridines tetrahydrobiopterin and dihydrobiopterin can be oxidized to biopterin by treatment of KI/I$_2$ under acidic conditions. To selectively determine the amount of tetrahydrobiopterin, a second series of samples is oxidized at alkaline pH. Under these conditions, dihydrobiopterin is

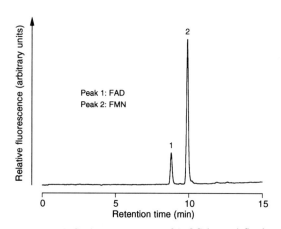

FIG. 2   HPLC chromatogram of NOS-bound flavins.

converted to biopterin, whereas tetrahydrobiopterin loses the side chain at C-6 and is oxidized to pterin. Thus, the difference between the amount of biopterin formed by oxidation at acidic pH (i.e., total biopterin) versus biopterin formed at alkaline pH (i.e., total biopterin minus tetrahydrobiopterin) corresponds to the amount of tetrahydrobiopterin in the sample.

Solutions and Buffers

Solution A: 0.1 M I_2 dissolved in an aqueous solution of 0.5 M KI (stable for at least 1 year at room temperature in the dark)

1 M HCl

1 M NaOH

0.2 M ascorbic acid (prepared freshly)

Buffer B: 20 mM NaH_2PO_4 (adjusted to pH 3.0 with 85% H_3PO_4) in 5% (v/v) methanol

Procedure

For acidic oxidation, solution A is mixed with an equal volume of 1 M HCl, and 10 μl of this freshly prepared reagent is added to 50 μl of a solution containing 5–50 pmol NOS. A second series of samples is subjected to alkaline oxidation by treating 50-μl aliquots with 10 μl of a freshly prepared mixture of solution A and 1 M NaOH (1:1, v/v). The amount of the I_2/KI solution required for complete oxidation of pteridines largely depends on the amount of reducing agents present in the samples. The method has been worked out for NADPH-free NOS solutions containing 12 mM 2-mercapto-ethanol. Sufficient I_2/KI has been added when a brownish-yellow color is still observable after the 1-hr oxidation period and neutralization. Tubes are kept at ambient temperature in the dark for 1 hr prior to neutralization of acidic and alkaline samples with 5 μl of 1 M NaOH and 5 μl of 1 M HCl, respectively. Subsequently, excess iodine is destroyed by the addition of 5 μl of 0.2 M ascorbic acid. The amount of ascorbic acid must be increased when the sample does not become colorless after addition of the reductant. Owing to the limited stablity of biopterin, it is recommended to analyze samples within a few hours by injection of 30 μl of the sample onto the reversed-phase column described above. Pteridines are eluted isocratically with buffer B at a flow rate of 1 ml/min and detected by fluorescence at excitation and emission wavelengths of 350 and 440 nm, respectively. Run times are routinely set to 15 min without any additional column regeneration required for at least 50 samples. The method is calibrated by acidic oxidation of authentic tetrahydrobiopterin under strictly identical conditions. Biopterin calibration curves are linear in the range from 10 nM to 1 μM (0.2–20 pmol per 30-μl injection).

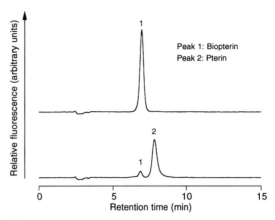

FIG. 3   HPLC chromatogram of NOS-bound tetrahydrobiopterin. *Top*: acidic oxida-toin; *bottom*: alkaline oxidation.

Figure 3 shows HPLC chromatograms obtained by treatment of purified recombinant rat brain NOS (40 μg/ml) with I_2/KI at acidic (upper trace) and alkaline pH (lower trace). Under acidic conditions, biopterin is the sole reaction product (Peak 1, 6.9 min, Fig. 3), and a total biopterin content of 0.44 ± 0.01 equivalent per subunit (mean $\pm$ SEM, $n = 3$) is calculated from the peak area. Under alkaline conditions, two peaks are observed, which correspond to biopterin (peak 1, 6.9 min, Fig. 3) and tetrahydrobiopterin-derived pterin (peak 2, 7.9 min, Fig. 3). The amount of alkali-stable biopterin (derived from dihydrobiopterin) is calculated as 0.031 ± 0.001 equivalent per subunit (mean $\pm$ SEM, $n = 3$). Accordingly, this particular NOS preparation contained 0.41 ± 0.025 mol tetrahydrobiopterin per mole of protein subunit.

Acknowledgments

We acknowledge the excellent technical assistance of Eva Leopold. This work was supported by Grants P 10098 (B.M.), P 10573 (K.S.), and P 9685 (E.R.W.) of the Fonds zur Förderung der Wissenschaftlichen Forschung in Austria and the Deutsche Forschungsgemeinschaft (SFB 366) in Germany.

References

1. O. W. Griffith and D. J. Stuehr, *Annu. Rev. Physiol.* **57,** 707 (1995).
2. P. F. Chen, A. L. Tsai, and K. K. Wu, *J. Biol. Chem.* **269,** 25062 (1994).
3. M. K. Richards and M. A. Marletta, *Biochemistry* **33,** 14723 (1994).

4. D. S. Bredt, P. M. Hwang, C. E. Glatt, C. Lowenstein, R. R. Reed, and S. H. Snyder, *Nature* (*London*) **351,** 714 (1991).
5. P. Klatt, B. Heinzel, M. John, M. Kastner, E. Böhme, and B. Mayer, *J. Biol. Chem.* **267,** 11374 (1992).
6. P. Klatt, M. Schmid, E. Leopold, K. Schmidt, E. R. Werner, and B. Mayer, *J. Biol. Chem.* **269,** 13861 (1994).
7. B. Mayer, P. Klatt, E. R. Werner, and K. Schmidt, *J. Biol. Chem.* **270,** 655 (1995).
8. B. Mayer and E. R. Werner, *Naunyn-Schmiedeberg's Arch. Pharmacol.* **351,** 453 (1995).
9. C. Harteneck, P. Klatt, K. Schmidt, and B. Mayer, *Biochem. J.* **304,** 683 (1994).
10. D. R. Groebe, A. E. Chung, and C. Ho, *Nucleic Acids Res.* **18,** 4033 (1990).
11. B. Mayer, P. Klatt, E. R. Werner, and K. Schmidt, *Neuropharmacology* **33,** 1253 (1994).
12. D. S. Bredt and S. H. Snyder, *Proc. Natl. Acad. Sci. U.S.A.* **87,** 682 (1990).
13. B. Mayer, M. John, and E. Böhme, *FEBS Lett.* **277,** 215 (1990).
14. J. T. Kindt, A. Woods, B. M. Martin, R. J. Cotter, and Y. Osawa, *J. Biol. Chem.* **267,** 8739 (1992).

[12] Prokaryotic Expression, Purification, and Characterization of Intact and Modular Constructs of Neuronal Nitric Oxide Synthase

Bettie Sue Siler Masters, Kirk McMillan, Linda J. Roman, Pavel Martasek, and Jonathan Nishimura

Introduction

The ability to isolate and purify the various isoforms of nitric oxide synthase (NOS) has enabled investigators to determine the molecular nature of enzymes that produce nitric oxide (NO·) from the monooxygenation of L-arginine to L-citrulline. The varied functions of NO· are vital to the maintenance of hemodynamic homeostasis by endothelial cells, to the formation of ultimately cytotoxic species in phagocytic cells (macrophages) that kill bacteria, fungi, and even tumor cells, and to the production of intracellular signals and/or retrograde neurotransmitters in neuronal cells. Because these diverse functions are of physiological and, often, pathological significance, it has become important to understand how the three major isoforms are regulated physiologically and what unique properties of each could lend themselves to the design of potent and specific inhibitors. The constitutive neuronal (nNOS; NOS-I) and inducible (macrophage; iNOS; NOS-II) isoforms were the first to be purified to the extent that the heme constituency could be unequivocally ascertained (1–4), and this revelation led to an entirely new examination of the mechanism by which these enzymes perform the incorporation of molecular oxygen into the substrate L-arginine to form the oxygenated products L-citrulline (5) and NO· (5–8). The requirement for (6R)-5,6,7,8,-tetrahydro-L-biopterin (BH$_4$) had been established earlier by Tayeh and Marletta (9) and Kwon *et al.* (10) for iNOS and by Bredt *et al.* (11) and Mayer *et al.* (12) for nNOS. The assumption had been made, by analogy to the aromatic amino acid hydroxylases (13), that BH$_4$ would be involved in the oxygenation mechanism. This role for BH$_4$ has been under considerable debate, and the predominant mechanisms suggest that heme is both necessary and sufficient for both hydroxylation steps, namely, the formation of N^ω-hydroxy-L-arginine (NHA) and, subsequently, L-citrulline plus NO·. On the other hand, there is strong evidence for a regulatory and conformational role for BH$_4$ in the various nitric oxide synthases (14, 15).

Methods in Neurosciences, Volume 31

The fact that there remained a number of uncertainties regarding the structure–function aspects of the nitric oxide synthases prompted our laboratory to embark on a series of studies beginning with efforts to purify large quantities of the intact rat neuronal isoform from stably transfected human kidney 293 cells (3), followed by the dissection of the protein into functional heme protein and flavoprotein domains. Modules of the heme domain were also created by molecular cloning techniques, permitting expression in appropriate strains of *Escherichia coli* by McMillan and Masters (16) and Nishimura *et al.* (17), respectively. While pursuing these experiments, Roman *et al.* (18) were successful in expressing the intact neuronal nitric oxide synthase in *E. coli* by utilizing a combination of techniques, including the coexpression of molecular chaperonins in protease-minus *E. coli* strains. Simultaneously, Ortiz de Montellano and co-workers (19) reported the prokaryotic expression of neuronal NOS in DH5α cells and subsequent purification and characterization of the heme environment by phenylation of the prosthetic heme. This chapter describes the techniques developed in our laboratory in detail.

Prokaryotic Expression and Purification of Intact Neuronal Nitric Oxide Synthase

Introduction

The study of full-length neuronal nitric oxide synthase (nNOS) has been hampered by the small amounts of active enzyme that could be isolated from cerebellar tissue [9 μg pure protein from 18 rat brains (20)], cell culture (8–10 mg nNOS from 12 liters of human kidney 293 cells; K. McMillan and B. S. S. Masters, unpublished observations, 1992), or baculovirus systems [30 mg nNOS from 1 liter (approximately seventeen 225-cm^2 flasks) (21)]. We have developed a quick, easy, and inexpensive expression system for nNOS in *E. coli* that produces 20–24 mg of enzyme per liter of cells. The enzyme preparation described in the following protocol is replete with heme and flavins but is about 90% deficient in pterin. However, the enzyme can be reconstituted with BH$_4$, if desired, and is also an excellent source of BH$_4$-deficient enzyme for use in the investigation of the role of BH$_4$ in catalysis. It is expected that other isoforms of NOS, including those from different species, will also be expressible in *E. coli*, although the expression systems may require some modifications. The expression system described herein will greatly facilitate the study of structure–function relationships in NOS, which, in turn, will aid in drug design and development. It will also be useful for the screening of site-directed mutants, which may give further insight into the mechanism of NO· production.

Cloning

The full-length rat cerebellar isoform of NOS is overexpressed in *E. coli* using the nNOSpCW plasmid, which contains the entire cDNA sequence of nNOS in pCW, a vector that has been used successfully for the expression of cytochromes P450 (22, 23). The nNOS cDNA is originally obtained in the Bluescript SK(−) plasmid (pNOS, provided by Drs. Snyder and Bredt at Johns Hopkins Medical School, Baltimore, MD) and is moved into the pCW plasmid. The initial 1215 nucleotides of pNOS, from the ATG start site to the *Nar*I restriction site, are amplified by the polymerase chain reaction (PCR) to incorporate an *Nde*I restriction site (CATATG) straddling the start codon, but no changes in the amino acid sequence are designed, except that an alanine is incorporated as the second residue, directly following the ATG start codon. The upstream primer is 5′ TCATCAT<u>CATATG</u>GCTGAAGA-GAACACGTT 3′. An additional seven nucleotides are included at the 5′ end to facilitate the cleavage of the *Nde*I site, which does not cleave very well with overhangs of 1, 2, 3, or 6 nucleotides (New England Biolabs, Beverly, MA). The return primer is 5′ CATGCTT<u>GGCGCC</u>AT 3′, which spans a *Nar*I restriction site in the nNOS cDNA sequence. The ends are digested with *Nde*I and *Nar*I, and this piece is ligated to the *Nar*I/*Xba*I fragment of nNOS cDNA (3529 nucleotides, which encompass the remaining 75% of the nNOS cDNA) and the *Xba*I/*Nde*I fragment of pCW$_{ori+}$ (4970 nucleotides, containing the ampicillin resistance gene), both of which have been gel-purified (GeneClean II, Bio101, La Jolla, CA). The resulting ligation product, nNOSpCW (Fig. 1), is transformed initially into *E. coli* JM109 supercompetent cells (Stratagene, La Jolla, CA), and the correct construct is isolated. The plasmid DNA, isolated via alkaline lysis minipreparation (24), is cotransformed with pGroELS, which encodes the *E. coli* chaperonins GroEL and GroES and confers chloramphenicol resistance, into *E. coli* BL21 protease-deficient cells ([*lon*], *ompT*) via electroporation. The nNOSpCW plasmid is approximately 10 kb long, and pGroELS is approximately 7 kb long.

Expression

Escherichia coli BL21 cells (1 ml of an overnight culture) containing nNOS-pCW and pGroELS are grown in 1 liter of modified Terrific Broth (20 g yeast extract, 10 g bactotryptone, 2.65 g KH_2PO_4, 4.33 g Na_2HPO_4, and 4 ml glycerol) containing 50 μg/ml ampicillin and 34 μg/ml chloramphenicol at 37°C, with shaking at 250 rpm. At an OD_{600} of 1.0–1.4, protein expression

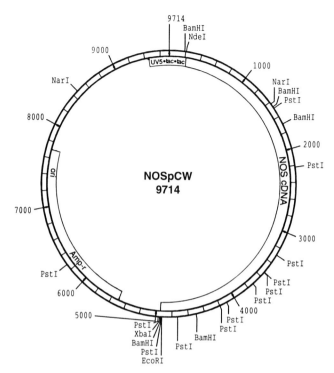

FIG. 1   Plasmid vector, nNOSpCW, for expression of intact neuronal nitric oxide synthase.

of nNOS and the chaperonins is induced with 0.5 mM isopropyl-β-D-thiogalactopyranoside (IPTG). Lower concentrations of IPTG (0.1 and 0.25 mM) seem to induce protein production just as well. The heme and flavin precursors, δ-aminolevulinic acid and riboflavin, are also added to the culture medium at this time to final concentrations of 450 and 3 μM, respectively. Then ATP (1 mM final concentration), a required cofactor for the chaperonins, is also added. The flasks are moved to room temperature ($\sim$25°C) and shaken at 250 rpm in the dark, either in a dark room or in flasks covered with aluminum foil, to protect the light-sensitive riboflavin. The cells are harvested at 40–44 hr postinduction, and the cell paste is frozen at −80°C until purification. If the cells are harvested 1 day postinduction, instead of at 2 days, the spectrally detectable NOS protein is greatly decreased (i.e., one-fifth of the final yield is seen at 18–20 hr).

Protein Purification

The cells from 1 liter of culture (~10 g wet weight) are resuspended in 30 ml of resuspension buffer [100 mM Tris-HCl, pH 7.4, 1 mM EDTA, 1 mM dithiothreitol (DTT), 10% glycerol, 1 mM phenylmethylsulfonyl fluoride (PMSF), and 5 μg/ml leupeptin/pepstatin]; if frozen cells are used, the cell paste is first thawed on ice. The cells are lysed by pulsed sonication [a total of 4 min: 2.5 sec on, 0.5 sec off at 80% power, large probe, Fisher Scientific (Pittsburgh, PA) Model 550] and the cell debris is removed by centrifugation at 150,000 g for 70 min.

The supernatant is applied to a 2',5'-ADP-Sepharose 4B column (~6 ml, Pharmacia, Piscataway, NJ) equilibrated in buffer B (50 mM Tris-HCl, pH 7.4, 0.1 mM EDTA, 0.1 mM dithiothreitol, 10% glycerol, and 100 mM NaCl). The column is extensively washed with at least 10 column volumes of buffer B, followed by an additional wash with buffer B containing 500 mM NaCl. The protein is eluted with buffer B containing 500 mM NaCl and 5 mM 2'-AMP, and appears to be approximately 75% pure after this step. An alternate method, yielding enzyme that appears to be over 80% pure, involves loading the supernatant in the presence of 1 mM 2'-AMP and no NaCl, washing with buffer B without NaCl, washing with buffer B containing 500 mM NaCl, and finally eluting, as in the first method, with buffer B containing 500 mM NaCl and 5 mM 2'-AMP. Fractions are pooled by color (yellow to yellow-brown, depending on the concentration of nNOS present), appearance on a Coomassie-stained polyacrylamide gel, and/or the presence of a heme peak at about 400 nm as viewed spectrophotometrically. The pooled fractions are concentrated using a Centriprep 30 (Amicon, Beverly, MA) and washed by repeated dilution and concentration to lower the NaCl concentration below 100 mM. Reconstitution with BH$_4$, if maximally active enzyme is desired, is performed after protein concentration by adding L-arginine and BH$_4$ to final concentrations of 500 and 250 μM, respectively, and incubating overnight at 4°C. Because BH$_4$ is light-sensitive, this incubation should be performed in the dark. All manipulations up to this point have been performed at 4°C.

Final purification is achieved by FPLC at room temperature using ion-exchange chromatography (Mono Q column; Pharmacia). The enzyme is applied to the column equilibrated in buffer B, pH 7.8, the column is washed with this same buffer, and the flow-through is monitored at 280 nm until the absorbance returns to baseline. The nNOS is eluted at about 200 mM NaCl in a 0–500 mM NaCl gradient in buffer B, pH 7.8. The nNOS-containing fractions are pooled as described above, by color, polyacrylamide gel electrophoresis (PAGE) analysis, and presence of heme. The purified nNOS is concentrated and stored at −80°C. This storage procedure does not call for dialysis and so the final preparation contains approximately 200 mM NaCl. If a preparation without NaCl is desired, the buffer can be altered by dialysis, but it must contain at least 10% glycerol and BH$_4$ for enzyme stability. If stored properly, the enzyme will lose no more than 10% of its activity with each freeze–thaw cycle. The best method of storing the enzyme is in aliquots of at least 10 μM containing over 10% glycerol at $-80°$C so that repeated freezing and thawing are prevented.

The yields of nNOS throughout purification, as measured by CO difference spectra, are as follows. The cytosolic extract per liter of cells contains 100–150 nmol nNOS (16–24 mg; 100% yield), the 2′,5′-ADP-Sepharose 4B column pool contains 50–100 nmol enzyme (8–16 mg; ~65% yield), and the Mono Q column pool contains 25–50 nmol enzyme (4–8 mg; ~30% yield). The specific activity of the purified nNOS reconstituted with BH$_4$, as measured at 25°C by the conversion of ^{3}H-labeled L-arginine to L-citrulline (25, 26) and by the NO capture assay (3, 20), is approximately 450 nmol/min/mg, and all spectral characteristics measured are identical to those of nNOS isolated from eukaryotic cell systems (18). As isolated, 10% of the E. coli-expressed nNOS contains a pterin moiety, and turnover numbers of 75 and 200 nmol/min/mg without and with BH$_4$ in the assay mixture, respectively, are obtained. After incubation with BH$_4$ and L-arginine, as described, about 70% of the enzyme contains BH$_4$, and turnover numbers are 190 and 435 nmol/min/mg in the absence and presence of BH$_4$, respectively.

Prokaryotic Expression of Heme (Oxygenase) Domain of Neuronal Nitric Oxide Synthase Using pCW$_{ori+}$ Expression Vector

The expression of the heme domain of nNOS in E. coli has been described in detail by McMillan and Masters (16). The ability to express this heme domain (and the two flavin domain constructs) had been suggested to us by the pioneering studies from the Fulco laboratory (27) on the Bacillus megaterium P450 BM-3 fatty acid ω-hydroxylase, which also consists of FAD- and FMN-containing and heme-containing domains in the same polypeptide chain. It was also possible, using this cloned and expressed heme domain, to demonstrate that cysteine-415 is the heme–thiolate ligand which confers the property of the visible absorption of the reduced, CO difference spectrum (λ_{max} ~450 nm) shared with over 300 expressed cytochromes P450. This heme domain construct produced high levels of the expressed protein in E. coli JM109 cells, utilizing the pCW$_{ori+}$ vector of Gegner and Dahlquist (28).

Prokaryotic Expression and Purification of Modules of Oxygenase Domain of Neuronal Nitric Oxide Synthase Using Glutathione *S*-Transferase Fusion Protein Constructs

Introduction

Further analysis of the oxygenase domain of nNOS (amino acids 1–721) has been provided impetus from the following observations. First, the comparison of nNOS with two other known isoforms, endothelial NOS (eNOS) and inducible NOS (iNOS), reveals that nNOS differs in length as a result of its N-terminal 220 amino acids. The presence of this N-terminal module has suggested that these first 220 amino acids may behave as an independent module or domain, playing a role in protein–protein interaction in lieu of cell–cell interactions or intercellular junctions as described for other proteins exhibiting the GLGF motif(s) and sharing some degree of homology (29). Second, the role of BH_4 in the enzymatic conversion of L-arginine to NO· has not been established, but it has been suggested to serve in either a redox or allosteric role. The proposal that BH_4 binding might occur in the region of amino acid residues 558–721 of nNOS was based on the structural similarity of this part of nNOS to dihydrofolate reductase (DHFR) (30). Using homology-based modeling, it has been possible to predict the structure of a 150-residue stretch just N-terminal of the calmodulin-binding site in neuronal NOS.

To address the aforementioned, we use a fusion protein strategy to investigate the modularity of the oxygenase domain of nNOS, using PCR-amplified DNA coding for different parts of the nNOS protein ligated to the pGEX-4T1 plasmid. The pGEX plasmids are designed for inducible, high-level expression of genes or gene fragments as fusions with *Schistostoma japonicum* glutathione *S*-transferase (GST) (31). Fusion proteins are easily purified from bacterial lysates by affinity chromatography using glutathione-Sepharose 4B. The thrombin or factor Xa protease recognition site for cleaving the desired protein from the fusion product is located immediately upstream from the multiple cloning site on the pGEX plasmids, although for many applications, such as investigating substrate binding, it has been unnecessary to remove the tag. The fusion protein approach has been used successfully for purification and further characterization of a large variety of proteins or modules derived from them. Such fusion proteins offer increased likelihood of expression, as well as a convenient tag for purification by affinity chromatography.

Cloning

The original nNOS cDNA in the Bluescript SK(−) plasmid was mentioned in a previous section, describing prokaryotic expression and purification of

TABLE I   Description of Primers Used in Preparation of Constructs

Construct	Amino acids of nNOS	Corresponding nucleotides of primers in nNOS sequence[a]	
		Upstream	Downstream
Fusion proteins used to study protein–nNOS interactions			
nNOS$_{69}$	1–69	349–368 (*Bam*HI)	537–555 (*Xho*I)
nNOS$_{99}$	1–99	349–368 (*Bam*HI)	628–645 (*Xho*I)
nNOS$_{70-220}$	70–220	556–573 (*Bam*HI)	991–1008 (*Xho*I)
GLGF	1–220	349–368 (*Bam*HI)	991–1008 (*Xho*I)
Fusion proteins used to address substrate binding and role of BH$_4$			
Module III[b]	220–721	1006–1023 (*Bcl*I)	2491–2508 (*Xho*I)
Module II[b]	220–558	1006–1023 (*Bcl*I)	1982–2020 (*Xho*I)
DHFR[b]	558–721	2020–2037 (*Bam*HI)	2491–2508 (*Xho*I)

[a] Restriction site used in primer.
[b] Reference 17.

the entire nNOS. Table I details the modules that are being expressed as fusion products with glutathione *S*-transferase. For all the constructs, plasmid pGEX-4T1 (Pharmacia, Uppsala, Sweden) was digested by *Bam*HI and *Xho*I and gel purified. To amplify DNAs coding for GLGF and DHFR motifs, a *Bam*HI restriction site was included in the upstream primer. Because the cDNAs coding for modules II and III contain *Bam*HI restriction site(s), a *Bcl*I restriction site was included in the upstream primers for the amplification of DNA encoding this part of the protein. *Bam*HI and *Bcl*I share identical overhangs, which were used to ligate to the *Bam*HI site of the plasmid. In all downstream primers, the *Xho*I site was used. The DNAs produced by PCR were purified, cleaved by *Bam*HI/*Bcl*I and *Xho*I, and ligated with pGEX-4TI. The resulting ligation product was used to transform *E. coli* DH5α cells.

Expression

Routinely, a single fresh colony of transformed *E. coli* DH5α cells is grown to mid-log phase in modified Terrific Broth (see above) supplemented with 1% D-glucose and ampicillin (50 μg/ml) at 30°C. Lower temperature and glucose supplementation are believed to affect the ratio of soluble protein to insoluble inclusion bodies. Next, IPTG is added to a final concentration of 0.1 mM, and the culture is incubated with shaking at 24°C for 2 hr. The cells are harvested by centrifugation at 4000 g for 15 min and frozen at -80°C until used.

Preparation of Fusion Proteins

Thawed cells are suspended in 1/50 of the original culture volume of phosphate-buffered saline (PBS)/0.5% Triton X-100 containing protease inhibitors (2 μg/ml leupeptin, 2 μg/ml pepstatin, 1 mM EDTA, 1 mM PMSF) and lysozyme (1 mg/ml). The cells are lysed by mild sonication, and the crude extract is centrifuged twice at 17,000 g for 20 min. Subsequent purification of fusion proteins in the supernatant is performed using a glutathione-Sepharose 4B column (Pharmacia). The clarified lysate is placed on a column, which is washed with 5–10 volumes each of PBS/0.5% Triton X-100, then PBS, and the GST fusion proteins are eluted with 10 mM reduced glutathione (Sigma, St. Louis, MO) in 50 mM Tris-HCl (pH 8.0 before glutathione addition). Yields are sufficient even though certain fusion proteins are poorly soluble. On the basis of analysis of preparations by sodium dodecyl sulfate–polyacrylamide gel electrophoresis (SDS–PAGE) and subsequent staining with Coomassie blue, the purity of the fusion proteins is estimated to be between 70 and 90% (Fig. 2). For the purposes of our studies, involving

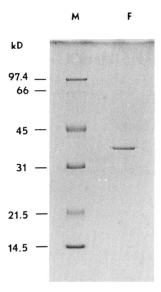

FIG. 2   Analysis by SDS–PAGE of GST–rat neuronal NOS 1–99 fusion protein. Left-hand side (lane M) represents protein markers for molecular masses as indicated. Right-hand side (lane F) is the GST–rat neuronal NOS 1–99 fusion protein after elution from the glutathione-Sepharose column with 10 mM glutathione. Approximately 2 μg of fusion protein was loaded on a 12% gel and visualized by Coomassie blue staining.

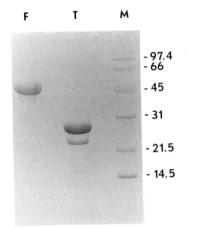

FIG. 3   Analysis by SDS–PAGE of GST–rat neuronal NOS 1–220 fusion protein and its thrombin digest. Right-hand side (lane M) represents protein markers for molecular masses as indicated. On the left-hand side (lane F) is the GST–nNOS 1–220 fusion protein after elution from the glutathione-Sepharose column by 10 mM glutathione. Approximately 6 μg of fusion protein was loaded on a 12% gel and visualized by Coomassie blue staining. In the middle (lane T) is the fusion protein after digestion with thrombin.

substrate binding assays for modules II, III, or DHFR, GST fusion constructs are very useful. If there is need for analysis of proteins without the GST tag, site-specific proteolysis of fusion proteins can be performed using thrombin (Fig. 3).

Troubleshooting

If basal expression of fusion proteins is obtained, the addition of glucose (~1%) may help. If the majority of the fusion protein is insoluble, then decreasing the growth temperature (30°C for growth without IPTG followed by induction with no more than 0.1 mM IPTG at 24°C) facilitates the production of sufficient yields of fusion proteins in soluble form. Although none of the modular constructs thus far produced in our laboratory have required such manipulations, the inclusion of molecular chaperonins (e.g., groEL and groES) by introduction on the same plasmid or cotransfected on another plasmid (see above) could facilitate the proper folding of the modules, provided their α-helical and/or β-sheet structure(s) have been preserved in the constructs.

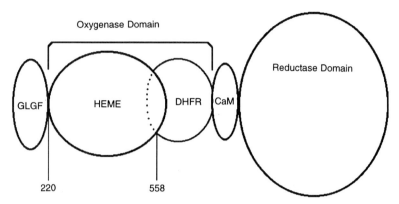

FIG. 4   Putative modular structure of neuronal nitric oxide synthase.

Summary

The diagram in Fig. 4 represents the possible molecular structure of rat neuronal nitric oxide synthase. This chapter addresses the production of modules and submodules from the oxygenase domain, including the GLGF-containing N terminus, of this enzyme for the purpose of determining binding sites and structure–function relationships. In addition, the ability to prepare adequate quantities of lower molecular weight domains which retain native structure provides the opportunity for biophysical characterization, including X-ray crystallography. The GLGF motif has generated much interest as a putative recognition–binding sequence for proteins interacting with dystrophin indirectly through syntrophins in skeletal muscle and, perhaps, vital proteins in other innervated tissues (32, 33).

References

1. K. A. White and M. A. Marletta, *Biochemistry* **31,** 6627 (1992).
2. D. J. Steuhr and M. Ikeda-Saito, *J. Biol. Chem.* **267,** 20547 (1992).
3. K. McMillan, D. S. Bredt, D. J. Hirsch, S. H. Snyder, J. E. Clark, and B. S. S. Masters, *Proc. Natl. Acad. Sci. U.S.A.* **89,** 11141 (1992).
4. P. Klatt, K. Schmidt, and B. Mayer, *Biochem. J.* **288,** 15 (1992).
5. N. S. Kwon, C. F. Nathan, C. Gilker, O. W. Griffith, D. E. Mathews, and D. J. Stuehr, *J. Biol. Chem.* **265,** 13442 (1990).
6. M. A. Marletta, *J. Biol. Chem.* **268,** 12231 (1993).
7. O. W. Griffith and D. J. Stuehr, *Annu. Rev. Physiol.* **57,** 707 (1995).
8. B. S. S. Masters, *Annu. Rev. Nutr.* **14,** 131 (1994).

9. M. Tayeh and M. A. Marletta, *J. Biol. Chem.* **264,** 19654 (1989).

10. N. S. Kwon, C. F. Nathan, and D. J. Stuehr, *J. Biol. Chem.* **264,** 20496 (1989).

11. D. S. Bredt, P. M. Hwang, C. E. Glatt, C. Lowenstein, R. R. Reed, and S. H. Snyder, *Nature (London)* **351,** 714 (1991).

12. B. Mayer, M. John, and E. Böhme, *FEBS Lett.* **277,** 215 (1990).

13. S. Kaufman, "The Enzymes" (P. D. Boyer and E. G. Krebs, eds.), Vol. 18, p. 217. Academic Press, Orlando, Florida, 1987.

14. K. J. Baek, B. A. Thiel, S. Lucas, and D. J. Stuehr, *J. Biol. Chem.* **268,** 21120 (1993).

15. P. Klatt, M. Schmid, E. Leopold, K. Schmidt, E. R. Werner, and B. Mayer, *J. Biol. Chem.* **269,** 13861 (1994).

16. K. McMillan and B. S. S. Masters, *Biochemistry* **34,** 3586 (1995).

17. J. Nishimura, P. Martasek, K. McMillan, J. C. Salerno, Q. Liu, S. S. Gross, and B. S. S. Masters, *Biochem. Biophys. Res. Commun.* **210,** 288 (1995).

18. L. J. Roman, E. A. Sheta, P. Martasek, S. S. Gross, Q. Liu, and B. S. S. Masters, *Proc. Natl. Acad. Sci. U.S.A.* **92,** 8428 (1995).

19. N. C. Gerbers and P. R. Ortiz de Montellano, *J. Biol. Chem.* **270,** 17791 (1995).

20. D. S. Bredt and S. H. Snyder, *Proc. Natl. Acad. Sci. U.S.A.* **87,** 682 (1990).

21. V. Riveros-Moreno, B. Hefferman, B. Torres, A. Chubb, I. Charles, and S. Moncada, *Eur. J. Biochem.* **230,** 52 (1995).

22. H. J. Barnes, M. P. Arlotto, and M. R. Waterman, *Proc. Natl. Acad. Sci. U.S.A.* **88,** 5597 (1991).

23. M. Nishimoto, J. E. Clark, and B. S. S. Masters, *Biochemistry* **32,** 8863 (1993).

24. J. Sambrook, E. F. Fritsch, and T. Maniatis, "Molecular Cloning: A Laboratory Manual," 2nd Ed., Cold Spring Harbor Laboratory, Cold Spring Harbor, New York, 1989.

25. M. Kelm and J. Schrader, *Circ. Res.* **66,** 1561 (1990).

26. E. A. Sheta, K. McMillan, and B. S. S. Masters, *J. Biol. Chem.* **269,** 15147 (1994).

27. L. O. Narhi and A. J. Fulco, *J. Biol. Chem.* **262,** 6683 (1987).

28. J. A. Gegner and F. W. Dahlquist, *Proc. Natl. Acad. Sci. U.S.A.* **88,** 750 (1991).

29. C. P. Ponting and C. Phillips, *Trends Biochem. Sci.* **20,** 102 (1995).

30. J. C. Salerno and A. J. Morales, "Biochemistry and Molecular Biology of Nitric Oxide, First International Conference" (L. Ignarro and F. Murad, eds.), p. 72. Abstract. Los Angeles, CA, 1994.

31. D. B. Smith and K. S. Johnson, *Gene* **67,** 31 (1988).

32. J. E. Brenman, D. S. Chao, H. Xia, K. Aldape, and D. S. Bredt, *Cell (Cambridge, Mass.)* **82,** 743 (1995).

33. B. S. Masters, L. J. Roman, P. Martasek, J. Nishimura, K. McMillan, E. A. Sheta, S. S. Gross, T. J. McCabe, W.-J. Chang, and J. T. Stull, "The Biology of Nitric Oxide Part 5" (J. Stamler, S. Gross, S. Moncada, and E. A. Higgs, eds.) in press. Portland Press, London, 1996.

[13] Generation of Monoclonal Antibodies to Endothelial Nitric Oxide Synthase

Jennifer S. Pollock and Masaki Nakane

Introduction

The generation of specific and sensitive antibody probes for endothelial nitric oxide (NO) synthase was necessary to facilitate the determination of the distribution of endothelial NO synthase under normal conditions and pathological conditions in various tissues. These antibody probes are also a valuable tool to aid in the ultrastructural localization of endothelial NO synthase (1). The intact protein was utilized as the antigen to facilitate production of a very specific probe. Monoclonal antibodies were generated as opposed to polyclonal antibodies because of the comparatively small amounts of antigen needed to produce monoclonal antibodies.

Monoclonal Antibody Production

Particulate bovine endothelial NO synthase is purified from cultured bovine aortic endothelial cells as previously described (2, 3). Purified fractions in buffer including 10 mM Tris-HCl (pH 7.4), 1 mM EDTA, 0.1% 2-mercaptoethanol, 100 mM NaCl, 10% glycerol, and 10 mM CHAPS are pooled and concentrated with Ultrafree-CL concentrators (Millipore, Bedford, MA) to approximately 0.5 mg/ml. As purity rather than activity of the protein is of interest, these fractions do not have exogenous bovine serum albumin added at the end of the purification protocol.

Approximately 10 μg of pure protein is mixed with Freund's complete adjuvant in equal proportions with a double-barreled syringe. When the mixture has a homogenous consistency, a maximum of 250 μl (5 μg/mouse) is injected into two BALB/c mice, each intraperitoneally. About 4 weeks later the mice are boosted with approximately 4 μg of protein with Freund's incomplete adjuvant intraperitoneally. An aliquot of serum is obtained from each mouse by capillary tube collection from the tail vein 10 days after the boost and is used to assess the production, specificity, and sensitivity of the antibodies by Western blotting techniques prior to fusion of the splenocytes.

Methods in Neurosciences, Volume 31

Approximately 8 μg of protein in phosphate-buffered saline (PBS) is then injected intraperitoneally. Four days later the spleen is removed, minced, and the red blood cells lysed. The resulting pellet of splenocytes is fused with mouse myeloma Sp 2/0 cells with polyethylene glycol and utilizing hypoxanthine/aminopterin/thymidine (HAT) medium (GIBCO, Grand Island, NY) for selection of the hybridomas. The cells are plated into 24-well tissue culture plates. Supernatant fractions are removed from the wells of the growing hybridomas within 1–2 weeks after the fusion and screened for the presence of specific anti-NO synthase antibody by enzyme-linked immunosorbent assay (ELISA) as described below. Hybridoma cells that screen positive for antibody production are selected and then cloned by limiting dilution twice. Specific antibody type and class is assessed by a typing kit (Amersham, Arlington Heights, IL). Quantitative analysis of the tissue culture supernatant is assessed by ELISA with type-specific antibodies. Ascites is produced by priming mice with pristane (Aldrich, Milwaukee, WI) 2 weeks prior to injecting approximately 1×10^6 monoclonal hybridoma cells/mouse. Mice are sacrificed and ascitic fluid collected within 1 month of injecting the cells. Specifically, hybridoma H32 produces approximately 56 μg antibody/ml of tissue culture supernatant, whereas the ascites fluid titers to approximately 100 times as concentrated as the tissue culture supernatan.

Enzyme-LInked Immunosorbent Assay

An enzyme-linked immunosorbent assay (ELISA) has been developed for purified bovine endothelial NO synthase and rat brain NO synthase to assess the specificity of the hybridomas at the first screen. Hybridomas that test positive for both endothelial and brain NO synthase are not selected, whereas hybridomas testing positive for endothelial NO synthase and negative for brain NO synthase are selected for further growth. The ELISA is performed in microtiter plates (Immulon-3, Dynatech). Separate plates are coated overnight at 4°C with 25 ng per well of purified rat brain or endothelial NO synthase in 50 mM Tris-HCl buffer, pH 8.0. This is the minimum amount of purified NO synthase that can be coated to obtain quantitative results. The plates are then washed with PBS and blocked with 2% bovine serum albumin in PBS at room temperature for 4 hr. Tween 20 does not enhance the results and is found to actually give a lower quantitative result, although the qualitative results are the same.

Aliquots (200 μl) of the cell culture supernatant from growing hybridomas are incubated in each plate for 2 hr at room temperature followed by four washes with PBS. The plates are then incubated with horseradish peroxidase-conjugated sheep anti-mouse immunoglobulin (1 : 6000, Amersham) for 1 hr at room temperature and washed four times with PBS. The positive wells are developed using o-phenylenediamine and hydrogen peroxide and the reaction is stopped with 1 N H$_2$SO$_4$. Plates are read at 490 nm in a microplate reader (Bio-Tek). All positive readings are then decoded, and the hybridomas are selected for continuing in growth cultures. The cultures are monitored weekly, and four final hybridomas are selected based on specificity between brain and endothelial NO synthase reactivity in the ELISA. Hybridomas H32, H210, H98, and H106 have been selected. H32 and H210 are typed to IgG$_{2a}$, and H98 and H106 are designated IgM. The IgM antibodies have not been characterized further because of the difficulties in working with IgM antibodies. In saturated cell cultures of H32 and H210, quantitative analyses have revealed that H32 generates 56 μg/ml of antibody whereas H210 generates only 3 μg/ml of antibody.

Western Blotting

Western blots are utilized to assess the specificity of the monoclonal antibodies to other tissue sources of NO synthase. Crude, partially purified, or purified enzyme preparations from various tissue and cell preparations are separated on 7.5% denaturing sodium dodecyl sulfate–polyacrylamide gels (SDS–PAGE). The proteins are then blotted onto nitrocellulose (Hybond-ECL, Amersham) by semidry electroblotting (Millipore) for 45 min. The blots are allowed to air dry for 1 hr or more and then blocked overnight at 4°C with 6% nonfat dry milk in Tris–saline buffer (40 mM Tris-HCl, pH 7.6, and 300 mM NaCl). The blots are washed twice in Tris–saline buffer and incubated with the primary antibody for 2 hr at room temperature. Incubation with the primary antibody could be extended for more sensitivity. Cell culture supernatant fractions from hybridoma H32 at a dilution of 1 : 1000 or the ascites fluid from hybridoma H32 at a dilution of 1 : 10,000 in 1% nonfat dry milk in Tris–Saline buffer are determined to be the optimum dilutions. The blots are washed four times in Tris–Saline buffer and incubated with horseradish peroxidase-conjugated sheep anti-mouse immunoglobulin antibody (1 : 3000, Amersham) for 1 hr at room temperature. Finally, the blots are washed four additional times, and the specific proteins are detected by enhanced chemiluminescence (ECL, Amersham). If background staining persists, the second antibody dilution is increased to 1 : 6000, and most of the problem is corrected.

References

1. J. S. Pollock, M. Nakane, L. D. K. Buttery, A. Martinez, D. Springall, J. M. Polak, U. Förstermann, and F. Murad, *Am. J. Physiol.* **34,** C1379 (1993).
2. J. S. Pollock, U. Förstermann, J. A. Mitchell, T. D. Warner, H. H. H. W. Schmidt, M. Nakane, and F. Murad, *Proc. Natl. Acad. Sci. U.S.A.* **88,** 10480 (1991).
3. J. S. Pollock and M. Nakane, *Methods Neurosci.* **31,** Chap. 10 (1996).

[14] Genetic Modulation of Inducible Nitric Oxide Synthase Expression

Edith Tzeng and Timothy R. Billiar

Introduction

Nitric oxide (NO) has a number of biological effects that vary over a wide spectrum, ranging from beneficial to deleterious. The precise action of this short-lived mediator is intimately linked to its cellular source as well as its rate of synthesis. The constitutively expressed NO synthase isoforms (NOS types 1 and 3) synthesize NO in low levels and are essential for hemodynamic regulation as well as for neurotransmission and memory formation (1, 2). However, elucidation of all the regulatory roles of these isoforms is still far from complete. Even more complex are the actions of the inducible NOS (NOS type 2) as a growing number of cells and tissues demonstrate the capacity to express this enzyme in response to different stimuli (1, 2). Beneficial actions of NOS-2-derived NO include maintaining organ perfusion during systemic or local inflammatory perturbations, promoting tumor and microbial killing, and scavenging free oxygen radicals. Conversely, supraphysiological levels of NO have been linked to refractory septic shock, tissue injury, and the pathophysiology of some autoimmune diseases such as juvenile diabetes mellitus, rheumatoid arthritis, and ulcerative colitis (3).

The potential neural effects of NO are many. Nitric oxide, as derived from NOS-1, has been demonstrated to be a neurotransmitter in the peripheral and central nervous systems (CNS) as well as in the gastrointestinal tract. There is evidence to support the role of NO in establishing synaptic connections (2) and long-term memory (4). Nitric oxide also appears to be a major regulator of vasodilator nerves of cerebral arterial smooth muscle and is important in the determination of cerebral perfusion (5). Less advantageous, NO has also been implicated in the pathogenesis of some neurodegenerative disorders including Alzheimer's and Huntington's diseases (6) as well as the dementia associated with acquired immunodeficiency syndrome (7, 8). Application of NOS inhibitor compounds have been beneficial in preventing the cytotoxic effects of N-methyl-D-aspartate (NMDA) administration and reduced cellular injury following ischemic hypoxic insults (6). The sources of NO in the central and peripheral nervous systems consist of a small population of neurons that express NOS-1 as well as astrocytes and microglial cells that have the capacity to express the inducible NOS-2 isoform (9).

Methods in Neurosciences, Volume 31

Extensive investigation is required to elicit the precise role of each of these sources of NO in neuronal function and pathophysiology.

As a consequence of the complexity of discerning the biological effects of NOS-2 expression in the nervous system, manipulating NOS-2 enzyme expression and function may permit a deeper understanding of the role of the enzyme in the CNS. A great effort is currently dedicated to the design of NOS-2-specific inhibitors for the clinical treatment of septic shock. These inhibitors will have to be highly specific because of the danger of simultaneously inhibiting the physiological function of the other NOS isoforms. In addition, the systemic administration of NOS inhibitors results in nonspecific reduction in NOS-2-dependent NO synthesis, including at sites that may benefit from high levels of NO. Perhaps the best mechanism to modulate NOS-2 activity is through the local regulation of enzyme expression, either increasing or decreasing the level of functional enzyme produced. Even more exciting would be to harness the NOS-2 enzyme to act at naive sites to exploit specific properties of NO. One such application, for example, is to increase NO synthesis at sites of cerebral ischemia to increase vasodilation and reduce vasospasm to support tissue perfusion and minimize infarct volume. Alternatively, the development of methods to locally inhibit NOS-2 expression, such as in various neurodegenerative diseases, may limit disease progression. To increase NOS-2 expression in a site-specific or cell-specific manner, gene transfer modalities are the most attractive. These modalities include both nonviral and viral methods and allow manipulation of the time frame over which increased NO synthesis is observed. To reduce NOS-2 expression locally, the use of antisense technology may prove to be the ideal method.

Gene Transfer

As early as the 1960s, it was known that exogenous genetic material could be taken up by mammalian cells and then expressed (10). However, the full impact of genetic transformation was not realized until the advent of recombinant DNA technology and the deciphering of the retroviral life cycle and reverse transcription. This led to the engineering of the first retroviral vectors in the early 1980s and opened the door for the treatment of genetic diseases (11). Since the early era of gene transfer, a number of new techniques have been developed that exploit both nonviral and viral mechanisms of genetic transformation. The initial applications for gene therapy involved correcting inherited metabolic disorders, namely, diseases stemming from the absence of a functional gene product. Targeted diseases included Lesch–Nyhan syndrome, adenosine deaminase deficiency, and Gaucher's disease, to name a few (11). The list of potential gene therapy applications continues to grow and includes the treatment of malignancies, infectious diseases, transplantation, and vascular complications. Beyond the clinical applications, the benefit of gene transfer technologies to basic science research is unlimited by permitting the study of disease pathogenesis and metabolic pathways in a controlled cellular system.

Vectors

A vector is defined as any genetic vehicle that facilitates the transfer and expression of genetic material into and within targeted cells. Vectors are typically composed of a mammalian enhancer/promoter complex, the genetic sequence being transferred, and a polyadenylation signal. Vectors may also include selectable markers that confer a survival advantage to cells expressing the gene products encoded by the vector. The simplest vector is the expression plasmid, which is the basis for nonviral gene transfer techniques such as electroporation, naked DNA transfer, calcium phosphate precipitation, and liposomes. Methods of gene transfer utilizing expression plasmid vectors are limited by transient gene expression, as the transferred plasmid is not replicated.

Another group of gene transfer vectors is the viral vector. These vectors are similar to the expression plasmids in that viral or mammalian promoters regulate the transcription of the genetic sequence of interest. However, they incorporate viral sequences necessary for the packaging of the vector into infectious viral particles. These viral vectors also retain characteristics of the parent wild-type virus from which they were derived. We discuss retroviral, adenoviral, and herpes simplex type I (HSV)-based vectors. Retroviruses (12) are RNA viruses whose life cycle involves genomic integration, permitting long-term and inheritable expression of a gene shuttled by this virus. The simplicity of retroviral genomes has allowed the development of replication-free, clinically safe retroviral vectors. Unfortunately, retroviruses preferentially target proliferating cells which limits its efficacy in postmitotic neurons and in *in vivo* applications. Adenoviral vectors (13) target both dividing and nondividing cells and are advantageous for *in vivo* gene transfer experiments. The adenoviral life cycle centers around a burst of viral replication that ultimately results in host cell lysis. Methods to minimize the production of cytotoxic adenoviral proteins are still being sought. Unlike the case for retrovirus, the large size of the adenoviral genome as well as the composition of the genome with overlapping coding sequences make deletional manipulations difficult. However, HSV (14), a relatively new addition to the viral vector family, has great potential for success as a gene transfer vehicle to the CNS. The natural target of HSV are peripheral neurons, and the virus can traverse synapses to neighboring cells. In neurons, HSV may enter a latent phase wherein the virus persists as a benign episomal entity in the cell nucleus. Genes may be inserted to replace the latent viral genes and can then be expressed for the lifetime of the cell. Although HSV is also a sizable virus, the HSV genomic composition is more amenable to genetic manipulation.

Vector Preparation

Because of the many expression plasmids and viral vectors that are commercially available or available through collaboration, describing the precise details of vector preparation is not feasible. For this reason, a broad description of how vector plasmids can be constructed is provided. For precise details of the molecular techniques, reference to Sambrook *et al.* (15) will be sufficient.

Expression plasmids consist of a bacterial plasmid backbone (such as pBR322, pBS) that permits bacterial amplification. Inserted into the plasmids are a variety of mammalian promoters, including those from cytomegalovirus (CMV) and Rous sarcoma virus (RSV) and retroviral long terminal repeats (LTR) that regulate the level of transcription of a gene inserted downstream. Also included in these plasmids are enhancer elements as well as polyadenylation signals. Between the promoter and the polyadenylation sequences have been inserted a segment of DNA that contains a multicloning region. The expression plasmid is linearized at the multicloning site by digestion with one or two of the restriction enzymes that can cut in that region. The desired cDNA is then inserted into this linearized plasmid. We constructed a human NOS-2 expression vector utilizing the expression plasmid pCIS (Genentech) (Fig. 1). pCIS has a multicloning region with the following enzyme restriction sites: *Xho*I, *Not*I, *Cla*I, *Hpa*I, and *Xba*I. The coding region of the human hepatocyte NOS-2 cDNA (16) is conveniently encompassed between *Xba*I sites. Therefore, the *Xba*I–*Xba*I iNOS fragment was isolated and ligated into pCIS that had been linearized at the *Xba*I site. If the fragment of DNA to be inserted does not have compatible DNA overhangs, compatible ends can be generated with DNA linkers. The orientation of the NOS-2 fragment in pCIS has to be determined through restriction digest mapping and only a plasmid containing NOS-2 in the sense orientation is isolated and purified.

Similarly, constructing a retroviral vector requires the insertion of the coding sequence of the NOS-2 cDNA into an appropriate plasmid containing the retroviral sequences. Retroviral plasmids are also available commercially as well as through collaboration. For example, we employed the retroviral

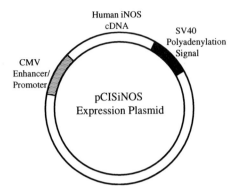

FIG. 1   Human NOS-2 expression plasmid. This mammalian expression vector was constructed using the human hepatocyte NOS-2 cDNA inserted into the pCIS expression plasmid. The cytomegalovirus (CMV) enhancer and promoter region drives expression of the NOS-2 gene in a constitutive manner.

vector called MFG (17). Essentially the entire viral coding sequence except for Ψ, the viral packaging signal, and the LTRs have been deleted from the parent Moloney murine leukemia virus. The cloning site of this plasmid is a 5' *Nco*I site and a 3' *Bam*HI site, located downstream from Ψ. The *Nco*I site incorporates the ATG translational start codon for the inserted gene and ensures a correct reading frame. An *Nco*I site was created in the human NOS-2 cDNA using polymerase chain reaction (PCR)-directed site mutagenesis, and a 3' *Bam*HI compatible site was created using *Bcl*I linkers. In addition, a selectable marker, such as the neomycin phosphotransferase (Neo) gene, can be inserted into the vector. We ligated the Neo gene linked to an internal ribosome entry site (IRES) into the *Bam*HI site just beyond the NOS-2 gene. The IRES permits translation of a second gene product from a single, polycistronic mRNA and has been shown to function superior to a second, internal promoter. This second viral construct was designated DFGiNOS (Fig. 2).

The complexity of creating adenoviral and herpes vectors is beyond the scope of this chapter and the capability of most researchers unfamiliar with these viruses. The large size of these viruses requires cellular recombination to generate infectious viral particles. To screen for successful recombinants and to purify high-titer viral supernatants requires some expertise. We recommend consulting groups specializing in these viral vectors to generate a particular NOS vector over attempting to isolate one independently.

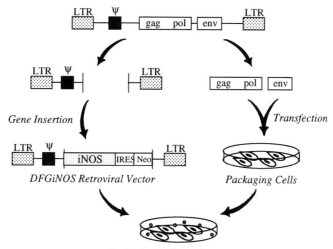

FIG. 2 Diagram of the creation of a retroviral vector carrying the human NOS-2 cDNA. The human hepatocyte NOS-2 (iNOS) cDNA was subcloned into a deletional mutant of the parent Moloney murine leukemia virus. This mutant retained the long terminal repeat (LTR), the viral promoter, as well as the Ψ packaging signal. Included in the retroviral construct is an internal ribosome entry site (IRES), which permits the efficient translation of a second gene product from a single viral transcript, and a neomycin phosphotransferase (Neo) gene to provide a selectable marker. The retroviral vector was designated DFGiNOS. Packaging cells that have been previously engineered with the deleted viral sequences were then transfected with DFGiNOS. The packaging cells are able to package the viral RNA species carrying Ψ with the viral envelope and core proteins that are available *in trans*. The packaged and infectious, but replication-deficient, DFGiNOS retroviruses are shed into the growth medium and can be collected.

Retrovirus Production by Packaging Cells

As listed in Table I, there are many cell lines capable of packaging the viral vectors into replication-deficient retroviral particles (18). Because the viral genes have been deleted from the retroviral plasmid, the products encoded by these genes have to be supplied by another source. Packaging cells are engineered to supply these products *in trans* (Fig. 2), possessing the missing viral sequences inserted into the cellular DNA through multiple previous transfections. Only RNA species carrying the Ψ signal are packaged into

TABLE I   Select Retrovirus-Packaging Cell Lines

Name	Host range[a]	Maximum titer[b]	Drug resistance[c]	Ref.
Ψ-2	Ecotropic	10^7	gpt	Mann et al. (1983)[d]
Ψ-AM	Amphotropic	2×10^7	gpt	Cone and Mulligan (1984)[e]
PA317	Amphotropic	4×10^7	tk	Miller and Buttimore (1986)[f]
PE501	Ecotropic	10^7	tk	Miller and Rosman (1989)[g]
ΨCRE	Ecotropic	10^6	hph, gpt	Danos and Mulligan (1988)[h]
ΨCRIP	Amphotropic	10^6	hph, gpt	Danos and Mulligan (1989)[h]
BOSC23	Ecotropic	10^7	hph, gpt	Pear et al. (1993)[i]

[a] Ecotropic, targets murine cells; amphotropic, targets many mammalian cell types.
[b] Highest reported titers in colony-forming units (cfu/ml).
[c] Drug resistance genes already present in these packaging cells due to prior DNA transfer procedures and selection. Selection for vectors carrying these markers should be avoided. gpt, Xanthine–guanine phosphoribosyltransferase; tk, herpes simplex virus thymidine kinase gene; hph, hygromycin phosphotransferase.
[d] R. Mann, R. C. Mulligan, and D. Baltimore, Cell 33, 153 (1983).
[e] R. D. Cone and R. C. Mulligan, Proc. Natl. Acad. Sci. U.S.A. 81, 6349 (1984).
[f] A. D. Miller and C. Buttimore, Mol. Cell Biol. 6, 2895 (1986).
[g] A. D. Miller and G. J. Rosman, BioTechniques 7, 980 (1989).
[h] O. Danos and R. C. Mulligan, Proc. Natl. Acad. Sci. U.S.A. 85, 6460 (1988).
[i] W. S. Pear, G. P. Nolan, M. L. Scott, and D. Baltimore, Proc. Natl. Acad. Sci. U.S.A. 90, 8392 (1993).

viral particles that are then shed from the cells. Packaging cell lines differ in host range, from ecotropic lines that target predominantly murine cells to amphotropic lines that target a wide range of hosts including human cells.

To generate virus, purified retroviral vector plasmid DNA can be transiently transfected into a packaging cell line using the calcium phosphate method. The method yields viral titers of 10^3 to 10^5 colony-forming units/ ml. An exception is the BOSC23 producer cells which generate virus with titers of 10^6 to 10^7. On day 1, plate retroviral packaging cells at 5×10^5 cells/ 6-cm tissue culture dish. On day 2, replace the culture medium with 4 ml of fresh medium. For each plasmid preparation to be transfected, combine the following: 25 μl of 2.0 M $CaCl_2$, 10 μg plasmid DNA, and sterile water to a total volume of 200 μl. In a dropwise fashion with constant agitation, mix this DNA–$CaCl_2$ solution in a polystyrene tube with 200 μl of HEPES-buffered saline: 280 mM NaCl, 10 mM KCl, 1.5 mM $Na_2HPO_4 \cdot 2H_2O$, 12 mM dextrose, 50 mM HEPES, pH 7.05. Incubate the mixture for 30 min at room temperature and then add it to the cells. Improved transfection efficiency can be obtained if chloroquine (100 μM) is simultaneously added to the cells. Allow the transfection to occur for 8–24 hr at 37°C in a 5% (v/v) CO_2 incubator. After the incubation, the medium is replaced with fresh medium. Virus production will begin within the first 24 hr. Virus-containing medium can be collected and centrifuged at 3000 g for 5 min at 4°C or filtered through a 0.45-μm sterile filter to remove cell debris. This viral supernatant can be used immediately to infect other cells or can be stored at −70°C for later use.

A more efficient method of generating virus is through the creation of a stable virus-producing cell line. The cell line producing the highest titer virus can be isolated and propagated or stored, providing a near endless supply of retroviral vector. A stable producer cell line can be generated by calcium phosphate cotransfection of the retroviral vector plasmid (10 μg) along with a plasmid carrying a selectable marker (0.5 μg) or by transfecting a retroviral vector with an internal selectable marker such as Neo. Forty-eight hours after transfection, the packaging cells are trypsinized and passaged 1:2. Neomycin (G418) is added to the culture medium at an initial dose of 750 μg/ml and tapered to 500 μg/ml when cell death begins. Only cells that express Neo will survive. Fresh medium and G418 should be provided every 3–4 days to remove dead cells. Colonies of G418-resistant cells will develop and can be individually picked and propagated. Colonies are visualized from the bottom of the culture dish and circled. The medium is removed and each colony can be scraped up with a pipette tip with a small drop of medium and transferred to a 12- or 24-well plate. Special care must be taken to minimize desiccation by performing the picking process quickly. The colonies are tested for NOS-2 expression, as a marker for retrovirus production, by assaying the supernatant for nitrite accumulation with the Griess reaction (19). The colony that produces the highest levels of nitrite, and presumably the highest virus production, is expanded to 10-cm plates with 10 ml of medium. Viral supernatant is collected every 24 hr once the cells reach confluence. This viral supernatant is filtered or centrifuged to eliminate cellular debris and can be used immediately or stored at −70°C.

A number of packaging cell lines are derived from a murine fibroblast progenitor. These fibroblasts lack the biosynthesis of tetrahydrobiopterin (BH_4), an essential cofactor for all NOS enzymes. The BH_4 behaves as an allosteric effector by maintaining a functional, dimeric quaternary configuration (20). To detect NOS activity in these cells, the cofactor must be supplemented in the medium either directly or through the addition of sepiapterin, a substrate that can be converted to BH_4 intracellularly. Doses of 10–100 μM are more than sufficient to maximize NOS activity. Although antiviral effects of NO have been reported (21), we did not detect any difference in viral titers generated by packaging cells exposed to NO synthesis versus cells maintained in a NOS inhibitor. Ideally, selecting a packaging cell line like ΨCRIP (22), a BH_4-deficient cell line, allows isolation of a high-titer producing colony without concerns for NO synthesis. For example, we created a ΨCRIP cell line stably producing the DFGiNOS retroviral vector.

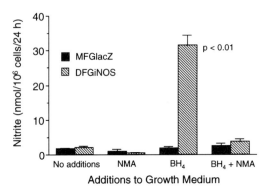

FIG. 3   Nitric oxide synthesis by ΨCRIP packaging cells transfected with DFGiNOS is dependent on supplemental tetrahydrobiopterin (BH_4), an essential cofactor for NOS enzymatic activity. In the absence of BH_4, these cells are unable to synthesize NO, as measured by nitrite accumulation. By providing exogenous BH_4, NOS activity is easily detected. Cells packaging a retrovirus carrying the β-galactosidase gene (MFGlacZ) never demonstrated any NOS activity. These data serve to emphasize the importance of BH_4 availability for the activity of a genetically transferred NOS enzyme.

As shown in Fig. 3, NO synthesis by these producer cells, as represented by NO_2^- levels, was intimately linked to BH_4 supplementation.

Gene Transfer Techniques

Calcium Phosphate-Mediated Transfection

Calcium phosphate-mediated transfection is one of the most widely utilized methods of DNA transfer. The exact mechanism of DNA uptake is not well understood, but it is widely accepted that DNA entry is through endocytosis. DNA–calcium phosphate coprecipitates enhance DNA uptake, and transfection efficiencies may approach 20% in certain cell types. Transfection efficiency is primarily determined by the concentrations of calcium phosphate and DNA to optimize the formation of DNA–calcium phosphate coprecipitates. This method of transfection is ideal for transient expression of a foreign DNA in the form of an expression plasmid. Transient transfection is associated with higher levels of gene product because multiple copy numbers of the expression plasmid are present within transfected cells. Gene transfer using calcium phosphate-mediated transfection into postmitotic chick embryo neurons and photoreceptors was successful with a 20–25% transfection efficiency (23).

The methodology is as previously described in the retrovirus production section. Special care should be taken to prepare the $CaCl_2$ and the precipitation buffers properly because errors can result in the formation of coarse DNA–calcium phosphate precipitates that are poorly endocytosed and are cytotoxic. Transfection efficiency may be improved by concurrent treatment of cells with chloroquine. Chloroquine is thought to act by inhibiting intracellular DNA degradation, permitting more DNA to reach the nucleus. The tolerance of various cell types to chloroquine varies and should be individually determined. A concentration of 100 μM, however, is well tolerated by most cells for a 3- to 5-hr period.

When employing this or any other method of transfection to express NOS-2 in foreign cell types, certain considerations must be made. High levels of NOS-2 expression and NO synthesis may have cytotoxic consequences. For this reason, including a NOS inhibitor, such as N^G-monomethyl-L-arginine (L-NMA), from transfection until just prior to testing for NOS-2 expression is recommended. As mentioned earlier in this chapter, certain cells lack BH_4 biosynthesis in the resting state and are unable to sustain the transferred NOS-2 activity in the absence of supplemental cofactor (Fig. 3). Even cells that exhibit NO synthesis activity following transfection may benefit from cofactor addition if the level of NOS-2 expression exceeds the BH_4 level within the cell. Repleting cellular BH_4 levels may be accomplished two ways. First, the addition of BH_4 (Dr. B. Shircks, Switzerland) directly to the culture medium at a concentration of 10–100 μM is sufficient. However, BH_4 is unstable when exposed to molecular oxygen and must be prepared and stored in the presence of an inert gas. Second, sepiapterin (Dr. B. Shircks, Switzerland) can also be used. Sepiapterin is a substrate for the salvage pathway of BH_4 biosynthesis. Unlike BH_4, sepiapterin is very stable and stock solutions can be stored and reused.

Liposome-Mediated Gene Transfer

Another commonly employed method of nonviral gene transfer is liposome-mediated transfection. This technique relies on the formation of cationic lipid–DNA complexes that facilitate DNA uptake into cells through membrane fusion. The polycationic lipids package and mask the negative charges of DNA, permitting its entry through the cytoplasmic membrane. In comparison to calcium phosphate, liposomes can be 5- to 100-fold more efficient. Commercially available liposome compounds include Lipofectin (GIBCO-BRL, Gaithersburg, MD), a 1 : 1 formulation of N-[1-(2,3-dioleyloxy)propyl]-

N,N,N-trimethylammonium chloride (DOTMA) and dioleoylphosphatidyl-ethanolamine (DOPE), and Lipofectamine (GIBCO-BRL), a 3 : 1 formulation of 2,3-dioleyloxy-N-[2-({2,5-bis[(3-aminopropyl)amino]-1-oxypentyl}amino) ethyl]-N,N-dimethyl-2,3-bis(9-octadecenyloxy)-1-propanaminium trifluo-roacetate (DOSPA) and DOPE. Some other lipopolyamines that have been reported to be effective in transfecting primary neuronal cultures in-clude dioctadecylamidoglycylspermine (DOGS), dioctadecylamidospermine (DOS), and dipalmitoylphosphatidylethanolamidospermine (DPPES) (24). A wide variety of nerve cells have been transfected with DOGS, ranging from central (cortex, striatum, septum hippocampus, hypothalamus, or cerebel-lum) to peripheral nerves (dorsal root ganglia) and astrocytes.

Regardless of the cationic lipid employed, the transfection protocol is the same. The factors that determine liposome transfection efficiency include cell density, relative concentrations of DNA to lipid compounds, and the length of transfection. These parameters must be optimized to determine the conditions that yield the highest level of gene expression for each cell type. The targeted cells are plated onto 6-well or 35-mm tissue culture dishes such that the cells reach 50–80% confluence within 24 hr (the ideal cell density will vary). On the day of transfection, dissolve 1–2 μg of expression plasmid DNA preparation in 100 μl of serum-free medium (without added antibiotics) for each well of cells to be transfected. In a separate set of tubes, dissolve 2–25 μl of the lipopolyamine in 100 μl of serum-free medium. Lipofectin is prepared as a 1 mg/ml solution, Lipofectamine as 2 mg/ml, and DOGS as 2.5 mg/ml. Combine the DNA and lipopolyamine solutions and incubate at room temperature for 10–45 min to allow DNA–lipid complexes to form. Wash the cells with serum-free medium just prior to transfection. For each transfection, add 0.8 ml of serum-free medium to the DNA–lipopolyamine mixture and overlay on the cells. Incubate the cells for 5–24 hr at 37°C in a (v/v) CO_2 incubator. After transfection, replace the DNA–lipopolyamine solution with growth medium (with serum) and incubate for 48–72 hr. Gene expression can be assayed following that time period.

Viral Infections

Of all the gene transfer techniques currently available, viral infection is the simplest once the viral vectors have been generated. For cells that are actively proliferating, the retroviral vector may be very effective. For cells that are postmitotic or for direct *in vivo* NOS gene transfer, either adenoviral or herpes simplex vectors are far superior and may be quite effective. All three viral methods have been utilized successfully for *in vitro* neural gene transfer (14, 25, 26).

To infect target cells with retrovirus, cells should be plated at a density that permits maximal cellular proliferation (50–70% confluence). Remove the medium from cells and add fresh or rapidly thawed viral supernatant. One milliliter of viral supernatant is sufficient to infect a 5-cm culture dish of cells. To augment the effectiveness of the infection process, the polycation Polybrene (Sigma, St. Louis, MO) is added at the time of infection at a concentration of 8 μg/ml. Incubate the cells with the virus for 3–8 hr at 37°C, occasionally swirling the dish to keep cells from desiccating. Following the incubation, remove the viral supernatant and replace with fresh medium. Expression of the virally shuttled gene can be detected as early as 24 hr after infection. We recommend adding an NOS inhibitor to the culture medium after infection until testing is anticipated to reduce any cytotoxicity of NO synthesis. If the retroviral vector was constructed to carry an internal selectable marker such as Neo, such as our DFGiNOS vector, G418 should be added to the infected cells 48 hr after infection. This will yield a population of cells that are all expressing the transferred NOS and Neo genes.

Adenoviral and herpes simplex infections are carried out in a similar fashion. An important difference is that infection with a great excess of either virus will be cytotoxic, resulting in cell lysis. Therefore, the multiplicity of infection (MOI, number of plaque-forming units/cell) that yields the greatest infection efficiency with the least toxicity must be determined for each cell type. MOIs between 10 and 100 are generally well tolerated. Most adenoviral and herpes viral supernatants have undergone purification and concentration to viral titers of 10^9–10^{10} pfu/ml. This purified supernatant is diluted to the appropriate MOI in a very small volume of serum-free medium (the presence of serum can reduce infectivity) and applied to the cells or injected directly into animals. *In vitro* infection can be 100% effective with an incubation time as short as 1 hr.

Other Transfection Techniques

There are a number of other techniques for delivering foreign DNA into cells. Two of these methods include electroporation (27) and biolistics (28). High-voltage electric shock was initially designed to create membrane pores or electropores for cell fusion. However, these pores were found to facilitate cellular entry of macromolecules ranging from DNA to proteins. Electroporation was then adapted to the introduction of exogenous DNA into a broad spectrum of cell lines. The effectiveness of electroporation is limited by several factors. The strength and duration of the electric pulse dictates the success of pore formation with high voltages resulting in irreversible cellular damage. Cell temperature determines the time lapse to pore closure, and improved DNA transfer is achieved if cells are maintained at 0°C. Finally, linear DNA conformation is associated with improved transfection. A number of electroporation instruments are commercially available and are accompanied by detailed protocols.

Particle bombardment has been shown to be effective for neuronal gene transfer (29). Biolistics was developed for transfection of plant cells, the cell walls of which proved to be a formidable barrier to conventional transfection modalities. Modern biolistics employs gold particles (1–3 μm) coated with recombinant DNA as the bullets and have been effective in transfecting primary mammalian neurons, neuronal cell lines, and brain slices. The depth of penetration of the gold bullets is dependent on the gold particle size and the energy used to propel the bullets. The newer gene guns rely on helium gas energy. Gene expression can be detected for 1–2 weeks. The older gene guns were bulky and inconvenient, but the newer helium-powered models are hand-held and can be used for gene delivery to cell culture systems as well as to animals.

Antisense Oligonucleotides

Inhibition of NOS-2 activity in targeted cells can be achieved with substrate inhibitors and antisense oligonucleotides. Substrate inhibitors, to date, lack the specificity to block NOS-2 activity independent of the other NOS isoforms. In that respect, antisense oligonucleotides offer a significant advantage. There is adequate sequence diversity between the three known NOS isoforms to use antisense technology to target a single NOS isoform adequately. Antisense oligonucleotides work by the formation of oligonucleotide–RNA hybrids that are then targets for ribonuclease H digestion. The use of antisense oligonucleotides to block gene production formation has been performed in a number of cell types as well as for a number of gene products including c-*myb*, protein kinase C, p53, tau, integrins, and NOS-2.

The basis for antisense oligonucleotide technology is that cells can internalize these small DNA species by endocytosis or pinocytosis (30). Large quantities of the oligonucleotides are required so that a small number will survive the intracellular transit and reach the nucleus intact. Serum and intracellular nucleases rapidly degrade these small DNA molecules. Fortunately, the advent of phosphorothioate (PS) and methyl phosphonate (MP) (Fig. 4) oligonucleotides, where the normal phosphodiester bonds have been altered, have rendered the oligonucleotides nuclease resistant and have been shown to improve internalization. The specificity of the oligonucleotides is vitally important and is dictated by the length of the molecules. An oligonucleotide between 15 and 17 bases long would theoretically have a unique sequence

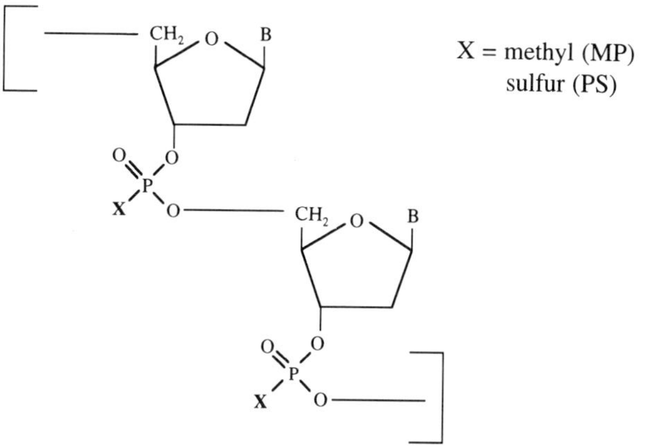

FIG. 4   Synthetic phosphorothioate (PS) and methyl phosphonate (MP) oligonucleotides utilized for antisense targeting. Replacement of the oxygen group on the phosphate in the phosphodiester bond with either a methyl or a sulfur group renders the oligonucleotide species insensitive to nuclease degradation. These compounds are much more useful for both research and clinical applications for this reason.

compared to the entire human genome. Although long sequences may intuitively appear to possess more specificity, they may actually increase nonsequence-specific mRNA degradation because of the greater length-dependent potential hybridization sites. Generally, oligonucleotides between 15 and 30 bases in length are designed. Automated oligonucleotide synthesis technology has enhanced the availability of oligonucleotides. Such synthesis facilities are generally available to most research laboratories.

Antisense oligonucleotides to iNOS should be designed to target a region in the NOS-2 mRNA that minimally cross-reacts with NOS-1 and NOS-3. Oligonucleotide design is vital to the success of antisense targeting. Special precautions should be taken to avoid sequences prone to internal secondary structure or that can form oligonucleotide dimers. Popular oligonucleotide target sites include the splice donor–acceptor site, the polyadenylation signal, the initiation codon region, and the 3' untranslated region. Once the sequence to be targeted has been selected, the antisense oligonucleotide should be synthesized as phosphorothioate oligonucleotides. Such PS oligonucleotides retain the net charge and aqueous solubility of phosphodiester oligonucleotides but are resistant to nuclease degradation. Sense oligonucleotides should also be synthesized to serve as experimental controls.

For cell culture administration, direct addition of the oligonucleotides to the culture medium has been shown to be successful. Targeted cells are passaged to the appropriate sized plates and allowed to attach for 24–48 hr in the required medium. At the time of the experiment, the medium is replaced with medium containing 1–50 μg/ml of either the antisense or sense oligonucleotide. Because PS oligonucleotides are resistant to nuclease digestion, the inclusion of serum in the medium is acceptable if this type of oligonucleotide is being utilized. If phosphodiester oligonucleotides are being used, serum-free conditions are encouraged. Incubate the cells with the oligonucleotide-containing medium for the duration of the experiment. The effectiveness of the antisense oligonucleotide at inhibiting NOS-2 expression can be tested by stimulating the cells to express NOS-2 with the optimal combination of cytokines and quantitating end product (nitrite and nitrate), NOS-2 protein, or mRNA at various concentrations of oligonucleotides. Because of nonsequence specific actions, sense oligonucleotides may also prove to have some inhibitory effect on NOS-2 expression. To assess the specificity of the antisense oligonucleotides for NOS-2, target cell expression of the native NOS-1 (if applicable) should also be examined. Liposomes can also be used to deliver the antisense oligonucleotides to cells in culture with as much as a 1000-fold enhancement in antisense oligonucleotide activity. Lipofectin or Lipofectamine can be used as described earlier. The optimal ratio of liposome to oligonucleotide as well as the length of incubation must be determined with each cell type targeted. When using liposomes, serum-free conditions should be employed with the oligonucleotide–liposome incubation. Once the oligonucleotide–liposome mixture has been removed, the effect of the antisense oligonucleotides is transient and is detectable for only a few days.

In vivo delivery of antisense oligonucleotides has been accomplished with both intraperitoneal and intravenous routes. Administration of the oligonucleotides through these routes required continual or periodic injection of the oligonucleotides in order for the inhibitory effects to persist. In mice, doses of 5–10 mg/kg/day of c-*myb* antisense oligonucleotides administered via an intravenous pump produced significant reduction in leukemic progression in *scid* mice (31). Higher doses (50–100 mg/kg/day) of protein kinase C antisense oligonucleotides administered intraperitoneally reduced mRNA levels by 70–90% (32). However, the detection of radiolabeled oligonucleotides in the brain was minimal with either of these techniques (33). Perhaps administering oligonucleotides directly into the brain parenchyma or in the cerebral spinal fluid will improve CNS targeting.

Another approach to antisense targeting utilizes all the gene transfer techniques described earlier to deliver antisense ribonucleotide sequences. Expression plasmids and viral vectors are constructed with the full length or a portion of the NOS-2 cDNA inserted in an antisense orientation. Transcription of these sequences results in antisense mRNA species. Antisense NOS-2 ribonucleotide sequences would, therefore, be synthesized for a transient period or long term depending on the gene delivery system employed. Unlike antisense oligodeoxynucleotides, repeated or constant infusion of the oligo-nucleotide will not be needed. Antisense RNAs have been reported to reduce neuroblast integrin (34) expression by 50% using retroviral delivery and reduced PC12 tau (35) expression by 20–40% following liposome-mediated transfection of an antisense tau expression plasmid.

References

1. C. Nathan and Q. Xie, *Cell (Cambridge, Mass.)* **78,** 915 (1994).
2. H. H. H. W. Schmidt and U. Walter, *Cell (Cambridge, Mass.)* **78,** 919 (1994).
3. S. M. Morris and T. R. Billiar, *Am. J. Physiol.* **266,** E829 (1994).
4. G. A. Böhme, C. Bon, M. Lemaire, M. Reibaud, O. Piot, J. M. Stutsmann, A. Doble, and J. C. Blanchard, *Proc. Natl. Acad. Sci. U.S.A.* **90,** 9191 (1993).
5. F. M. Faraci and J. E. Brian, Jr., *Stroke* **25,** 692 (1994).
6. T. M. Dawson, V. L. Dawson, and S. H. Snyder, *Ann. Neurol.* **32,** 297 (1992).
7. S. A. Lipton and H. E. Gendelman, *N. Engl. J. Med.* **332,** 934 (1995).
8. V. L. Dawson, T. M. Dawson, G. R. Uhl, and S. H. Snyder, *Proc. Natl. Acad. Sci. U.S.A.* **90,** 3256 (1993).
9. S. Murphy, R. L. Minor, Jr., G. Welk, and D. G. Harrison, *J. Neurochem.* **55,** 349 (1990).
10. E. R. M. Kay, *Nature (London)* **191,** 387 (1961).
11. T. Friedmann, *Nat. Genet.* **2,** 93 (1992).
12. A. D. Miller, *Curr. Top. Microbiol. Immunol.* **158**(Viral Expression Vectors), 1 (1992).
13. K. L. Berkner, *Curr. Top. Microbiol. Immunol.* **158**(Viral Expression Vectors), 39 (1992).
14. J. C. Glorioso, W. F. Goins, C. A. Meaney, D. J. Fink, and N. A. DeLuca, *Ann. Neurol.* **35,** 528 (1994).
15. J. Sambrook, E. F. Fritsch, and T. Maniatis, "Molecular Cloning: A Laboratory Manual." Cold Spring Harbor Laboratory, Plainview, New York, 1989.
16. D. A. Geller, C. J. Lowenstein, R. A. Shapiro, A. K. Nussler, M. Di Silvio, S. C. Wang, D. K. Nakayama, R. L. Simmons, S. H. Snyder, and T. R. Billiar, *Proc. Natl. Acad. Sci. U.S.A.* **90,** 3491 (1993).
17. G. Dranoff, E. M. Jaffee, A. Lazenby, P. Golumbeck, H. Levitsky, K. Brose, V. Jackson, H. Hamada, D. M. Pardoll, and R. C. Mulligan, *Proc. Natl. Acad. Sci. U.S.A.* **90,** 3539 (1993).
18. O. Dranos, *Methods Mol. Biol.* **5**(Practical Molecular Virology, Viral Vectors for Gene Expression), 17 (1991).
19. L. C. Green, D. A. Wagner, J. Glogowski, P. L. Skipper, J. S. Wishnok, and S. R. Tannenbaum, *J. Anal. Biochem.* **126,** 131 (1982).
20. E. Tzeng, T. R. Billiar, P. D. Robbins, M. Loftus, and D. J. Stuehr, *Proc. Natl. Acad. Sci. U.S.A.* **92,** 11771 (1995).
21. K. D. Coren, *J. Clin. Invest.* **91,** 2446 (1993).
22. O. Danos and R. C. Mulligan, *Proc. Natl. Acad. Sci. U.S.A.* **85,** 6460 (1988).

23. M. Werner, S. Madreperla, P. Lieberman, and R. Adler, *J. Neurosci. Res.* **25,** 50 (1990).
24. J. P. Loeffler and J. P. Behr, *in* "Methods of Enzymology" (R. Wu, ed.), Vol. 217, p. 599. Academic Press, San Diego, 1993.
25. E. J. Boiviatsis, M. Chase, M. X. Wei, T. Tamiya, R. K. Hurford, Jr., N. W. Kowall, R. I. Tepper, X. O. Breakefield, and E. A. Chiocca, *Hum. Gene Ther.* **5,** 183 (1994).
26. G. Le Gal La Salle, J. J. Robert, S. Berrard, V. Ridoux, L. D. Stratford-Perricaudet, M. Perricaudet, and J. Mallet, *Science* **259,** 988 (1993).
27. H. Potter *in* "Methods of Enzymology" (R. Wu, ed.), Vol. 217, p. 461. Academic Press, San Diego, 1993.
28. J. C. Sanford, F. D. Smith, and J. A. Russell, *in* "Methods of Enzymology" (R. Wu, ed.), Vol. 217, p. 483. Academic Press, San Diego, 1993.
29. D. C. Lo, A. K. McAllister, and L. C. Katz, *Neuron* **13,** 1263 (1994).
30. C. A. Stein and Y. C. Cheng, *Science* **261,** 1004 (1993).
31. M. Z. Ratajczak, J. A. Kant, S. M. Luger, N. Hijiya, J. Zhang, G. Zon, and A. M. Gevirtz, *Proc. Natl. Acad. Sci. U.S.A.* **89,** 11823 (1992).
32. N. M. Dean and R. McKay, *Proc. Natl. Acad. Sci. U.S.A.* **91,** 11762 (1994).
33. S. Agrawal, J. Temsamani, and J. Y. Tang, *Proc. Natl. Acad. Sci. U.S.A.* **88,** 7595 (1991).
34. D. S. Galileo, J. Majors, A. F. Horwitz, and J. R. Sanes, *Neuron* **9,** 1117 (1992).
35. B. Esmaeli-Azad, J. H. McCarty, and S. C. Feinstein, *J. Cell Sci.* **107,** 869 (1994).

[15] Expression of Constitutive Brain Nitric Oxide Synthase and Inducible Nitric Oxide Synthase by Cultured Neurons and Glia

Dahlia Minc-Golomb and Joan P. Schwartz

Introduction

Nitric oxide (NO) is a membrane-permeant molecule, with a short half-life, which is formed through the conversion of L-arginine to L-citrulline by the enzyme nitric oxide synthase (NOS). The isoforms of NOS fall into two major categories: (1) inducible NOS (iNOS), which is found in macrophages and neutrophils only following induction by cytokines or other treatments, an induction regulated at the level of transcription; and (2) constitutive Ca^{2+}-regulated NOS (cNOS), which includes two isoforms, neuronal (nNOS) and endothelial (eNOS) (1). As a messenger, NO is unique because it is a highly diffusible molecule that can pass through the cell membrane without the need for specific membrane structures such as synapses or receptors. Its primary signal transduction mechanism is through direct activation of the soluble guanylate cyclase.

In the central nervous system (CNS), NO serves as an important neuromodulator. In postsynaptic neurons that contain nNOS, the NO generated in response to synaptic signaling is thought to act as a retrograde messenger (2). Several lines of evidence further support the possibility that NO functions as a retrograde messenger for long-term potentiation (LTP) (3–5). Neuronal NO may also function as a mediator to couple cerebral blood flow to neuronal activity (6). In contrast to these beneficial effects, NO has also been implicated in pathological conditions in the brain, as indicated by a number of studies which have shown that either NO or L-arginine facilitates glutamate neurotoxicity in several brain regions (7, 8).

Because of the potential role of NO in both neuromodulation and neurotoxicity, regulation of expression of NOS isoforms in the CNS has attracted much scientific attention. Bredt *et al.* (9) described and localized the expression of nNOS in the brain and demonstrated that cerebellar granule cells (CGCs) contained the highest levels of nNOS of all neurons. We demonstrated that expression of nNOS is maintained when CGCs are put into primary culture (10) and that these neurons can be stimulated to express iNOS (11). In addition, we and others have shown that astrocytes can express both nNOS and iNOS (12).

In this chapter we present the basic methodologies for studying the expression of NOS in primary cultures of neurons and astrocytes derived from the CNS. The primary tool for analyzing gene expression from small numbers of cells is polymerase chain reaction (PCR). Although this approach is powerful for detecting a specific mRNA in a whole preparation, it cannot localize a specific gene transcript to a specific cell type. The technique of choice for this purpose is *in situ* hybridization: we describe the use of fluorescent *in situ* hybridization (FISH). In addition, we demonstrate the use of immunohistochemistry (IHC) to show expression of NOS protein. Furthermore, FISH and IHC can be combined to prove colocalization of an mRNA and its protein product in a single cell. Since methods to measure NO levels directly are rapidly evolving and improving (13), we have not included them in this text.

Preparation of Cell Cultures

Cerebellar Granule Cells

The following principles underlie the ability to establish relatively pure cultures of CGC neurons. First, the age of the animals selects for the desired type of neurons because CGCs in postnatal day 8 animals are still undifferentiated, allowing them to survive in culture, whereas the other neuronal types, such as Purkinje cells, have already differentiated by that time. Second, continuous exposure of the cultures to an antimitotic agent (cystosine arabinoside) from the first day *in vitro* (DIV 1) prevents the growth of nonneuronal dividing cells. Third, a depolarizing concentration of KCl, which is specifically obligatory for survival of CGC *in vitro* in contrast to other types of neurons, is employed. One final advantage of CGC cultures is that the cells differentiate with a comparable time course as is seen *in vivo*. As they differentiate, putting out a network of processes (illustrated in Fig. 1A), certain biochemical–pharmacological processes, such as γ-aminobutyric acid (GABA) receptors, become functional, whereas others, such as the type 1 somatostatin receptor, disappear. The cultures can be used at any stage during this time course, but the questions to be asked will necessarily differ.

The CGCs are prepared from 8-day-old rat pups as described by Levi *et al.* (14). Cerebellum is dissected, freed of meninges, and chopped into small pieces. Cells are dissociated by a combination of trypsinization (250 μg/ml) and trituration and, after a series of washes, are plated onto poly(L-lysine)-coated dishes in Eagle's basal medium (BME) supplemented with 10% heat-inactivated fetal bovine serum, 2 mM L-glutamine, 25 mM KCl, and 10 μg/ml gentamicin at a concentration of 2×10^6 cells/ml. Twenty-four hours

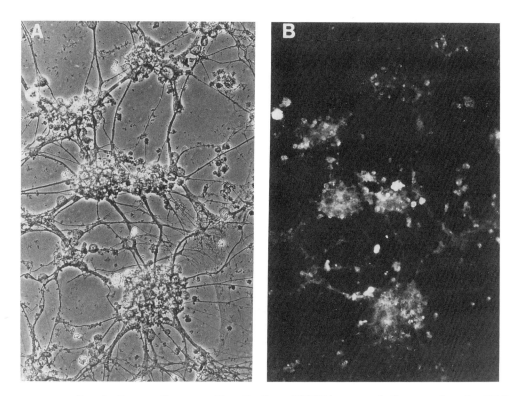

FIG. 1   Immunofluorescent localization of iNOS in rat cerebellar granule cells. CGC were incubated with LPS and IFN-γ for 4 hr prior to fixation. (A) Bright-field view of CGC at DIV12. (B) The same field as shown in (A) after immunohistochemical staining of CGC with antibody to iNOS. Magnification: $\times 400$.

later, 10 μM cytosine arabinoside is added to the cultures. The medium is not changed thereafter, although distilled water is added to maintain the volume. The cultures are maintained in a humidified CO_2 incubator in 95% air–5% CO_2 (v/v).

Immunohistochemical characterization of these cultures has shown that more than 95% of the cells are CGCs. The neuronal morphology of the CGC can be demonstrated by staining with either an antibody to neurofilament proteins, such as HO37, a monoclonal antibody which recognizes the 140- and 200-kDa forms of neurofilament or an antibody to neuron-specific enolase (15). The absence of microglia and astrocytes is verified by staining with antibodies to ED-1 and glial fibrillary acidic protein (GFAP), respectively.

Astrocyte Cultures

Astrocytes can be prepared relatively easily, in part because they survive more readily than differentiated neurons, and in part because they divide continuously once in culture and attach so well to the bottom of the dish that other contaminating cell types (microglia, oligodendrocytes, O_2A progenitors, and type 2 astrocytes) can physically be shaken off and removed. Astrocytes are prepared from different brain regions by dissecting cortex, striatum, cerebellum, or any other region individually. We prepare them from postnatal day 8 rat pups as described previously (16). After removal of the meninges, the tissue is dissociated mechanically in Leibowitz L-15 medium by passage through smaller and smaller needles, and the cells are washed and plated in Dulbecco's modified Eagle's medium (DMEM) containing 4.5 g glucose/liter and supplemented with 10% fetal bovine serum, 2 mM glutamine, 1 mM pyruvate, and 10 μg/ml gentamycin. When the culture reaches confluence (7–10 days), the flask is shaken overnight three times and the medium changed after each shake. After the third shake, the cells are trypsinized and replated in fresh medium containing 10 μM cystosine arabinoside for 24 hr, to kill off dividing fibroblasts. The medium is changed to remove cytosine arabinoside, and the cells are cultured until confluent again (7–14 days later), with medium changes every 2–3 days. Cells are then subcultured for experiments: at this point the cultures are 98% GFAP-positive type 1 astrocytes. A typical culture is illustrated in Fig. 2A. The cultures

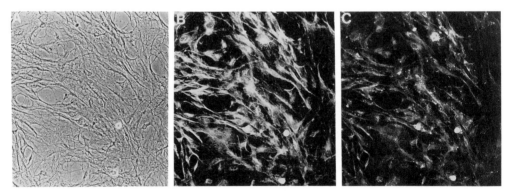

Fig. 2   A FISH analysis of expression of nNOS and iNOS in cortical astrocytes after exposure to LPS and IFN-γ for 4 hr. (A) Bright-field view of astrocytes. (B) Expression of nNOS mRNA in same field as (A). (C) Expression of iNOS mRNA in same field as (A). Magnification: ×200.

are maintained throughout in a humidified atmosphere of 90% air–10% CO_2, which selects against neurons.

RNA-Specific Polymerase Chain Reaction

The numbers of cells that can be obtained in primary cultures, particularly neurons, is rather limited. Therefore sensitive methods are required for the examination of gene expression in these cells. The reverse transcription–polymerase chain reaction (RT–PCR) provides high sensitivity but may lead to false-positive results. One of the main sources for such false-positives is the inadvertent amplification of fragments of genomic DNA contaminating the RNA sample. The genomic sequence for rat NOS isoforms is not yet available, so one cannot select primers that span an intron, which would obviate the problem. There are three other ways to prevent the problem created by genomic DNA contamination: (a) treat all RNA samples with RNase-free DNase; (b) use only poly(A)$^+$ RNA for reverse transcription; (c) use the technique of RNA-specific RT–PCR (RS-PCR) (17). The third approach is the simplest and least likely to result in differential loss of mRNAs.

The RNA is prepared by the single-step method (18), using a commercial solution (RNAzol, Tel-Test, Friendswood, TX). Reverse transcription is carried out using a specific primer which consists of a unique nucleotide sequence that does not occur in the genome at the 5′ end and an antisense sequence for the desired mRNA at the 3′ end. Other conditions include 50 mM Tris, pH 8.3, 40 mM KCl, 6 mM $MgCl_2$, 1 mM dithiothreitol, 20 units avian myeloblastosis virus (AMV) reverse transcriptase (RT), 500 μM of each deoxynucleoside triphosphate (dNTP), 1.0 μM primer, and 5 μg RNA. The reaction is carried out for 60 min at 42°C. Five percent of this product is amplified by a PCR reaction, in which the upstream primer is a sense sequence from upstream of the RT primer site and the downstream primer is the unique sequence on the 5′ end of the RT primer (without the antisense sequence). The reaction is carried out with 10 mM Tris, pH 8.3, 50 mM KCl, 1.5 mM $MgCl_2$, 200 μM of each dNTP, and 0.025 units AmpliTaq polymerase per microliter (GIBCO, Grand Island, NY). Then 22–30 cycles of reaction, consisting of 1 min at 94°C, 1 min at 59°C, and 1 min at 72°C, are carried out. Aliquots of the PCR products are electrophoresed on a 2% agarose gel, stained with ethidium bromide to visualize the PCR products, and transferred onto a nylon membrane (Nytran, Schleicher & Schuell, Keene, NH). Blots are baked in a vacuum oven at 80°C for 2 hr and then hybridized overnight with random primer-labeled cDNA probe. Negative controls that do not include AMV RT or RNA are included.

The following specific primers are used for the amplification of mRNAs for rat nNOS and iNOS:

1. Neuronal NOS RT primer (T30 tagged antisense oligonucleotide, the last 20 bases of which are complementary to the brain cNOS sequence from nucleotide 3040 to 3059, GenEMBL X59949):

5'-GAACATCGATGACAAGCTTAGGTATCGATATCCAGGAGGGTGTCCACCGC-3'

2. Neuronal NOS upstream PCR primer (nucleotides 2779–2800):

5'-AATGGAGACCCCCCTGAGAAC-3'

3. Downstream PCR primer for T30 tag

5'-GAACATCGATGACAAGCTTAGGTATCGATA-3'

4. Inducible NOS RT primer for iNOS mRNA (T30 tagged antisense oligo-nucleotide, the last 20 bases complementary to the iNOS sequence from nucleotide 3554 to 3573, GenEMBL M84373)

5'-GAACATCGATGACAAGCTTAGGTATCGATAGCAGTCTTTTCCTATGGGGC-3'

5. iNOS upstream PCR primer (nucleotides 3440–3461)

5'-GCTACCACATTGAAGAAGCTGG-3'

The RS-PCR analysis of RNA prepared from cortical, cerebellar, and striatal astrocytes for the gene for iNOS is shown in Fig. 3. Astrocytes are incubated for 4 hr with (+) or without (−) 10 μg/ml lipopolysaccharide (LPS; *Escherichia coli* serotype O128 : B12, Sigma Chemical, St. Louis, MO) and 100 U/ml mouse γ-interferon (IFN-γ, Genzyme, Cambridge, MA). The RNA is prepared and analyzed by RS–PCR as described. Figure 3 shows an ethidium bromide-stained gel, in which the specific products of the expected size, 164 bp, can only be detected in the samples from LPS–IFN-γ-treated cultures. Thus, as in macrophages, iNOS transcripts are detected in astrocytes only following cytokine induction. The same observation has been made in CGCs (11).

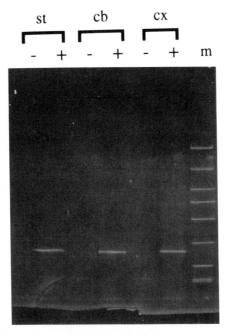

FIG. 3 The RS-PCR products of iNOS mRNA from cortical (cx), cerebellar (cb), and striatal (st) astrocytes, without (−) or with (+) exposure to LPS and IFN-γ for 4 hr. m, DNA size markers, corresponding to 1000, 700, 500, 400, 300, 200, 100, and 50 bp from the top (Research Genetics, Huntsville, AL).

Fluorescent *in Situ* Hybridization

As no primary culture is totally free of contaminating cells, the finding of a specific PCR-amplified product in RNA from such a culture cannot serve as definitive evidence for expression and existence within the expected cell type. Only morphological localization of the specific mRNA and/or protein in the cells is a direct proof for its expression by that specific cell type. To localize NOS mRNA to specific cell types, we use FISH. Utilization of oligonucleotide probes for FISH offers several advantages over the use of cDNA or RNA probes. Aside from eliminating the requirement for the cloned cDNA, one can synthesize probes designed to detect specific individual members of a gene family; in this study, we use probes which independently detect the nNOS and iNOS transcripts. Labeling the oligo-probes by tailing with biotin allows nonradioactive detection of the specific mRNA. Thus, the advantages of FISH are many: (a) one can combine

FISH with immunohistochemistry to colocalize the mRNA with its protein product; (b) the time required for visualization of the hybridization product is significantly shortened relative to radioactive methods; and (c) the resolution of FISH is higher than that of radioactive methods for *in situ* hybridization.

Neurons or astrocytes are cultured in poly(L-lysine)-coated glass chamber slides; the use of organic solvents in the procedure precludes the use of plastic. However, when neurons are grown on glass, the cell bodies of the neurons form dense aggregates and differ morphologically from those seeded on plastic. Following incubation with or without LPS–IFN-γ, the cells are fixed in 2% (w/v) paraformaldehyde for 10 min. The FISH procedure is based on the method described by Young *et al.* (19), as modified in our laboratory (11). Following fixation the slides are washed in phosphate-buffered saline (PBS) for 10 min, immersed in 0.1 M triethanolamine hydrochloride, 0.9% NaCl, 0.25% acetic anhydride for 10 min, washed in 2 × SSC (1 × SSC is 0.154 M NaCl, 0.015 M sodium citrate) for 1 min and dehydrated sequentially in 70, 80, 95, and 100% ethanol. The slides are transferred to chloroform for 5 min and washed in 100 and 95% ethanol for 1 min each.

To optimize the sensitivity of FISH, we use two different oligoprobes for each specific mRNA. The two nNOS-specific antisense oligonucleotides are based on the rat brain sequence and correspond to nucleotides 2779–2800 and 3040–3059 (Gen EMBL X59949). The two iNOS-specific antisense oligonucleotides are based on the rat macrophage sequence and correspond to nucleotides 3440–3461 and 3554–3573 (GenEMBL M84373). Each oligoprobe is biotinylated at both the 5′ end (using biotin–phosphoramidite) and the 3′ end (using biotin-controlled pore glass, Glen Research, Sterling, VA). The corresponding sense and inverse antisense oligomers serve as negative controls.

The slides are prehybridized for 4 hr at 37°C, in a solution containing 50% (w/v) formamide, 4× SSC, 1× Denhardt's solution (0.02% (w/v) bovine serum albumin, 0.02% (w/v) Ficoll, 0.02% (w/v) polyvinylpyrrolidone), 25 μg/ml tRNA, and 50 μg/ml salmon sperm DNA, and hybridized overnight at 37°C in the same solution to which 350 ng/ml each of two oligoprobes has been added. Following hybridization, the slides are washed four times in 2× SSC–50% formamide at 40°C for 15 min per wash, washed twice for 1 hr in 1× SSC at 25°C, rinsed in 70 and 95% ethanol for 5 min each, and air-dried. The slides are then incubated with streptavidin fluorescein (diluted 1 : 200 in PBS; Amersham, Arlington Heights, IL) for 20 min, followed by two washes with PBS. After mounting, the slides are examined in a Zeiss fluorescent microscope. An example of cortical astrocytes labeled by FISH for nNOS mRNA (Fig. 2B) and for iNOS mRNA expression following 4 hr exposure to LPS–IFN-γ (Fig. 2C) is shown in Fig. 2.

Immunohistochemistry

Immunohistochemistry is used for the localization of NOS protein in cell cultures. Cells are fixed in 2% (w/v) paraformaldehyde, followed by incubation for 10 min in 5% (v/v) acetic acid/95% (v/v) ethanol at −20°C. The antibodies are obtained and used as follows: anti-nNOS polyclonal antibody directed against a synthetic peptide corresponding to the amino terminus of rat brain nNOS (a gift from B. Mayer) is used at 1:2000 dilution; anti-iNOS monoclonal antibody (Transduction Laboratory, Lexington, KY) is used at 1:5 dilution. The primary antibodies are incubated with the cells at 4°C overnight. Incubation with secondary antibody diluted 1:100 (goat anti-rabbit IgG coupled to fluorescein or anti-mouse IgG coupled to rhodamine, Jackson ImmunoResearch Laboratory, West Grove, PA) is carried out for 1 hr at room temperature. After extensive washes in PBS, the cells are mounted in Gel-Mount (Biomeda, Foster City, CA) and visualized using a Zeiss fluorescent microscope. Figure 1B shows CGCs stained with iNOS antibody; Figure 1A represents the same field in phase contrast.

When cells are double-labeled by both FISH and IHC, the slide is first hybridized with the oligoprobes for FISH, washed, and then exposed to the IHC primary antibody for 1 hr at 37°C. The slide is then incubated with a combination of streptavidin–fluorescein and the secondary antibody coupled to Texas Red for 1 hr prior to mounting.

Antibodies used for IHC characterization of the cultures include the following: anti-GFAP polyclonal antibody (a gift from R. Lipsky, American Red Cross, Bethesda, MD) used at 1:500 dilution; HO-37 monoclonal antibody specific to the 140- and 200-kDa neurofilament proteins (a gift from V. M.-Y. Lee, Univ. Pennsylvania, Philadelphia, PA) used at 1:5 dilution; anti-NSE polyclonal antibody (Polysciences, Warrington, PA) used at 1:500; and anti-ED-1 monoclonal antibody (Serotec, Oxford, England) used at 1:400. The same fixation and immunohistochemical procedures were used as described above.

Conclusions

Use of a combination of primary CNS cultures together with several techniques for analysis of specific mRNAs and proteins has allowed us to study expression of two isoforms of nitric oxide synthase at the level of both mRNA and protein. We have demonstrated that neurons, the cerebellar granule cells, as well as astrocytes from three different brain regions (cortex, cerebellum, and striatum), express nNOS under control conditions. We have further shown that expression of iNOS can be induced in both types of neural cells by treatment with a combination of LPS and γ-interferon. Analysis by RS-PCR demonstrated the presence of mRNA specific for nNOS and for iNOS in RNA prepared from both types of cultures. Use of FISH allowed definitive localization of the mRNA to the specific cell type under study, and immuno-histochemistry demonstrated that not only was the specific mRNA expressed but its protein product was also present. All of these techniques have the requisite sensitivity and specificity needed for analyses of cultured cells: cell cultures in turn allow one to test various factors for their effects on NOS expression/activity in the absence of other cell types which might contribute other active agents that would confuse the analysis.

Acknowledgments

We thank the following individuals for gifts of cDNA probes or antibodies: D. S. Bredt, B. Mayer, and V. M.-Y. Lee. We also thank Dr. I. Tsarfaty for invaluable assistance with FISH.

References

1. S. Moncada, R. M. J. Palmer, and E. A. Higgs, *Pharmacol. Rev.* **43,** 109 (1991).
2. T. M. Dawson and S. H. Snyder, *J. Neurosci.* **14,** 5147 (1994).
3. T. J. O'Dell, R. D. Hawkins, and E. R. Kandel, *Proc. Natl. Acad. Sci. U.S.A.* **88,** 11285 (1991).
4. E. M. Schuman and D. V. Madison, *Science* **254,** 1503 (1991).
5. T. V. P. Bliss and G. L. Collingridge, *Nature (London)* **361,** 31 (1993).
6. H. H. W. Schmidt, T. D. Warner, and F. Murad, *Lancet* **339,** 986 (1992).
7. V. L. Dawson, T. M. Dawson, E. D. London, D. S. Bredt, and S. H. Snyder, *Proc. Natl. Acad. Sci. U.S.A.* **88,** 6368 (1991).
8. S. A. Lipton, Y. B. Choi, Z. H. Pan, S. Z. Lei, H. S. Chen, N. J. Sucher, J. Loscalzo, D. J. Singel, and J. S. Stamler, *Nature (London)* **364,** 626 (1993).
9. D. S. Bredt, P. M. Hwang, and S. H. Snyder, *Nature (London)* **347,** 768 (1990).
10. D. Minc-Golomb and J. P. Schwartz, *NeuroReport* **5,** 1466 (1994).
11. D. Minc-Golomb, I. Tsarfaty, and J. P. Schwartz, *Br. J. Pharmacol.* **112,** 720 (1994).
12. S. Murphy, M. L. Simmons, L. Agullo, A. Garcia, D. L. Feinstein, E. Galea, D. J. Reis, D. Minc-Golomb, and J. P. Schwartz, *Trends Neurosci.* **16,** 323 (1993).
13. T. Malinski and Z. Taha, *Nature (London)* **358,** 676 (1992).
14. G. Levi, F. Aloisi, M. T. Ciotti, W. Thangnipon, A. Kingsbury, and R. Bálazs, *in* "A Dissection and Tissue Culture Manual of the Nervous System" (A. Shahar, J. deVellis, A. Vernadakis, and B. Haber, eds.), p. 211. Alan R. Liss, New York, 1989.
15. T. Taniwaki, S. P. Becerra, G. J. Chader, and J. P. Schwartz, *J. Neurochem.* **64,** 2509 (1995).

16. J. P. Schwartz and D. J. Wilson, *Glia* **5,** 75 (1992).
17. A. R. Shuldiner, K. Tanner, C. A. Moore, and J. Roth, *BioTechniques* **11,** 760 (1991).
18. P. Chomczynski and N. Sacchi, *Anal. Biochem.* **162,** 156 (1987).
19. W. S. Young, E. Mezey, and R. E. Siegel, *Neurosci. Lett.* **70,** 198 (1986).

[16] Structural Characterization of Human Neuronal Nitric Oxide Synthase Gene: Methodologic Approach to a Complex Transcription Unit

Anthony H. Cheung and Philip A. Marsden

Introduction

Nitric Oxide Synthase Genes

Nitric oxide synthases (NOS) catalyze the 5-electron oxidation of L-arginine to produce L-citrulline and the free radical nitric oxide (NO) (1, 2). The NOS enzymes are best characterized as apoenzymes. The enzymatic catalysis of L-arginine requires dimerization of enzyme monomer units and numerous essential cofactors: tetrahydrobiopterin, calmodulin, heme, flavins, NADPH, and molecular oxygen (3–6). There are three members of the human nitric oxide synthase family identified to date (7, 8). The neuronal, inducible, and endothelial constitutive NOS isoforms localize to human chromosomes 12q24.2, 17q11.2-12, and 7q35-36, respectively (9–13). Though as a family these genes are structurally related in terms of exon–intron organization and are undoubtedly derived by gene duplication from a common ancestral gene, the NOS enzymes are restricted in their cellular expression and mechanisms of cellular regulation in manners unique for each isoform. Data indicate that the only structurally related gene in the human genome is cytochrome P450 reductase, localized to 7q11.2 (7, 14).

Complex versus Simple Genes and Approaches to Complex Genes

This chapter describes the considerations and some of the techniques that are relevant and reliable in characterizing complex transcription units. With advances in molecular cloning techniques and cloning vectors, rapid progress has been made in the structural characterization of complex genes. In many ways the current focus of studies, though based on structural characterization of the genomic DNA, now includes isolation and characterization of informative DNA markers and determination of mechanisms of genomic regulation implicated in transcriptional regulation and alternative splicing

Methods in Neurosciences, Volume 31

patterns, many of which are cell-specific. Advances in the human genome project have motivated a desire to integrate unique transcription units with the human genetic map in general. Specifically, an interest is evolving as to neighboring transcription units or intragenic transcription units. Data presented in Table I indicate characteristic features of complex versus simple transcription units. This dichotomy is often useful in designing the most robust approach to structural characterization of a gene. A simple gene, that is to say one whose total transcriptional length is small, is composed of minimal numbers of introns and exons and shows little evidence for the existence of pseudogenes, might be simply approached using λ phage-derived bacteriophage cloning methods (Table I). These vectors are capable of accommodating inserts of 15 kilobases (kb) (i.e., EMB3, λ FIXII). The approach to a complex transcription unit is dependent on a detailed understanding of mRNA structure [i.e., 5'-RACE (rapid amplification of cDNA ends) and 3'-RACE performed first] and can utilize cloning vectors ranging from cosmid vectors (35–45 kb insert sizes) to P1-bacteriophage and P1-artificial chromosomes (PAC) (75–125 kb insert sizes, respectively) and yeast artificial chromosomes (YAC) (0.3–1.0 megabase insert size). At present preparing a gridded array of YAC or P1 clones and screening with standard Southern hybridization colony lift methods or a PCR cloning strategy are the most efficient, reliable methods for cloning large contigs of any genome. For a smaller laboratory, commercial services offer attractive time-saving screening services.

TABLE I Simple versus Complex Genes

Characteristic	Simple	Complex
Size of gene	Small	Large
Size of mRNA	Small and uniform	Large and varied
Number of exons	Few	Many
Exon usage	Simple	Complex (splicing variants)
Size of exons	Conform to average exon size of 130–140 nt	Marked variation in size with respect to average exon size
First exon	Single	Multiple
Last exon	Single poly(A)site	Multiple poly(A)sites
Pseudogenes	Absent	Present
Internal transcription units	Absent	Present
Repetitive elements	Absent	Present
Allelic variants	Absent	Present
Transcriptional regulation	Simple promoter	Complex single promoter with distal enhancers and/or silencers

Methods and Discussion

Cloning and Sequencing Large Molecular Weight Contigs

For projects seeking to define neighboring contiguous genomic regions or genes of undocumented complexity, a YAC offers potential advantages in terms of longer term project design. The YACs can accept insert sizes varying from 0.3 to 1.0 megabases (Mb). A representative vector would be pYAC-4 in the host *Saccharomyces cerevisiae* AB1380. Representative YAC libraries often contain partial *Eco*RI digests of human diploid genomic DNA previously size fractionated with pulsed-field gel electrophoresis (PFGE). Chimerism of clones, that is, ligations of noncontiguous portions of the human genome, represents a common artifact of YAC cloning vectors. With this methodology the YAC is manipulated against a background of *S. cerevisiae* genomic DNA. Given that the genomic complexity of *S. cerevisiae* approaches 1.4×10^7 base pairs (bp) it is evident that a YAC insert of 1.0×10^6 bp represents approximately 1/15 of the DNA isolated. It is technically challenging to separate large amounts of YAC DNA from the host background.

To isolate genomic fragments of the human neuronal NOS gene, localized to 12q24.2, a human genomic DNA YAC library is screened with standard Southern blot hybridization methods using a cDNA probe representing 5'-UTR sequences and amino acids 1–423 (exon 2–6) of human neuronal NOS. A 0.9-Mb YAC clone is isolated and subsequently mapped by the method of partial enzymatic digestion with the rare cutting restriction enzymes *Not*I, *Mlu*I, and *Sph*I on a CHEF DRIII mapping system (Bio-Rad, Hercules, CA) (9, 10). Southern transfers are done as described (9). Blots are probed with a *Pvu*II/*Bam*HI digest of pBR322 to produce centromeric and acentromeric end-specific pYAC-4 vector probes, respectively. This useful method allows isolation of either end of the YAC given that pYAC-4 is a pBR322 derivative. Yeast clones are maintained on *ura⁻*, *trp⁻*-deficient AHC agar to avoid recombination events and deletion–insertion rearrangements. Each arm of the YAC cloning vector is required for the AB1380 yeast host to grow on this nutritionally deficient substrate.

Southern blots are then probed with varying regions of the human neuronal NOS cDNA to determine whether all exons are present in the YAC. Having defined a partial long-range restriction enzyme map and defining whether the entire gene is present on the YAC, a library of subclones is prepared to produce vectors useful for daily manipulation (i.e., sequencing, pure DNA preparation).

Preparation of High Molecular Weight DNA from Yeast

1. AHC medium, 100 ml, is inoculated with a 48-hr miniculture (4 ml) and incubated (30°C, 250 rpm, 20 hr).

2. Yeast cells are collected (2000 g, 22°C, 10 min), resuspended, and washed in 25 ml of sterile water and spun again (2000 g, 22°C, 10 min).

3. Yeast cells are pooled in 3.5 ml SCEM buffer–2-mercaptoethanol before the addition of 10 μl of 100 units/μl lyticase (Sigma, St. Louis, MO) (37°C, 2 hr occasional gentle mixing) to remove the yeast wall.

4. Pour the yeast solution down the wall of a 250-ml flask containing 3 ml of lysis buffer.

5. Add 4 ml of lysis buffer to the original tube, mix, and again pour the yeast solution down the wall of the flask.

6. Add 350 μl proteinase K (20 mg/ml), incubate (37°C, 1 hr), and then cool to 22°C.

7. Prepare a sucrose step gradient in a 38-ml centrifuge tube (Nalgene 3410-2539): 3 ml of 50% (w/v) sucrose, 12 ml of 20% (w/v) sucrose, and 12 ml of 15% (w/v) sucrose. Layer lysed yeast gently onto the gradient.

8. Load tubes into Beckman SW28 rotor (Beckman Instruments, Palo Alto, CA) and centrifuge (125,000 g, 22°C, 3 hr).

9. Collect 5 ml of the 20–50% sucrose interface which contains the yeast chromosomal DNA and dialyze (4 liters Tris/EDTA (TE), pH 8.0, 4°C, 24 hr). Change dialyzate, and repeat for a further 8 hr.

10. Concentrate DNA, approximately twofold, using granular sucrose in a sterile petri dish. Rinse off granular sucrose with TE, pH 8.0.

11. Dialyze again (4 liters TE, pH 8.0, 16 hr, 4°C), transfer yeast chromosomal DNA to a sterile tube, and store at 4°C.

Whenever handling high molecular weight DNA always use wide bore pipettes and tips to minimize DNA shearing; freezing should also be avoided for this reason.

Required Solutions

    SCEM buffer: 1 M sorbitol, 0.1 M sodium citrate, 10 mM EDTA; add 20.8 μl of fresh 14.4 M 2-mercaptoethanol to 10.0 ml of SCEM buffer

    Lysis buffer: 0.3 M Tris-HCl, pH 8.0, 3% (v/v) N-lauroylsarcosine, 0.2 M EDTA

    TE, pH 8.0: 10 mM Tris, pH 8.0, 1.0 mM EDTA

    AHC medium: 1.7 g yeast nitrogen base (without amino acids and ammonium sulfate) (Difco Laboratories, Detroit, MI), 5 g ammonium sulfate, 10 g casein hydrolyzate (vitamin-free) (Difco Laboratories), 50 ml adenine sulfate dihydrate (4.4. mg/ml); add water to 900 ml, adjust to pH 5.8, autoclave, and add 100 ml of sterile 20% (w/v) glucose [to prepare AHC agar follow the recipe for AHC medium, but use only 10 ml adenine sulfate dihydrate (4.4 mg/ml) and add 20 g agar (Difco Laboratories), autoclave, add 100 ml sterile 20% (w/v) glucose after cooling and before pouring]

Making Sublibraries of High Molecular Weight Yeast DNA

Partial Sau3A Digest

Digest conditions are optimized with an analytical partial Sau3A digest to yield 35–50 kb size fragments of yeast DNA for cloning into cosmid vectors.

1. The analytical restriction enzyme digest of 5 μg yeast DNA is performed in a final volume of 200 μl. Prewarm mixture to 37°C for 5 min prior to addition of 2 μl diluted (0.05 U/μl) Sau3A (Boehringer Mannheim, Mannheim, Germany).

2. At 0, 5, 10, 15, 20, and 30 min remove 30 μl of reaction mix.

3. Arrest Sau3A digest by adding 0.5 μl of 0.5 M EDTA, pH 8.0, equilibrate at 65°C for 10 min, and then phenol/chloroform (1 : 1)(v/v) extract twice with final chloroform extraction. Avoid vigorous pipetting or vortexing.

4. Analyze digest and control (i.e., uncut yeast DNA) with PFGE (0.5× TBE, 2–20 switch time, 4.5 V/cm, 14°C, Bio-Rad CHEF DRIII). Stain gel postrun with ethidium bromide. Determine optimal digest conditions to prepare 35- to 50-kb fragments.

5. For preparative digest the analytical digest can be scaled up 5- to 10-fold. Following restriction enzyme inactivation DNA fragments are ethanol precipitated using standard approaches and lyophilized.

Ligation into Cosmid Vector

Ligation of DNA fragments into a vector will also yield unwanted DNA fragment chimeras. Calf intestinal alkaline phosphatase (CIAP) treatment limits the ligation of noncontiguous fragments.

1. Lyophilized DNA is gently resuspended in sterile water (15 min, 37°C) and treated with 1 μl of 1 U/μl CIAP (GIBCO-BRL Life Technologies, Gaithersburg, MD) (final volume 100 μl, 37°C, 1 hr).

2. Inactivate by adding 3 μl of 0.5 M EDTA (68°C, 10 min) and then phenol/chloroform (1 : 1)(v/v) extract twice with final chloroform extraction.

3. Ethanol precipitate DNA, lyophilize, and resuspend in 20 μl of sterile water (15 min, 37°C).

4. Ligate partial *Sau*3A digest/CIAP-treated DNA into the *Bam*HI site of SuperCosI (Stratagene, La Jolla, CA) and package with Gigapack XL (Stratagene) extract according to the manufacturer's directions.

Screening and Sequencing YAC Sublibraries

Titration of the cosmid library derived from partial *Sau*3A digests of YAC bearing *S. cerevisiae* clones should ideally contain greater than 10^5 independent recombinants. Practically, libraries with more than 10^4 independent clones have been screened by Southern hybridization successfully without the need of amplification or risk of clonal deletion. Although the cosmid library may be screened with a random primer-labeled Alu repeat sequence to select human genomic cosmid clones from the yeast background, this is not routinely necessary. Oligonucleotides (18- to 25-mers) derived from cDNA and/or genomic sequences are used to screen cosmid libraries using YAC-derived DNA as a positive control. T4 polynucleotide kinase (PNK) (New England Biolabs, Beverly, MA) and [γ^{32}P]ATP (NEN, specific activity 3000 Ci/mmol) end labeling of oligonucleotides is a rational approach. Primers arrayed every 300–500 nucleotides can be used for initial screens. Using human cDNAs as a positive control will help exclude oligonucleotides designed from cDNA sequences that cross intron–exon boundaries and therefore do not bind to genomic DNA. If such is the case, the oligonucleotide should be moved 30–50 nucleotides 5' or 3'. The goal of this approach is to develop an overlapping contig of the cosmid clones that bind oligonucleotides every 300–500 nucleotides in a 5' to 3' direction over the full length of the relevant cDNA. Particular attention should not be directed toward 3'-untranslated regions (UTR) as these often contain large exons and can be simply isolated with primers every 1–2 kb. For example, the last exon of human neuronal nitric oxide synthase gene measured 2160 nucleotides in length (15).

A similar approach can be employed with P1 or PAC clones as starting genomic material. P1 bacteriophage clones can accept insert sizes of 75–85 kb, and newer PACs can accept insert sizes up to 120 kb (16, 17). A common cloning vector would be a pAd10-SacBII. The P1 and PAC clones are therefore one-tenth the size of YAC clones. The P1/PAC clones uncommonly exhibit chimerism and are not complicated by the host background. It is relatively straightforward to separate P1 episomal DNA from bacterial genomic DNA using a modified protocol (GenomeSystems, St. Louis, MO) for Qiagen-Tip 500 columns (Qiagen, Chatsworth, CA). For manipulating P1/PAC clones the method of choice in our laboratory is to prepare a partial *Sau*3A digest of some 5- to 10-kb fragments cloned into the *Bam*HI site of pBluescript (Stratagene). DNA fragments of the P1/PAC vector are purified by preparative agarose gel electrophoresis with subsequent β-agarase digestion. Treatment of partial *Sau*3A digested P1/PAC DNA with CIAP is not required prior to ligation. However, researchers should be wary of *Sau*3A chimeras in subsequent clonal analysis. It is important to develop overlapping restriction enzyme contigs to exclude potential clonal chimeras.

Sequence Analysis of Human Neuronal Nitric Oxide Synthase Gene

The reported average exon size of eukaryotic exons is 137 nucleotides. A rational approach to oligonucleotide design includes 18–25 mers every 150 nucleotides on both the top and bottom strands of the cDNA. Staggering the location of the primers every 75 nucleotides helps exclude crossing intron splice junctions with any given oligonucleotide. It is helpful to design primers with computer-based screening approaches (e.g., Oligo version 4.0, National Biosciences, Plymouth, MN) and optimize melting temperatures such that primer pairs can be used in polymerase chain reaction (PCR) amplifications to aid in intron sizing. Sequence analysis with standard methods seeks to define diversions in sequence between cDNA and genomic sequences. Intron sizing can be accomplished with a combination of PCR amplification of genomic DNA, restriction enzyme mapping of cosmid or plasmid subclones, and/or direct sequence analysis. As an initial approach to genomic characterization it is not practical to sequence introns greater than 500 nucleotides in length.

Determination of Transcription Initiation and Termination Sequences

Characterization of transcription initiation requires mapping of the initiation site of RNA polymerase II. It is prudent to initiate such studies with the technique of 5'-rapid amplification of cDNA ends (RACE). This accomplishes two goals. It defines with unidirectional PCR exon 1 of the gene of interest. A representative 5'-RACE approach incorporates a gene-specific antisense primer to prepare first-strand cDNA and two series of PCRs following tailing of first-strand cDNA synthesis with terminal deoxyribonucleotide transferase. Two 5' primers are used initially, one of which incorporates a poly(T) tail with a 5'-multiple cloning site. Subsequent rounds of PCR use two gene-specific antisense primers, nested in a 5' → 3' direction. The PCR products are cloned using a sense–primer multiple cloning site 5' end and a convenient 3'-restriction enzyme site known to exist in the cDNA of interest. The 5'-RACE technique represents a reasonable method for determination of transcription initiation sites if the location of the initial first-strand synthesis primer is within 250 nucleotides of the 5'-cap site of the mRNA. The potential to define alternate promoters or alternate splicing at the 5′ end of the mRNA transcripts is an appealing component of this relatively sensitive method.

5′-Rapid Amplification of cDNA Ends Protocol

A modified 5′-RACE protocol is used to characterize further the 5′ termini of mRNA transcripts for human neuronal NOS derived from normal human brain total cellular RNA (18). Total cellular RNA (5 mg) is reverse transcribed with an antisense gene-specific primer (5′-CCA TGG TAA CTA GCT TCC-3′) by Moloney murine leukemia virus (MMLV) reverse transcriptase (GIBCO-BRL Life Technologies). Template RNA is removed by RNase H (Pharmacia Biotech, Uppsala, Sweden), and first-strand cDNA product is tailed with dATP by terminal deoxynucleotide transferase (GIBCO-BRL Life Technologies). First-round PCR amplification is performed in a total volume of 100 μl for 35 cycles with primer annealing at 55°C (Perkin-Elmer Cetus 480 thermocycler) (Perkin-Elmer Cetus, Emeryville, CA) (sense generic primers 5′-GAC TCG AGT CGA CGA ATT CAA T$_{17}$-3′, 2.5 pmol; 5′-GAC TCG AGT CGA CGA ATT CAA-3′, 25 pmol; antisense gene-specific primer 5′-TCT CTA AGG AAG TGA TGG TTG-3′, 25 pmol). A second round of amplification is performed for 35 cycles with primer annealing at 57°C in a total volume of 100 μl (sense generic primer 5′-GAC TCG AGT CGA CGA ATT CAA-3′, 25 pmol; antisense gene-specific primer 5′-GGC TGT GTC TAG AAG TGA CG-3′, 25 pmol). The PCR products are digested with EcoRI/XbaI, subcloned into pBluescript I SK(−), and subjected to DNA sequence analysis.

Primer extension analysis with independent antisense transcripts labeled at the 5′ terminus with [γ-^{32}P]ATP (NEN, specific activity 3000 Ci/mmol) and T4 PNK (New England Biolabs), defines transcription initiation with nucleotide resolution. Primer extension products should be size-fractionated against a genomic sequencing ladder prepared with the dideoxy-Sanger sequencing method using the identical primer for accurate size comparison. Both 5′-RACE and primer–extension analysis are based on reverse transcriptase (RT) and are subject to limitations of RT extension in regions of secondary RNA structure. S1 nuclease analysis and/or RNase protection represent RT-independent methods to corroborate transcription initiation sites identified with 5′-RACE analysis and 5′-primer extension analysis (18). Similar approaches are utilized to define the 3′ terminus of mRNA transcripts. The 5′-RACE and 3′-RACE techniques revealed structurally distinct ends of mRNA transcripts for human neuronal NOS (15, 19). This diversity, especially at the 5′-end, underscores the necessity to isolate cDNA clones that include regions in or near the mapped transcription initation site. Regions 25 nucleotides longer than cloned cDNA fragments may well contain exonic sequences derived from genomic regions considerably 5' with respect to downstream exons.

Genetic Markers

Dispersed repetitive sequences are characteristic of eukaryotic DNA and provide the potential for allelic DNA polymorphism. Microsatellites consisting of variable numbers of mono-, di, tri-, and tetranucleotide repeats represent a major class of tandemly repeated sequences that are evenly distributed in the human genome and readily amplified with PCR. The dinucleotide pair $(dC-dA)_n$ represents a subclass of tandem repeated DNA segments that are present every 10–20 kb in the human genome and are of great diagnostic utility. The extent of allelic polymorphism is positively correlated with the size of the $(CA)_n$ repeat. Cosmid libraries can be screened with a $[\gamma^{32}P]ATP$ (NEN, specific activity 3000 Ci/mmol) end-labeled $(CA)_{15}$ oligonucleotide to identify cosmid clones expressing dinucleotide repeats of marked complexity. Following this a complete *Sau*3A digest of positive cosmid clones can be used to prepare a library cloned into the *Bam*HI site of a prokaryotic cloning vector (e.g., pBluescript, Stratagene) to define sequences flanking CA repeats for design of gene-specific oligonucleotide primers.

Screening with a CA Oligonucleotide

Oligonucleotides of choice are as follows: $(AC)_{11}AA$, $(AC)_{11}AG$, $(AC)_{11}AT$, $(AC)_{11}CC$, $(AC)_{11}CG$, and $(AC)_{11}CT$.

1. Spot cosmid-bearing bacteria colonies, in a grid fashion, onto a sterile nitrocellulose filter laid on top of LB agar containing 100 μg/ml ampicillin (Sigma) and incubate (37°C, 12–16 hr).

2. Make replica filters for processing: place filters with colonies facing upward onto 10% sodium dodecyl sulfate (SDS)-soaked Whatman (Clifton, NJ) 3MM paper for 1 min, transfer onto denaturing solution-soaked 3MM paper for 1 min, put on neutralization solution-soaked 3MM paper for 1 min, wash twice in neutralization bath for 5 min, rinse in 2× SSC, and then UV cross-link the filter.

3. Prehybridize filters (2–3 hr, 55°C) in hybridization solution (6× SSC, 5× Denhardt's solution, 0.5% SDS, and 0.1 mg/ml sonicated salmon sperm DNA).

4. Probe filters with $[\gamma^{32}P]ATP$ (NEN, specific activity 3000 Ci/mmol) end-labeled $(CA)_{11}$ oligonucleotides (55°C, 2 hr)

5. Wash filters with 6× SSC–0.5% SDS as follows: quick rinse (100 ml, 55°C), wash (100 ml, 45 min, 55°C)

6. Orient filters and expose to Kodak (Rochester, NY) XAR film ($-80°C$, 2–18 hr)

Genomic Amplification of Human Peripheral Blood Genomic DNA to Assess Extent of DNA Allelic Polymorphism in Normal Individuals

An intragenic microsatellite has been located in the human neuronal NO synthase gene (Fig. 1). The intragenic microsatellite corresponding to a dinucleotide $(dC-dA)_n$ repeat is characterized using PCR methodology and the following primers: 5'-flanking sense primer 5'-CCT GCG TGG CTA CTA CAT TC-3', antisense primer 5'-AGA CGT CGC AAC CCT CAT TA-3' (annealing temperature T_a of 52°C). T4 PNK (New England Biolabs, Beverly, MA) and $[\gamma^{32}P]ATP$ (NEN, specific activity 3000 Ci/mmol) are used to end-

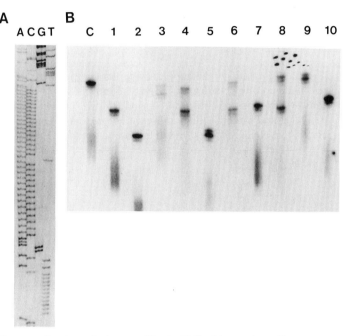

FIG. 1 Characterization of microsatellite in 5'-flanking region of human neuronal NO synthase gene. (A) Sanger $[^{35}S]dATP$-labeled dideoxynucleotide sequencing ladder derived from a genomic clone with a reverse antisense primer. (B) Autoradiograph of $[^{32}P]ATP$ end-labeled PCR amplification of control genomic plasmid subclone used to generate the sequencing ladder (labeled C) or peripheral blood genomic DNA from 10 normal individuals (labeled 1–10). Different sized PCR products represent unique alleles in homozygous (i.e., 1, 2, 7, and 10) or heterozygous individuals (i.e., 3, 4, 5, 6, 8, and 9).

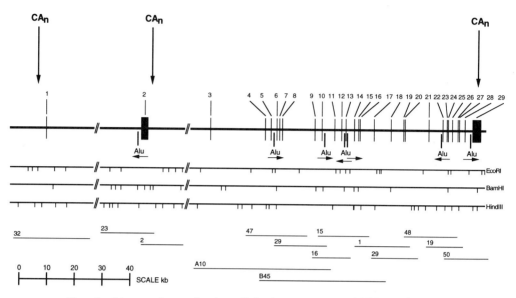

FIG. 2   Structural organization of the human neuronal NO synthase gene. Exons are numbered and indicated by boxes. Introns and 5' and 3'-flanking regions are indicated by lines. *Eco*RI, *Bam*HI, and *Hin*dIII restriction sites are indicated. Bacteriophage and cosmid clones representing human genomic sequences are depicted. Repetitive elements found in the human neuronal NO synthase gene are illustrated. CA_n, dinucleotide $(dC-dA)_n$ sequence repeat; *Alu*, *Alu* 7 SL-derived retroposon repeat.

label 10 picomol of sense primer, and 1×10^5 cpm of primer is used for PCR amplification of 50 ng of human peripheral mononuclear leukocyte genomic DNA in a volume of 50 μl for 35 cycles then analyzed via electrophoresis on a 6% denaturing polyacrylamide gel (Fig. 1). Primer pairs should be optimized initially with respect to PCR conditions (annealing temperatures, Mg^{2+} requirements, and absence of nonspecific genomic amplification) and analyzed with agarose gel electrophoresis.

Summary

The approaches presented in this chapter represent a rational approach to the genomic characterization of a complex transcription unit. Analysis of the human neuronal nitric oxide synthase gene at 12q24.2 revealed that the gene contains 29 exons spanning approximately 160 kb of genomic DNA and is present as a single copy in the haploid human genome (15). Translation initiates in exon 2 and terminates in exon 29, respectively (15). The structural organization of the human neuronal nitric oxide synthase gene is shown in Fig. 2. Comparison with the human endothelial and inducible nitric oxide synthases reveals a remarkable degree of conservation of exon size and intron splice location; marked diversity in intron sizes was not unexpected.

Acknowledgments

Philip A. Marsden is a recipient of a Medical Research Council of Canada (MRCC) Scholarship and is supported by a Medical Research Council of Canada operating grant (MT-12565) and a Heart and Stroke Foundation of Canada operating grant.

References

1. L. J. Ignarro, *Annu. Rev. Pharmacol. Toxicol.* **30,** 535 (1990).
2. S. Moncada, R. M. Palmer, and E. A. Higgs, *Pharmacol. Rev.* **43,** 109 (1991).
3. D. S. Bredt and S. H. Snyder, *Proc. Natl. Acad. Sci. U.S.A.* **87,** 682 (1990).
4. D. S. Bredt, C. D. Ferris, and S. H. Snyder, *J. Biol. Chem.* **267,** 10976 (1992).
5. K. J. Baek, B. A. Thiel, S. Lucas, and D. J. Stuehr, *J. Biol. Chem.* **268,** 21120 (1993).
6. K. A. White and M. A. Marletta, *Biochemistry* **31,** 6627 (1992).
7. D. S. Bredt, P. M. Hwang, C. E. Glatt, C. Lowenstein, R. R. Reed, and S. H. Snyder, *Nature (London)* **351,** 714 (1991).
8. C. Nathan and Q. W. Xie, *J. Biol. Chem.* **269,** 13725 (1994).
9. P. A. Marsden, H. H. Heng, S. W. Scherer, R. J. Stewart, A. V. Hall, X. M. Shi, L. C. Tsui, and K. T. Schappert, *J. Biol. Chem.* **268,** 17478 (1993).
10. W. Xu, P. Gorman, D. Sheer, G. Bates, J. Kishimoto, L. Lizhi, and P. Emson, *Cytogenet. Cell Genet.* **64,** 62 (1993).
11. P. A. Marsden, H. H. Heng, C. L. Duff, X. M. Shi, L. C. Tsui, and A. V. Hall, *Genomics* **19,** 183 (1994).
12. N. A. Chartrain, D. A. Geller, P. P. Koty, N. F. Sitrin, A. K. Nussler, E. P. Hoffman, T. R. Billiar, N. I. Hutchinson, and J. S. Mudgett, *J. Biol. Chem.* **269,** 6765 (1994).
13. L. J. Robinson, S. Weremowicz, C. C. Morton, and T. Michel, *Genomics* **19,** 350 (1994).
14. D. J. Stuehr and M. Ikeda-Saito, *J. Biol. Chem.* **267,** 20547 (1992).
15. A. V. Hall, H. Antoniou, Y. Wang, A. H. Cheung, A. M. Arbus, S. L. Olson, W. C. Lu, C. L. Kau, and P. A. Marsden, *J. Biol. Chem.* **269,** 33082 (1994).

16. J. C. Pierce and N. L. Sternberg, *in* "Methods in Enzymology" (R. Wu, ed.), Vol. 216, p. 549. Academic Press, San Diego, 1992.
17. N. Sternberg, *Proc. Natl. Acad. Sci. U.S.A.* **87,** 103 (1990).
18. P. A. Marsden, K. T. Schappert, H. S. Chen, M. Flowers, C. L. Sundell, J. N. Wilcox, S. Lamas, and T. Michel, *FEBS Lett.* **307,** 287 (1992).
19. J. Xie, P. Roddy, T. K. Rife, F. Murad, and A. P. Young, *Proc. Natl. Acad. Sci. U.S.A.* **92,** 1242 (1995).

[17] Transcriptional Control of Endothelial Nitric Oxide Synthase Gene Expression

William C. Sessa

Introduction

Endothelial nitric oxide synthase (eNOS) is expressed in a variety of cell types inclusive of endothelial cells, hippocampal neurons, cardiac myocytes, and epithelial cells. Although once considered a constitutive "housekeeping gene," evidence suggests that expression of the eNOS gene may be activated via transcriptional mechanisms. The hemodynamic forces of shear stress and cyclic strain *in vitro* (1, 2) and chronic exercise *in vivo* (3) increase eNOS mRNA levels in endothelial cells, and the induction by cyclic strain is via transcriptional activation. Additionally, subconfluent endothelial cells express two to three times more eNOS mRNA than do confluent cells (4). Therefore, to begin to elucidate such mechanisms, the analysis of eNOS mRNA levels, transcription rate, and promoter–reporter gene analysis in endothelial cells have been useful tools in understanding the factors that influence eNOS transcription.

Northern Blot Analysis

The easiest way to determine if the expression of eNOS is influenced by a stimulus is to perform Northern blot analysis. Species specific eNOS cDNA probes for human, bovine, and rat eNOS are available (5–7) that selectively hybridize to the eNOS transcript, but not to iNOS or nNOS transcripts. In our laboratory, the bovine cDNA probe hybridizes selectively to a single transcript in RNA extracts from human, rat, or dog endothelium (~4.4 kb) and to two transcripts from extracts prepared from certain rat nervous system tissues (3, 8).

RNA Isolation, Electrophoresis, and Hybridization

Total cytoplasmic RNA or poly(A) RNA can be isolated from tissues or cultured cells to determine the level of eNOS expression. Total RNA can be extracted from cultured endothelial cells, vascular smooth muscle cells, and fresh tissues by guanidine isothiocyanate extraction and cesium chloride gradient fractionation by standard molecular biology techniques (9). We have used TRIzol, a one-step reagent for the isolation of RNA, with results equivalent to those seen with cesium chloride-purified RNA. The RNA [2–10 μg of total or 0.5–1 μg of poly(A)$^+$ RNA] is then denatured by heating (65° C) for 10 min in 50% (v/v) formamide/4.4 M formaldehyde and electrophoresed through a 1.2% agarose gel containing 2.2 M formaldehyde in Northern buffer [composed of 20 mM 3-(N-morpholino)propanesulfonic acid MOPS, 5 mM sodium acetate, and 1 mM EDTA, pH 7.0] for 2–3 hr. The formaldehyde–agarose gel is then equilibrated in transfer buffer (10X SSPE) for 1 hr, transferred to a nylon membrane in the same buffer (Magna NT, Nucleobond), and hybridized to ^{32}P-labeled eNOS cDNA by standard methods. Blots are washed twice in 2× SSC/0.1% sodium dodecyl sulfate (SDS) for 15 min at room temperature, twice in 0.3× SSC for 15 min at room temperature, then washed twice at high stringency in 0.3× SSC/0.1% SDS at 65°C; blots are then rinsed in 0.1 M phosphate buffer, pH 7.0, dried, and exposed to X-ray film in the presence of an intensifying screen for 2–10 days. Typically, with 5 μg of total RNA isolated from cultured bovine aortic endothelial cells (BAEC) a robust, 4.4-kb transcript is seen after a 2-day exposure. To ascertain if equal amounts of RNA were loaded per lane, examine the samples on a ethidium bromide-stained gel or reprobe the membrane with a cDNA from a constitutively expressed gene, such as glyceraldehyde-phosphate dehydrogenase (GAPDH) or von Willebrand's factor (a constitutively expressed gene in endothelial cells). Relative levels of eNOS can then be quantitated by laser densitometry of the autoradiographs or by phosphorimaging of the digitized areas.

Using Northern blotting, we have been able to determine 2- to 5-fold increases in eNOS expression in endothelial extracts from exercising dogs and cultured BAEC exposed to cyclic strain. Other groups have used Northern analysis to demonstrate increases in gene expression in proliferating BAEC (4) and BAEC treated with inhibitors of protein kinase C (10). The advantages of Northern blot analysis are that it is relatively simple and specific when sufficient quantities of RNA are available from cells or tissues. The disadvantages are potential artifacts due to hybridization of the cDNA probe to related mRNAs, the difficulty in quantification of small changes in gene expression, and the amount of RNA needed to detect a single-copy gene. Alternative approaches for determining eNOS mRNA levels are nuclease protection assays and reverse transcriptase–polymerase chain reaction (RT-PCR). Nuclease protection assays are extremely sensitive, specific, and quantitative. It is the method of choice if Northern blotting cannot discriminate between two related mRNA species. An RT-PCR assay is useful to determine the absence or presence of a particular mRNA. Competitive RT-

PCR is absolutely necessary if quantitation of PCR products is to be attempted. However, even with competitive PCR, there is no control over the reverse transcriptase reaction with different template RNAs. Most importantly, interpretation of all measurements for eNOS gene expression by the above techniques must be tempered by the inability to discern if stimulation or inhibition of expression is transcriptionally based or due to posttranscriptional mechanisms.

Nuclear Runoff Assays

If RNA analysis methods demonstrate differences in eNOS gene expression it is necessary to discern if the nuclear transcription rate is being affected or if the mRNA is being stabilized or destabilized. In brief, to quantify transcriptional rate isolated nuclei are incubated with radiolabeled deoxynucleotides to generate labeled nascent mRNAs which are then hybridized to immobilized full-length cDNAs or probes with 5' or 3' untranslated regions as described (11). Side-by-side comparison of nuclei isolated from control or treated cells allows the investigator to determine the transcriptional rates.

To isolate nuclei, cultured endothelial cells are washed with PBS, trysinized, and pelleted under sterile conditions. Nuclei are extracted in buffer containing 10 mM Tris-HCl, pH 7.4, 10 mM NaCl, 3 mM MgCl$_2$, 0.5% (v/v) Nonidet P-40 (NP-40) with protease inhibitors [1 μg/ml leupeptin, 5 μg/ml aprotinin, and 0.5 mM phenylmethylsulfonyl fluoride (PMSF)] and are recovered by centrifugation at 500 g for 5 min at 4°C. Nuclei are then resuspended in runoff buffer containing 35% glycerol, 10 mM Tris-HCl, pH 8.0, 5 mM MgCl$_2$, 80 mM KCl, 0.1 mM EDTA, and 0.5 mM dithiothreitol (DTT), 0.8 U RNasin, 4 mM each of dATP, dGTP, and dCTP, and 200 μCi [α-^{32}P]dUTP (3000 Ci/mmol, Amersham, Arlington Heights, IL) and incubated at 26°C for 30 min to generate labeled transcripts. Nuclei are then digested with RNase-free DNase I (10 μg) at 26°C for 5 min followed by incubation with proteinase K (10 mg in buffer containing 5% SDS, 50 mM EDTA, and Tris-HCl, pH 8.0) at 37°C for 30 min. Nascent, labeled transcripts are extracted with guanidium isothiocyanate, precipitated with 2-propanol, and dissolved in buffer containing 50 mM Tris-HCl, pH 8.0, 150 mM NaCl, and 1 mM EDTA. The incorporation of label into RNA derived from control or treated nuclei is quantified in a β-counter, and equal counts are then hybridized overnight at 65°C to equal amounts of linearized, alkaline-denatured eNOS or GAPDH cDNAs that were previously immobilized by UV cross-linking onto nylon or nitricellulose membranes. Membranes are washed as outlined under the section on Northern blotting and transcriptional rates determined by densitometry of autoradiographs or by phophorimaging.

Nuclear runoff assays are extremely informative; however, they are time-consuming and technically difficult. A rate-limiting factor is the number of cells needed to obtain sufficient amounts of high-quality nuclei. A good rule of thumb is to characterize the transcription of eNOS in nuclei isolated from control cells, then to scale up in order to run nuclei from control and treated cells side by side. Also, the amount of labeled transcript is usually very small, so the addition of trace amounts of carrier yeast tRNA during extraction and precipitation will improve yields. If data from runoff assays show no differences in the transcriptional rate between control and treated cells, then RNA stability measurements are warranted as previously described for the destabilization of eNOS transcript by tumor necrosis factor (TNF) treatment of endothelial cells (12).

Promoter–Reporter Gene Assays

If RNA stability studies and nuclear runoff data demonstrate that transcription of the endogenous gene is stimulated or inhibited, the use of promoter–reporter gene assays is a convenient and sensitive way to (1) examine if the nucleotide sequence upstream of the transcriptional start site can act as a cellular promoter; (2) identify the important cis-DNA elements within a promoter region; and (3) elucidate the trans-acting factors that are necessary and sufficient for gene expression. Promoter–reporter gene constructs have been used to characterize the important DNA elements of the iNOS and eNOS promoters (13–15). Because gene expression is a complex process, driven by promoter regions of genes in concert with multiple basal transcription factors, intronic enhancers, and specific cellular factors, reporter gene assays should always be interpreted cautiously and used in conjunction with measurements of the endogenous gene.

Polymerase Chain Reaction Cloning of Endothelial Nitric Oxide Synthase Promoter and Construction of Luciferase Reporter Gene

We have cloned the promoter for the human eNOS gene and have generated luciferase (LUC) reporter gene constructs using the following methodology (15). Based on the most 5′ sequence of the human eNOS gene, we have designed two oligonucleotide primers (5′-ATCTGATGCTGCC-3′) and 3′ downstream prior to the initiator methionine codon (5′-GTTACTGTGCGT-3′) to amplify a 1600-bp fragment by PCR using *Taq* polymerase (Promega, Madison, WI) with human genomic DNA (50 ng) as template. Denaturation, annealing, and elongation temperatures are 94°, 60°, and 72°C for 1 min each for 30 cycles, respectively. The PCR product is gel purified and subcloned into the TA cloning vector for PCR products (Invitrogen, San Diego, CA). Alkaline-denatured double-stranded plasmid DNA is sequenced on both strands and is virtually identical to the previously published sequences (16, 17).

The eNOS PCR product (F1) is subcloned into the *KpnI/XhoI* sites of the luciferase reporter gene vector, pGL2-Basic (Promega). This vector contains several unique polycloning sites in front of the luciferase gene product (LUC) followed by a poly(A) signal for stabilization of the transcribed product. We then generate deletion mutants to test if specific regions of the promoter are important for transcriptional activation. For the generation of deletion mutants, the following forward primers [5′ to 3′ notation, each with a *KpnI* site (underlined) preceded by AA] are used in a PCR reaction with the full-length promoter (−1600 to +22) as template.

F2: AAGGTACCACAGCCCGTTCCTTC (nt −1189)
F3: AAGGTACCCCGTTTCTTTCTTAAACT (nt −1033)
F4: AAGGTACCCTGCCTCAGCCCTAG (nt −779)
F5: AAGGTACCGAGGTGAAGGAGAGA (nt −494)
F6: AAGGTACCGTGGAGCTGAGGCTT (nt −166)

with the reverse primer (with a *XhoI* site at the 3′ end)

R1-CTCGAGGTTACTGTGCGTCCACTCT (nt +22 to +4)

The PCR products are gel purified, digested, and subcloned into the *KpnI/XhoI* sites of the luciferase reporter gene vector.

Transfection of Bovine Aortic Endothelial Cells with Endothelial Nitric Oxide Synthase Promoter Constructs

The BAEC are cultured in Dulbecco's modified Eagle's medium (DMEM) containing 10% heat-inactivated fetal bovine serum, 25 mM L-glutamine, 100 units/ml penicillin, and 100 μg/ml streptomycin sulfate (all from GIBCO-BRL, Gaithersburg, MD) into 6-well plastic tissue culture plates (C-6). We typically use BAEC between passages 3 and 6 for transfections. For transfection of BAEC, cells (60–70% confluent) are preincubated in OptiMEM medium (GIBCO-BRL) for 30 min at 37°C. Endothelial eNOS promoter plasmid DNAs (2 μg) and a plasmid containing simian virus 40 (SV40)-driven β-

galactosidase (1 μg, to normalize for transfection efficiency) are mixed with Lipofectamine (GIBCO-BRL, 6 μl/well) and incubated for 20 min at room temperature. The lipid-coated DNA is then added to each well containing 2 ml of OptiMEM medium and incubated overnight. The next day, the medium is removed and replaced with complete medium for an additional 24 hrs. The BAEC are then lysed (400 μl) and extracts centrifuged to remove unbroken cells and debris. Extracts are then used for measurement of luciferase (10 μl) or β-galactosidase activities (40 μl). Luciferase activity is measured at least three times in duplicate using a Berthold luminometer, and β-galactosidase is measured spectrophotometrically (at 420 nm) by the generation of o-nitrophenol from the substrate, o-nitrophenyl-β-D-galactopyranoside. All data are normalized as relative light units/β-galactosidase activity.

Comments

Transfection of the promoterless pGL2-Basic vector into BAEC yields very low levels of LUC activity (200–400 arbitrary light units per 10 μl of lysate), whereas transfection with the eNOS F1 promoter construct gives a 25- to 35-fold increase in LUC activity over background levels. Deletion analysis demonstrated that the regions between F3 and F4 and F5 and F6 are important for basal transcription in BAEC. Lipofectamine is our agent of choice when transfecting BAEC in C-6 well culture plates. Transfection efficiency and reproducibility are very good based on intra- and interassay variability; however, when larger numbers of cells are needed, we transfect 150-mm dishes of subconfluent BAEC utilizing the less expensive calcium phosphate method with good results. A critical factor with transient reporter gene assays is the quality and amount of supercoiled DNA. Typically if we are comparing the full-length construct to different deletion constructs, we perform simultaneous plasmid preparations and estimate the amount of supercoiled DNA on a 1% agarose gel in Tris–EDTA–acetic acid (1× TAE) buffer. If the amounts of supercoiled DNA look similar, we quantify the DNA spectrophotometrically and transfect. Another consideration when using LUC reporter gene assays is the time necessary to reach peak and stable luciferase activity. In BAEC, we typically measure luciferase activity 24 hr after transfection. Normalization of transfection efficiency by cotransfection with an additional reporter gene (β-galactosidase or human growth hormone) using a constitutively active promoter [early SV40 or cytomegalovirus (CMV) promoters] is essential for reliable results. We have tested both SV40- and CMV-driven β-galactosidase reporter plasmids in BAEC and have found that the CMV-based vector gives a much more robust increase in β-galactosidase activity than does the SV40 vector.

A major concern when using promoter–reporter constructs in a heterologous expression system is that transcription initiation may occur at alternative transcriptional start sites compared to that observed with the endogenous gene. Such a result would lead the investigator to study the promoter in an "unphysiological" context. To confirm that the promoter–reporter transgene utilizes the same initiation site as the endogenous gene, primer extension analysis of RNA isolated from cells transiently transfected with the reporter gene should be performed.

Site-Directed Mutagenesis

By deletion analysis, it is possible to determine positive or negative regulatory regions of a promoter construct as described. However to assess directly if an element is necessary for transcription of the reporter gene, deletion analysis must be coupled with site-directed mutagenesis, electrophoretic mobility gel shift, and supershift assays to characterize the putative trans-activating factor. For example, based on our results with the human eNOS promoter the Sp1 element is necessary for promoter activity because mutation of the Sp1 in the context of all LUC constructs (F3, F5, and F6) inhibited normalized LUC activity by 80–90%. The method herein describes site-directed mutagenesis of the Sp1 site using recombinant PCR with two rounds of amplification as previously described (18). The sequence of the eNOS Sp1 motif (−103) is 5′-GGATAGGGCCCGGGCGAGG-3′ with the hexanucleotide of GGGCCC being the core sequence. The PCR primers were designed to introduce an **A** for the first **C** in the motif and a **T** for the third **C** in the motif. Therefore, the PCR primers for the Sp1 mutation (mutations of wild-type sequence appear in bold) are (5′ to 3′ notation) GGATAGGG**A**CTGGG-CGAGG (sense) and CCTCGCCC**A**G**T**CCCTATCC (antisense). In brief, sense and antisense primers with the corresponding mutations were synthesized and incubated in separate reaction tubes with F3, F5, and F6 LUC as templates and one outside complementary primer from upstream or downstream of the mutation site, thus yielding two subfragments that each contained the appropriate mutation. Subfragments were end-filled and annealed, and a second round of PCR was performed using two outside primers. The PCR product was isolated and subcloned into the *Kpn*I/*Xho*I sites of pGL2-Basic as above.

Analysis of Nuclear Binding Proteins

Deletion analysis followed by site-directed mutagenesis will confirm if a cis-element is necessary for transcriptional activity, presumably because the cis-element binds an important transcription factor. However, this cannot be assumed because mutagenesis may interfere with higher DNA structure, not nuclear binding of a specific transcription factor. Two ways to determine if a cis-element can bind nuclear factors is via electrophoretic mobility shift assays (EMSA) and supershift assays. In brief, nuclei are isolated from BAEC and incubated with a ^{32}P-labeled oligonucleotide probe encompassing a putative cis-element (i.e., Sp1 site). DNA–protein complexes are then resolved on native gels and specific interactions with specific nuclear proteins can be determined.

Preparation of Nuclear Extracts

Nuclear extracts are prepared essentially as described by Dignam et al. (19) BAEC (T75 flasks) are rinsed with cold PBS, and cells trypsinized, and pelleted. Cell pellets are washed twice with cold PBS, lysed by the addition of 500 μl of lysis buffer (10 mM HEPES, 1.5 mM MgCl$_2$, 10 mM KCl, 0.5% NP-40 containing 1 μg/ml leupeptin, 5 μg/ml aprotinin, 1 mM DTT, and 0.5 mM PMSF), and are incubated at 4°C for 10 min. Lysates are centrifuged, and nuclei are resuspended in buffer containing 20 mM HEPES, 420 mM NaCl, 1.5 mM MgCl$_2$, 0.2 mM EDTA, and 25% glycerol with the above protease inhibitors and incubated at 4°C for 30 min. Nuclei are centrifuged (10,000 rpm) for 10 min at 4°C, and an equal volume of the supernatant is added to a nuclear homogenization buffer containing 20 mM HEPES, 100 mM KCl, 0.2 mM EDTA, 20% glycerol with protease inhibitors; extracts are stored at −80°C. Extracts are sonicated and clarified prior to use.

Electrophoretic Mobility Shift Assays and Supershift Assays

Probes are prepared by annealing complementary strands of the oligonucleotides overnight followed by purification on Sephadex G-25. Radioactive probes are end-labeled with [γ-^{32}P]ATP using T4 polynucleotide kinase (New England Biolabs, Beverly, MA) and purified prior to use. Typically specific activities are 10^5 cpm/ng DNA. Nuclear extracts (4–10 μg) from BAEC are incubated in binding buffer [25 mM HEPES (pH 7.5), 50 mM KCl, 1 mM DTT, 10 μmM ZnSO$_4$, 0.2 mg/ml bovine serum albumin (BSA), 10% glycerol, and 0.1% NP-40] containing 1 μg of poly(dI) · poly(dC) (Sigma, St. Louis, MO) at room temperature for 10 min and then the ^{32}P-labeled oligonucleotide probe (~5000 cpm or 50 fmol) is added for an additional 10 min in a total reaction volume of 15 μl. In competition studies, excess wild-type or mutant oligonucleotides are added in 50-fold molar excess prior to the addition of the ^{32}P-labeled probe. The wild-type Sp1 probe (upper strand) is 5′-GGATAG-

GGGCGGGGCGAGG-3', and the mutant probe is 5'-GGATAGGGACTGG-GCGAGG-3'.

To determine the composition of nuclear proteins that bind to Sp1-specific oligonucleotide sequences, supershifting of the DNA–protein complexes is performed. For supershift experiments, either nonimmune or Sp1 antisera (Santa Cruz Biotech, Santa Cruz, CA) is incubated overnight at 4°C with DNA–nuclear protein complexes, prior to electrophoresis. Authentic Sp1 (1 ng, Promega) is used to determine the specificity of the Sp1 antisera. Nuclear DNA–protein complexes are resolved on 6% nondenaturing polyacrylamide gels containing 7.5% glycerol in 0.25% Tris–borate–EDTA buffer. Dried gels are exposed to Kodak (Rochester, NY) XAR film for autoradiography.

Comments

The combination of site-directed mutagenesis followed by EMSA and super-shifting provides a relative easy way to determine (1) if a cis-element is necessary, (2) if the element can specifically bind nuclear proteins, and (3) the preliminary identification of the proteins that bind to the cis element. Utilizing the above methodology, we have determined that the Sp1 element in the eNOS promoter is necessary for gene expression and that Sp1 and other nuclear proteins can bind to the Sp-1 site.

Acknowledgments

I thank Ms. Rong Zhang for excellent technical assistance in this project. W.C.S. is supported, in part, by grants from the National Institutes of Health (R29-HL51948) and The Patrick and Catherine Weldon Donaghue Medical Research Foundation. The Molecular Cardiobiology Program at Yale is supported by Lederle Pharmaceuticals.

References

1. K. Nishida, D. G. Harrison, J. P. Navas, A. A. Fisher, S. P. Dochery, M. Uematsu, R. M. Nerem, R. W. Alexander, and T. J. Murphy, *J. Clin. Invest.* **90,** 2092 (1992).
2. M. A. Awolesi, W. C. Sessa, and B. E. Sumpio, *J. Clin. Invest.* **96,** 1449 (1995).
3. W. C. Sessa, K. Pritchard, N. Seyedi, J. Wang, and T. H. Hintze, *Circ. Res.* **74,** 349 (1994).
4. J. Arnal, J. Yamin, S. Dockery, and D. G. Harrison, *Am. J. Physiol.* **267,** C1381 (1994).
5. S. P. Janssens, A. Shimouchi, T. Quertermous, D. B. Bloch, and K. D. Bloch, *J. Biol. Chem.* **267,** 14519 (1992).

6. W. C. Sessa, J. K. Harrison, C. M. Barber, D. Zeng, M. E. Durieux, D. D. D'Angelo, K. R. Lynch, and M. J. Peach, *J. Biol. Chem.* **267,** 15274 (1992).

7. Partial eNOS clone from rat kidney, GenBank accession U02534.

8. W. C. Sessa, J. K. Harrison, D. R. Luthin, J. S. Pollock, and K. R. Lynch, *Hypertension (Dallas)* **21,** 934 (1993).

9. F. M. Ausubel, R. Brent, R. E. Kingston, D. D. More, J. G. Seidman, J. A. Smith, and K. Struhl (eds.) "Current Protocols in Molecular Biology." Wiley, New York, 1990).

10. Y. Ohara, H. S. Sayegh, J. J. Yamin, and D. G. Harrison, *Hypertension (Dallas)* **25,** 415 (1995).

11. M. E. Greenberg and E. B. Ziff, *Nature (London)* **311,** 433 (1984).

12. M. Yoshizumi, M. A. Perella, J. C. Burnett, and M. E. Lee, *Circ. Res.* **73,** 203 (1993).

13. Q. W. Xie, Y. Kashiwabara, and C. Nathan, *J. Biol. Chem.* **269,** 4705 (1994).

14. R. C. Venema, K. Nishida, R. W. Alexander, D. G. Harrison, and T. G. Murphy, *Biochim. Biophys. Acta* **1218,** 413 (1994).

15. R. Zhang, W. Min, and W. C. Sessa, *J. Biol. Chem.* **270,** 15320 (1995).

16. P. A. Marsden, H. Q. Heng, S. W. Scherer, R. J. Stewart, A. V. Hall, X. M. Shi, L. C. Tsui, and K. T. Schappert, *J. Biol. Chem.* **268,** 17478 (1993).

17. L. J. Robinson, S. Weremowicz, C. C. Morton, and T. Michel, *Genomics* **19,** 350 (1994).

18. R. Higuchi, B. Krummel, and R. K. Saiki, *Nucleic Acids Res.* **16,** 7351 (1988).

19. J. D. Dignam, R. M. Lebovitz, and R. G. Roeder, *Nucleic Acids Res.* **11,** 1475 (1983).

[18] Site-Directed Mutagenesis to Probe Endothelial Nitric Oxide Synthase Posttranslational Modifications

Jianwei Liu and William C. Sessa

Introduction

Nitric oxide (NO) is synthesized from L-arginine by the enzyme NO synthase (NOS). Nitric oxide mediates diverse physiological processes such as cardiovascular homeostasis, host defense mechanisms, learning, and memory. Three isoforms of NOS, neuronal (nNOS), cytokine-inducible (iNOS), and endothelial (eNOS), share approximately 50% amino acid sequence homology, indicative of a NOS gene family (1).

Although NOSs have similar cofactor requirements and can exist as soluble or membrane-associated proteins (2), the lipid modifications of eNOS are a unique feature compared to the other isoforms of NOS. Both cytosolic and membrane-bound forms of eNOS are N-myristoylated (3, 4), but only membrane-bound eNOS is palmitoylated (4, 5). Endothelial NOS is a Golgi-associated protein, and the lipid modifications are necessary for its localization (6).

Our understanding of the importance of N-myristoylation in the membrane association of eNOS has been greatly facilitated by site-specific mutagenesis. Mutation of glycine-2, the myristoylation site, to alanine inhibits myristoylation of eNOS, converts the membrane-bound eNOS into a cytosolic form (7, 8), and abolishes palmitoylation of the protein (4, 5). Therefore, precise functional role(s) of either modification remain to be clarified. We have successfully used site-directed mutagenesis to identify palmitoylation sites in eNOS, thus allowing us to study the functional roles of myrisotylation and palmitoylation in the activation and localization of the enzyme.

General Comments

The replacement of specific amino acid residues in proteins using site-directed mutagenesis allows for a rationally designed, systematic, and quantitative investigation of protein structure–function relationship. Site-specific mutations can be introduced into cloned DNA by several highly efficient and straightforward methods. Specific mutations are typically introduced by hybridization of a synthetic oligonucleotide primer containing the desired mutation to either single- or double-stranded wild-type DNA. Methods differ in the way the mutagenized strands are synthesized *in vitro*. The most reliable methods require single-stranded DNA as template for site-directed mutagenesis. Single-stranded DNA can be obtained by cloning the DNA fragment of interest into the phage vector, M13. However, the foreign DNA cloned into the bacteriophage is unstable and thus requires the investigator to subclone from M13 into plasmid vectors for subsequent propagation. The use of phagemids circumvent such subcloning steps and are genetically stable. Phagemids replicate as plasmids; however, in the presence of viral helper phage, single-stranded DNA is secreted into the medium.

The use of PCR (polymerase chain reaction) methodology to generate mutants does not require single-stranded DNA template and eliminates the screening for mutant clones. However, the most troublesome problem associated with PCR methods is the high frequency of second-site mutations. Regardless of the methods employed for mutagenesis, it is usually preferable to resequence the DNA to ensure that only the desired mutation has been achieved. We routinely use the phagemid vector pALTER-1 (Promega, Madison, WI) for site-directed mutagenesis. The pALTER vector contains an inactivated ampicillin resistance gene, which can be restored by annealing an ampicillin repair oligonucleotide primer during the mutagenesis reaction, thus providing an antibiotic selection for mutants. Here, mutation of eNOS palmitoylation site(s) is exemplified to describe the site-directed mutagenesis procedures we routinely employ.

Determination of Potential Palmitoylation Sites

Unlike N-myristoylation, no consensus sequence has been identified yet for protein cysteine palmitoylation. To identify the potential palmitoylation site(s) of eNOS, bovine aortic endothelial cells are labeled with [^{3}H]myristic acid or [^{3}H]palmitic acid. The labeled eNOS is purified by immunoprecipitation using a specific eNOS monoclonal antibody and partially digested with endoproteinase Glu-C (V8 protease). V8 protease mapping demonstrates similar labeled mapping patterns between [^{3}H]palmitic acid-labeled and [^{3}H]myristic acid-labeled eNOS peptides, implying that the palmitoylation site(s) must be close to the N-terminal myristoylation site of eNOS. The largest palmitoylated peptide fragment includes the first six cysteine residues as potential palmitoylation sites.

Design of Mutagenic Oligonucleotides

The full-length eNOS cDNA in Bluescript (Stratagene, La Jolla, CA) plasmid (9) is digested with *Hin*dIII/*Xba*I and subcloned into the corresponding sites of the mutagenesis vector, pALTER-1 as illustrated in Fig. 1. The single-stranded DNA produced from this phagemid is used as template for site-directed mutagenesis. Therefore, the mutagenic oligonucleotide primer must be complementary to the sense eNOS strand. Approximately, 8–10 perfectly matched nucleotides at the 5' terminal and 7–9 at the 3' terminal are required for formation of a stable hybrid and for primer extension. A new restriction

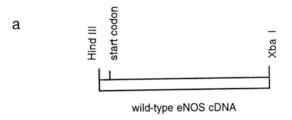

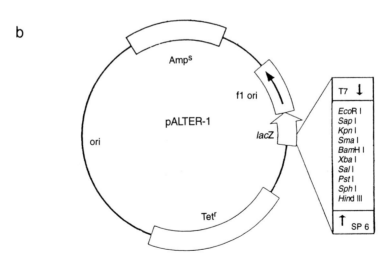

FIG. 1   Subcloning of eNOS into the mutagenesis vector pALTER-1. Endothelial NOS in Bluescript KS was digested with *Hin*dIII/*Xba*I (a) and subcloned into the same sites of pALTER-1 (b).

site is often introduced with a silent mutation (i.e., one that does not change the translated amino acid) into the mutagenic oligonucleotide primer to facilitate screening by restriction analysis. Here, we mutate the first cysteine codon (TGC) in the amino terminus into serine (TCC) as an example, and to facilitate selection of the mutant by restriction analysis, a novel *Avr*II site (CCtAGG) not found in wild-type eNOS is introduced to the mutagenic oligonucleotide primer. Thus, the mutagenic primer (antisense) is 5'-CCCAG-CCCtAGGCCGgAGGGGGGCC-3'. The 5' terminal of the oligonucleotide is then phosphorylated with T4 polynucleotide kinase for subsequent extension.

Site-Directed Mutagenesis

Mutagenesis is performed using standard DNA recombinant methods (10).

Preparation of Single-Stranded Template

Make a 5-fold dilution of an overnight culture of *Escherichia coli* containing the recombinant pALTER plasmid. After 30 min, infect the culture with helper phage R408 and let incubate overnight at 37°C. After overnight culture, precipitate the phage with polyethylene glycol (PEG) 8000; the single-stranded DNA can be purified using phenol–chloroform extraction and ethanol precipitation.

Synthesis of Mutant Strand

The phosphorylated ampicillin repair oligonucleotide primer is used to restore ampicillin resistance to the mutant strand. The molar ratios of primer to template 5:1 (ampicillin repair oligonucleotide primer) and 40:1 (the phosphorylated mutagenic oligonucleotide primer), respectively, are used in annealing. The oligonucleotide primers are hybridized to single-stranded template by incubation at 75°C for 3 min and allowing the reaction to slowly cool to room temperature. The hybridized primer is then extended with T4 DNA polymerase and ligated with T4 DNA ligase to form a closed, double-stranded plasmid.

Transformation and Verification

Synthesized DNA is first transformed into a repair-deficient *E. coli* strain such as BMH71-18 (Promega). After an overnight culture in the presence of ampicillin, isolated DNA is transformed into a repair-competent strain of *E. coli* such as JM109. When mutagenesis is performed by this method, a large population of the colonies (often more than 80%) contain phagemid carrying the desired mutation. Therefore, it is not necessary to screen colonies by hybridization with the ^{32}P-labeled mutagenic oligonucleotide probe. Instead, mutants are identified by *Avr*II digestion and further confirmed by DNA sequencing with Sequenase kit (U.S. Biochemical, Cleveland, OH).

The mutated fragment is recovered by restriction digestion of plasmid DNA and then used to replace the homologous segment of the wild-type cDNA. The mutant and wild-type eNOS cDNAs are subcloned into the mammalian expression vector, pcDNA3 (containing the neomycin-resistant gene) for expression and characterization of the protein in mammalian cells.

Expression of Mutants Transiently in COS-7 or Stably in HEK 293 Cells

A few eukaryotic proteins have been expressed efficiently in prokaryotic hosts. However, many eukaryotic proteins synthesized in bacteria fold incorrectedly or inefficiently and thus exhibit low specific activities and are insoluble. In addition, biological activities of many eukaryotic proteins require posttranslational modifications, some of which are not performed in bacteria. Because of these problems, many methods have been developed to introduce cloned DNA into mammalian cells. For rapid analysis of the mutant protein, we rely on transient transfection of COS-7 by the DEAE–dextran method. Usually, 60–80% confluent COS-7 cells are transfected with 15 μg of plasmid DNA per 1×10^6 cells by the DEAE–dextran procedure. Then 48–72 hr posttransfection, the transfected cells are used for Western blot analysis or NOS activity assays. However, for production of enough enzyme for kinetic studies or immunocytochemistry, stable cell lines expressing the cloned cDNA are required. We have transfected human embryonic kidney cells (HEK 293) with wild-type or mutant eNOS cDNAs cloned in the vector pcDNA3 (15 μg of plasmid DNA per 1×10^6 cells in 100-mm dishes) according to the calcium phosphate precipitation protocol. Transfected cells are selected in the presence of G418 (800 μg/ml) for 2–3 weeks. At this time, G418-resistant colonies are isolated and proliferated in complete DMEM containing maintenance concentrations of G418 (250 μg/ml). In preliminary experiments, we find that HEK 293 cells do not contain the mRNA for eNOS or express NOS activity.

Determination of Modification, Activities, and Localization of Mutant

Transfected cells are characterized in regard to incorporation of [^{3}H]palmitic acid, NOS activity, membrane association, and subcellular localization of the mutants. Radiolabeling of mutants with [^{3}H]palmitic acid is the most straightforward way to test if the mutated cysteine residue is a palmitoylation site. However, a valid concern frequently raised about mutant or chimeric proteins is the possibility that an amino acid replacement may alter protein conformation and subsequently affect its functional capacity. Therefore, kinetic analysis of the activities of wild-type or mutant eNOS is performed based on the conversion of [^{3}H]arginine into [^{3}H]citruline to test if the mutation affects substrate or cofactor recognition by the enzyme. Cell fractionation is performed to investigate the importance of palmitoylation in membrane association. To examine the importance of eNOS palmitoylation on its localization, confocal laser microscopy is used to study the localization of the palmitoylation mutant.

Acknowledgments

This work is supported, in part, by grants from the National Institutes of Health (R29-HL51948 and F32-HL09224) and The Patrick and Catherine Weldon Donaghue Medical Research Foundation. The Molecular Cardiobiology Program at Yale is supported by Lederle Pharmaceuticals.

References

1. W. C. Sessa, *J. Vasc. Res.* **31,** 131 (1994).
2. C. Nathan and Q. Xie, *J. Biol. Chem.* **269,** 13725 (1994).
3. J. Liu and W. C. Sessa, *J. Biol. Chem.* **269,** 11691 (1994).
4. J. Liu and W. C. Sessa, *Biochemistry* **34,** 12333 (1995).
5. L. J. Robinson, L. Busconi, and T. Michel, *J. Biol. Chem.* **270,** 995 (1995).
6. W. C. Sessa, G. Garcia-Cardena, J. Liu, A. Keh, J. S. Pollock, J. Bradley, S. Thiru, I. M. Braverman, and K. M. Desai, *J. Biol. Chem.* **270,** 17641 (1995).
7. W. C. Sessa, C. M. Barber, and K. R. Lynch, *Circ. Res.* **72,** 921 (1993).
8. L. Busconi and T. Michel, *J. Biol. Chem.* **268,** 8410 (1993).
9. W. C. Sessa, J. K. Harrison, C. M. Barber, D. Zeng, M. E. Durieux, D. D. D'Angelo, K. R. Lynch, and M. J. Peach, *J. Biol. Chem.* **267,** 15274 (1992).
10. F. M. Ausubel, R. Brent, R. E. Kingston, D. D. Moore, J. G. Seidman, J. A. Smith, and K. Struhl (eds.), "Current Protocols in Molecular Biology." Wiley, New York, 1990.

Section III

Assessment of Nitric Oxide-
Mediated Functions at Cell
and Organ Levels

[19] Nitric Oxide and Endothelial Regulation of Vascular Tone

Francesco Cosentino and Thomas F. Lüscher

Introduction

Nitric oxide (NO), an unstable gas, the by-product of automobile exhaust, electric power stations, and lightning, is enzymatically synthesized in a wide variety of species and tissues and is the mediator of biological functions as diverse as vascular signaling, neurotransmission in the brain, and defense against pathogens (1–8). The crucial discovery by Furchgott and Zawadzki (9) that the endothelium regulates the tone of the underlying vascular smooth muscle not only stimulated research activity worldwide but truly revolutionized cardiovascular sciences. Now, endothelium-dependent regulation of vascular tone (local regulation of blood flow), platelet function (antithrombotic mechanisms), and mitogenesis has become a key component of cardiovascular physiology.

Evidence continues to accumulate on the importance of paracrine substances formed in the vascular endothelium for regulation of the vascular system. These substances include nitric oxide and prostacyclin, which are both vasodilators and potent inhibitors of platelet function, and an unidentified endothelium-derived hyperpolarizing factor. By contrast, endothelial cells can also produce vasoconstrictors and growth promoters, such as thromboxane A_2 (TXA_2), prostaglandin H_2, oxygen-derived free radicals, endothelin, and angiotensin II (Fig. 1). Determination of the mechanisms governing the formation and release of these substances in different blood vessels as well as the maintenance of the balance between them is important for understanding their role in normal and pathological conditions (10). In this chapter we will summarize the current understanding of the role of nitric oxide as a regulator of vascular integrity.

Biosynthesis of Nitric Oxide

Although the effect of exogenously administered nitrates on the vasculature has been studied for decades, the first clue that endothelial cells released a substance that could cause vasodilatation was found only recently. The demonstration in 1980 of the phenomenon of endothelium-dependent relax-

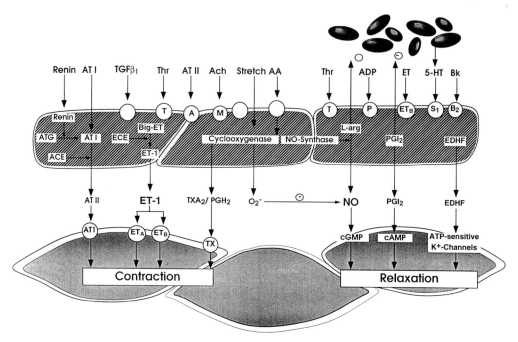

FIG. 1   Schematic diagram showing endothelium-derived vasoactive factors produced in blood vessels.

ations and of the release of endothelium-derived relaxing factor (EDRF) (9) led to a search for the chemical identity of this factor. Within a few years, it was discovered that this EDRF was nitric oxide (11, 12).

Nitric oxide is synthesized enzymatically from the amino acid L-arginine by different isoforms of nitric oxide synthase (NOS) in a number of species including humans. Many advances in the understanding of the L-arginine/nitric oxide pathway have come from molecular studies of NOS isoforms. The isozymes have been purified, characterized, cloned, and sequenced (13–19). Chromosomal mapping of the genes encoding NOSs hints at the existence of three distinct gene products (20). Two of these NOS isozymes are constitutively expressed in neurons (nNOS) and vascular endothelial cells (eNOS), whereas expression of a third isoform, inducible NOS (iNOS), is induced after stimulation with lipopolysaccharide and/or cytokines. To understand the role of NOS in vascular homeostasis, discrimination between the various isoforms of NOS may be important.

Constitutive NOS synthesizes nitric oxide within seconds in response to ligand–receptor coupling events at the cell surface and displays a strict dependence on Ca^{2+} and calmodulin (21–23). Alternatively, agonist-independent nitric oxide release also contributes to vascular tone. Both shear stress (24) and deformation of endothelium (25), due to pulsatile flow in blood vessels, stimulate nitric oxide release through poorly characterized mechanisms.

The activity of inducible NOS under basal conditions is negligible, but after stimulation a massive increase in activity occurs within 2 to 4 hr (26–29). Besides macrophages, a number of other nucleated mammalian cells, including hepatocytes, neurons, neutrophils, endothelial cells, and vascular smooth muscle cells, are capable of expressing inducible NOS (30–32). Once expressed, the inducible isoform, which is Ca^{2+}-independent, generates large amounts of nitric oxide over an extended time (33). In humans, iNOS has been found to be involved in endotoxemic shock (34). Indeed, induction of NOS may lead to excessive production of nitric oxide with subsequent impairment of arterial contractility, similar to the situation described in peripheral arteries during development of septic shock.

The different NOS isoforms catalyze the same reaction and exhibit similar biochemical properties. The reaction is a two-step redox process. L-Arginine is first hydroxylated to the intermediate N^G-hydroxy-L-arginine, which is not released in significant quantities from the enzyme but immediately undergoes oxidative cleavage to yield nitric oxide and L-citrulline (35, 36) (Fig. 2). The activity of both constitutive and inducible NOS and the formation of nitric oxide are dependent on the presence of tetrahydrobiopterin as a cofactor (37–42). However, the exact role of tetrahydrobiopterin in the synthesis of nitric oxide is not known. Current experimental evidence hints at a dual mode of action of tetrahydrobiopterin, involving both an allosteric effect on the NOS protein and a participation as a reactant in L-arginine oxidation (39, 41). In porcine and human vascular endothelial cells, inhibition of tetrahydrobiopterin synthesis reduces formation of nitric oxide induced by activation of constitutive enzyme with calcium ionophore A23187 or bradykinin (43, 44). These studies suggest that in cultured endothelial cells, optimal concentrations of tetrahydrobiopterin are essential for Ca^{2+}-dependent production of nitric oxide.

Molecular cloning of NOS revealed close amino acid sequence homology between NOS and cytochrome P450 reductase (13). This similarity, together with the fact that cytochrome P450 reductase is known to form superoxide anion, suggested that NOS may also, at least under certain conditions, generate reduced oxygen species (45). Furthermore, biochemical studies demonstrated that activation of purified NOS at suboptimal concentrations of tetrahydrobiopterin leads to production of oxygen-derived radicals (36, 46–48). The intriguing concept of a dysfunctional endothelial NOS as a source of hydrogen peroxide (H_2O_2) was confirmed in intact blood vessels depleted of

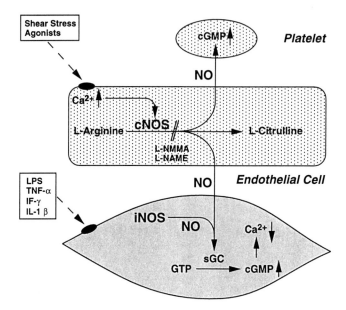

Vascular Smooth Muscle Cell

FIG. 2   The L-arginine/nitric oxide pathway in the blood vessel wall: endothelial cells form nitric oxide (NO) from L-arginine via the activity of the constitutive nitric oxide synthase (cNOS), which can be inhibited by analogs of the amino acid such as N^G-monomethyl-L-arginine (L-NMMA) and N^G-nitro-L-arginine methyl ester (L-NAME). Nitric oxide activates soluble guanylate cyclase (sGC) in vascular smooth muscle and platelets, and it causes increases in cyclic 3′,5′-guanosine monophosphate (cGMP), which mediates relaxation and platelet inhibition, respectively. Shear stress and receptor-operated agonists stimulate the release of NO. In addition, vascular smooth muscle cells can form NO via the activity of an inducible form of NO synthase (iNOS). LPS, Lipopolysaccharide; TNF-α, tumor necrosis factor-α; IF-γ, γ-interferon; IL-1β, interleukin-1β.

tetrahydrobiopterin (45). As H_2O_2 is a potent vasodilator it is conceivable that it may replace nitric oxide as a relaxing factor. However, prolonged increased concentrations of H_2O_2 may be harmful for endothelial cells as well as smooth muscle cells, leading to oxidative damage (49). Such NOS-catalyzed generation of oxygen radicals may represent an important mechanism underlying endothelial dysfunction and oxidative vascular injury (45). The first step of tetrahydrobiopterin biosynthesis involves activation of GTP cyclohydrolase I (40, 50). Undoubtedly, the regulation of tetrahydrobiopterin synthesis and its influence on the production of endothelial nitric oxide and oxygen-derived free radicals deserve further investigations.

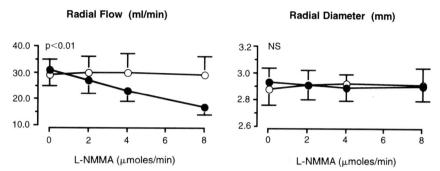

FIG. 3   Basal release of nitric oxide in human peripheral arteries. Mean radial blood flow, mean internal diameter during the control period (○) and after infusion of increasing doses of L-NMMA (●). Values are means ± SEM. Modified from Ref. 51.

Physiological Roles of L-Arginine/Nitric Oxide Pathway

The vasculature is in a constant state of active dilation mediated by nitric oxide (Fig. 3) (51). Endothelial cells continuously release small amounts of nitric oxide, producing a basal level of vascular smooth muscle relaxation. However, the expression of constitutive enzyme and the release of nitric oxide can be enhanced above basal levels after receptor stimulation by different agonists (Fig. 4) (51). Nitric oxide is a very small lipophilic molecule that can rapidly diffuse through biological membranes into underlying smooth muscle cells to cause relaxation and into the lumen of the blood vessel wall to inhibit platelet function. The binding of the heme group of guanylate cyclase causes an immediate and profound increase in catalytic activity, resulting in conversion of guanosine 5'-triphosphate to cyclic monophosphate guanosine 3',5'(cyclic GMP), which in turn causes relaxation and inhibition of platelet function by decreasing intracellular Ca^{2+} levels (52–55) (Fig. 2).

Regulation of Vascular Tone

Both *in vitro* and *in vivo* studies demonstrated that basal and stimulated production of nitric oxide in endothelial cells plays a key role in regulation of vascular tone. The strongest evidence comes from results obtained in studies with N^G-substituted analogs of L-arginine, which are potent and selective inhibitors of nitric oxide synthesis (2, 56). L-Arginine analogs such as N^G-monomethyl-L-arginine (L-NMMA), N^G-nitro-L-arginine methyl ester

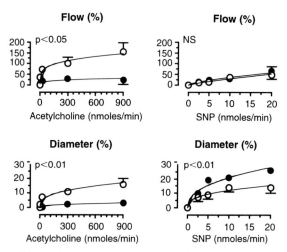

FIG. 4   Nitric oxide and response to acetylcholine and sodium nitroprusside in humans. Percent change from baseline of mean radial blood flow and mean internal diameter after infusion of increasing doses of acetylcholine and sodium nitroprusside in the control state (○) and after L-NMMA (●). Values are means ± SEM. From Ref. 51.

(L-NAME), and N^G-nitro-L-arginine (L-NA) cause endothelium-dependent contractions in a number of isolated arteries. These contractions are mediated by the inactivation of basal production of nitric oxide (2, 57, 58). The inhibitory effect of these compounds is prevented by L-arginine (but not D-arginine). This stereospecific inhibition indicates that L-arginine analogs compete with L-arginine for the nitric oxide synthase active site to prevent production of nitric oxide. Intravenous injections of L-NMMA into anesthetized rabbits resulted in an immediate and substantial rise in blood pressure, which can be reversed by L-arginine (59). The blood pressure elevating and vasoconstrictive effects of L-NMMA have now been demonstrated in a number of species including humans (60–62). Similar results have been obtained with the other known nitric oxide synthase inhibitors (63). These inhibitors have no intrinsic constrictor activity on vascular smooth muscle. Their activity is entirely endothelium-dependent and results from the inhibition of endogenous vasodilation. Thus it appears that the endothelial generation of nitric oxide by NOS is involved in the maintenance of normal blood flow and pressure.

In the porcine coronary circulation, L-NMMA inhibits the relaxations in response to serotonin, but only slightly those to bradykinin (64). Because similar effects are obtained by other inhibitors of the action of nitric oxide such as hemoglobin and methylene blue, endothelial cells appear to release a relaxing factor distinct from nitric oxide (65). Prostacyclin can be ruled out because indomethacin does not affect the responses to bradykinin (64). Furthermore, the observation that in small canine cerebral arteries endothelium-dependent relaxations to bradykinin are resistant to inhibitors of nitric oxide formation or cyclooxygenase (66) strongly confirm this conclusion. Studies suggest that bradykinin (and acetylcholine) cause not only an endothelium-dependent relaxation but also endothelium-dependent hyperpolarization of vascular smooth muscle (67–70). These data would be compatible with the concept that endothelial cells release a biochemically unidentified substance that has the capacity to hyperpolarize vascular smooth muscle cells via ATP-dependent potassium channels (71, 72). Hyperpolarization of vascular smooth muscle cells is associated with a decreased sensitivity to vasoconstrictor substances and may also contribute to vasodilator responses induced by prostacyclin and nitric oxide.

Inhibition of Platelet Function

Normally, platelets circulate in an inactivated state in the circulation. This may be at least in part due to the fact that endothelial cells continuously release inhibitors of platelet function such as nitric oxide and prostacyclin. Both mediators increase cyclic GMP and cyclic AMP in platelets and thereby prevent platelet adhesion and aggregation (52, 55, 73–75). In addition, platelets themselves possess an L-arginine/nitric oxide pathway that modulates the reactivity of the cells to aggregatory stimuli (76).

At sites where platelets are activated, they immediately release active mediators such as serotonin, TXA_2, ATP/ADP, platelet-derived growth factor, and transforming growth factor β-1 which may interact with the endothelial receptors. Indeed, it was demonstrated that in isolated human coronary and internal mammary arteries, aggregating platelets cause endothelium-dependent relaxations that are mediated by nitric oxide (77, 78). The platelet-derived mediator primarily responsible for the stimulation of nitric oxide formation is ADP, although serotonin may contribute under certain conditions (79, 80). In contrast to normal arteries, arteries devoid of endothelial cells or with dysfunctional endothelium markedly contract to aggregating platelets (78). Platelet-induced vasoconstriction is primarily mediated by serotonin and thromboxane A_2, which activate specific receptors on vascular smooth muscle cells (78, 81).

When platelets are stimulated, the coagulation cascade also is activated, leading to the formation of thrombin. Thrombin is an enzyme that is responsible for the formation of fibrin from fibrinogen. In addition to its activity in the coagulation cascade, thrombin can act as a vasoactive factor both directly and by modulating the effects of other vasoactive factors, particularly those released from platelets. In isolated vessels with intact endothelium, thrombin induces vasorelaxation through the generation of both nitric oxide and prostacyclin; this in turn also prevents platelet activation. On the other hand, in vessels denuded of endothelium thrombin induces profound platelet activation, and in turn the release of TXA_2, which causes marked contraction and further platelet activation (82, 83). Nitric oxide has also been shown to inhibit leukocyte activation both *in vitro* and *in vivo* (84, 85). Thus, nitric oxide appears to be even involved in regulating the interactions between leukocytes and the vascular endothelium.

Regulation of Vascular Growth

Mechanical removal of the endothelium by balloon cathether and dysfunction of endothelial cells invariably result in migration and proliferation of smooth muscle cells (10, 86–88), suggesting that the endothelium normally has a net inhibitory influence on these responses. Indeed, in normal vessels the endothelium appears to synthesize primarily substances like nitric oxide (89, 90) and heparan sulfate (91) that inhibit growth of smooth muscle cells. Furthermore, it has been demonstrated that nitric oxide inhibits angiotensin II-induced migration of rat smooth muscle cells (92). This is the first report providing direct evidence that nitrovasodilators may exert beneficial effects on vascular structure (Fig. 5) (92). However, under certain conditions, the endothelium can also generate growth promoters such as platelet-derived growth factor, basic fibroblast growth factor, and endothelin-1 (93–96). Thus the secretion of growth inhibitors or promoters by endothelial cells as well as their capacity to inhibit circulating blood cells modulates vascular structures (97).

Conclusions

In 1987, only a few articles were written about endogenously produced nitric oxide, but since then more than 3000 have been published. Nitric oxide has been shown to be an endogenous messenger that plays various physiological roles. Our discussion focused on the role of endothelial L-arginine/nitric oxide pathway in the maintenance of vascular homeostasis. The achievements of the last years have clearly demonstrated that the endothelium, due to its

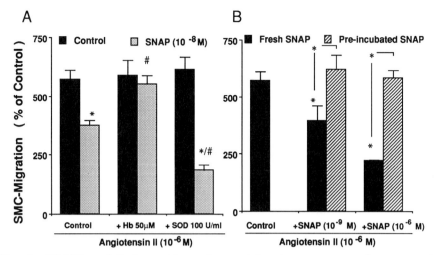

FIG. 5 (A) Effect of S-nitroso-N-acetyl($\pm$)penicillamine (SNAP) (10^{-8} M) on 10^{-6} M AII-induced SMC migration in the presence and in the absence of hemoglobin (50 μM) or superoxide dismutase (SOD, 100 U/ml). *Significant inhibition of AII-induced migration by SNAP; #Significantly different from SNAP-induced inhibition of migration ($p < 0.05$) in absence of hemoglobin or SOD. (B) inhibitory effect of SNAP (10^{-6} and 10^{-9} M), freshly prepared and preincubated for 48 hr on AII-induced SMC migration. *Significantly different from AII-induced migration in absence of SNAP. From Ref. 92.

strategic anatomical position between the circulating blood and vascular smooth muscle, is a primary candidate as the mediator and target of cardiovascular disease and atherosclerosis in particular (98). Functional alterations of the endothelial L-arginine/nitric oxide pathway may be important in cardiovascular disease, because a depressed activity of this protective mechanism would lead to impaired relaxation (and, in turn, vasoconstriction and reduced local blood flow) and also be associated with reduced antithrombotic properties of the endothelial layer.

The realization that tetrahydrobiopterin may regulate the availability of EDRF–nitric oxide has given further insights into the regulation of vascular tone and refocused attention on the complex chemistry associated with NOS catalysis. We anticipate that identification of signal transduction pathways involved in the control of gene expression and activity of constitutive NOS will provide a basis for understanding how the endothelial cells lose their protective functions and become the source of substances with vasoconstrictor, proaggregatory, and promitogenic activity.

References

1. C. J. Lowenstein and S. H. Snyder, *Cell (Cambridge, Mass.)* **70,** 705 (1992).
2. S. Moncada, R. M. Palmer, and E. A. Higgs, *Pharmacol. Rev.* **43,** 109 (1991).
3. C. Nathan, *FASEB J.* **6,** 3051 (1992).
4. T. F. Lüscher, *Br. J. Clin. Pharmacol.* **34** (Suppl 1), 29S (1992).
5. M. A. Marletta, *Trends Biochem. Sci.* **14,** 488 (1989).
6. L. J. Ignarro, *Annu. Rev. Pharmacol. Toxicol.* **30,** 535 (1990).
7. T. F. Lüscher, *Clin. Exp. Hypertens.* **12,** 897 (1990).
8. J. S. Stamler, D. J. Singel, and J. Loscalzo, *Science* **258,** 1898 (1992).
9. R. F. Furchgott and J. V. Zawadzki, *Nature (London)* **288,** 373 (1980).
10. T. F. Lüscher and F. C. Tanner, *Am. J. Hypertens.* **6,** 283S (1993).
11. T. M. Griffith, D. H. Edwards, M. J. Lewis, A. C. Newby, and A. H. Henderson, *Nature (London)* **308,** 645 (1984).
12. R. M. J. Palmer, A. G. Ferrige, and S. Moncada, *Nature (London)* **327,** 524 (1987).
13. D. S. Bredt, P. H. Hwang, C. Glatt, C. Lowenstein, R. R. Reed, and S. H. Snyder, *Nature (London)* **351,** 714 (1991).
14. W. C. Sessa, J. K. Harrison, and C. M. Barber, *J. Biol. Chem.* **267,** 15274 (1992).
15. S. Lomas, P. A. Marsden, G. K. Li, P. Tempst, and M. Thomas, *Proc. Natl. Acad. Sci. U.S.A.* **89,** 6348 (1992).
16. S. P. Janssens, A. Shimouchi, T. Ouertermous, D. B. Bloch, and K. D. Bloch, *J. Biol. Chem.* **267,** 14519 (1992).
17. Q. W. Xie, H. J. Cho, and J. Calaycay, *Science* **256,** 225 (1992).
18. C. J. Lowenstein, C. S. Glatt, D. S. Bredt, and S. H. Snyder, *Proc. Natl. Acad. Sci. U.S.A.* **89,** 6711 (1992).
19. C. R. Lyons, G. J. Orloff, and J. M. Cunningham, *J. Biol. Chem.* **267,** 6370 (1992).
20. C. Nathan and Q. W. Xie, *J. Biol. Chem.* **269,** 13725 (1994).
21. M. W. Radomski, R. M. J. Palmer, and S. Moncada, *Proc. Natl. Acad. Sci. U.S.A.* **87,** 10043 (1990).
22. U. Forstermann, J. S. Pollock, and H. H. H. W. Schmidt, *Proc. Natl. Acad. Sci. U.S.A.* **88,** 1788 (1991).
23. C. F. Nathan and D. J. Stuehr, *J. Natl. Cancer Inst.* **82,** 726 (1990).
24. G. M. Rubanyi, J. C. Romero, and P. M. Vanhoutte, *Am. J. Physiol.* **250,** 41145 (1986).
25. D. La Montagne, U. Pohl, and R. Busse, *Circ. Res.* **70,** 123 (1992).
26. R. Iyengar, D. J. Stuehr, and M. A. Marletta, *Proc. Natl. Acad. Sci. U.S.A.* **84,** 6369 (1987).
27. D. J. Stuehr and M. A. Marletta, *Proc. Natl. Acad. Sci. U.S.A.* **82,** 7738 (1985).
28. J. C. Drapier and J. B Hibbs, Jr., *J. Immunol.* **141,** 2407 (1988).
29. A. Ding, C. F. Nathan, and D. J. Stuehr, *J. Immunol.* **141,** 2412 (1988).
30. R. Busse and A. Mulsch, *FEBS Lett.* **275,** 87 (1990).
31. R. G. Knowles, M. Merret, M. Salter, and S. Moncada, *Biochem. J.* **270,** 833 (1990).
32. R. G. Knowles, M. Salter, S. L. Brooks, and S. Moncada, *Biochem. Biophys. Res. Commun.* **172,** 1042 (1990).

33. S. Moncada and A. Higgs, *N. Engl. J. Med.* **329,** 2002 (1993).
34. S. Moncada, *Schweiz. Rundsch. Med. Prax.* **82,** 1154 (1993).
35. D. J. Stuehr, N. S. Kwon, C. F. Nathan, and D. W. Griffith, *J. Biol. Chem.* **266,** 6259 (1991).
36. P. Klatt, K. Schmidt, G. Uray, and B. Mayer, *J. Biol. Chem.* **268,** 14781 (1993).
37. M. A. Tayeh and M. A. Marletta, *J. Biol. Chem.* **264,** 19654 (1989).
38. N. S. Kwon, C. F. Nathan, and D. J. Stuehr, *J. Biol. Chem.* **264,** 20496 (1989).
39. J. Giovanelli, K. L. Campos, and S. Kaufmann, *Proc. Natl. Acad. Sci. U.S.A.* **88,** 7091 (1991).
40. S. Kaufmann, *Annu. Rev. Nutr.* **13,** 261 (1993).
41. K. J. Baeck, B. A. Thiel, S. Lucas, and D. J. Stuehr, *J. Biol. Chem.* **268,** 21120 (1993).
42. S. S. Gross and L. Levi, *J. Biol. Chem.* **267,** 25722 (1992).
43. K. Schmidt, E. R. Felmayer, B. Mayer, B. Wachter, and W. R. Kukowetz, *Biochem. J.* **281,** 297 (1992).
44. G. Werner-Felmayer, E. R. Werner, D. Fuchs, and A. Hausen, *J. Biol. Chem.* **268,** 1842 (1993).
45. F. Cosentino and Z. S. Katusic, *Circulation* **91,** 139 (1995).
46. B. Mayer, M. John, B. Heinzel, E. R. Werner, H. Wachter, G. Schultz, and E. Böhme, *FEBS Lett.* **288,** 187 (1991).
47. B. Heinzel, M. John, P. Klatt, E. Böhme, and B. Mayer, *Biochem. J.* **281,** 627 (1992).
48. S. Pou, W. S. Pou, D. S. Bredt, S. H. Snyder, and G. M. Rosen, *J. Biol. Chem.* **267,** 24173 (1992).
49. Z. S. Katusic and F. Cosentino, *News in Physiological Sciences* **9,** 64 (1994).
50. C. A. Nichol, G. K. Smith, and D. S. Duch, *Annu. Rev. Biochem.* **54,** 729 (1985).
51. R. Joannides, V. Richard, W. E. Haefeli, L. Linder, T. F. Lüscher, and C. Thuillez, *Hypertension* **26,** 327 (1995).
52. R. Busse, A. Lückhoff, and E. Bassenge, *Naunyn Schmiedeberg's Arch. Pharmacol.* **336,** 566 (1987).
53. R. M. Rapoport, M. B. Draznin, and F. Murad, *Nature (London)* **306,** 174 (1983).
54. M. W. Radomski, R. M. J. Palmer, and S. Moncada, *Br. J. Pharmacol.* **92,** 181 (1987).
55. M. W. Radomski, R. M. J. Palmer, and S. Moncada, *Proc. Natl. Acad. Sci. U.S.A.* **87,** 5293 (1990).
56. K. Ishii, B. Chang and J. F. Korwin, *Eur. J. Pharmacol.* **176,** 219 (1990).
57. Z. S. Katusic and P. M. Vanhoutte, *in* "Endothelium-Derived Relaxing Factors" (G. M. Rubanyi and P. M. Vanhoutte, eds.), p. 95. Karger, Basel, 1990.
58. F. Cosentino, J. C. Sill, and Z. S. Katusic, *Am. J. Physiol.* **264,** H413 (1993).
59. D. D. Rees, R. M. J. Palmer, and S. Moncada, *Proc. Natl. Acad. Sci. U.S.A.* **86,** 3375 (1989).
60. P. Vallance, J. Collier, and S. Moncada, *Lancet,* 997 (1989).
61. P. Vallance, J. Collier, and S. Moncada, *Cardiovasc. Res.* **23,** 1053 (1989).
62. A. Calver, J. Collier, and S. Moncada, *J. Hypertens.* **10,** 1025 (1992).
63. D. D. Rees, R. M. J. Palmer, R. Schultz, H. F. Hodson, and S. Moncada, *J. Pharmacol.* **101,** 746 (1990).

64. V. Richard, F. C. Tanner, M. Tschudi, and T. F. Lüscher, *Am. J. Physiol.* **259,** H1433 (1990).
65. P. M. Vanhoutte, *Nature (London)* **327,** 459 (1987).
66. Z. S. Katusic, *Am. J. Physiol.* **262,** H1557 (1992).
67. T. Shibano, J. Codina, L. Birnbaumer, and P. M. Vanhoutte, *Am. J. Physiol.* **267,** H207 (1994).
68. M. Feletou and P. M. Vanhoutte, *Br. J. Pharmacol.* **93,** 515 (1988).
69. K. Komori, R. R. Lorenz, and P. M. Vanhoutte, *Am. J. Physiol.* **255,** H207 (1988).
70. P. M. Vanhoutte, *Circulation* **87**(Suppl. V), V9 (1993).
71. M. Tare, H. C. Parkington, H. A. Coleman, T. O. Neild, and G. J. Dusting, *Nature (London)* **346,** 69 (1990).
72. N. B. Standen, J. M. Quayle, and N. W. Davies, *Science* **245,** 177 (1989).
73. M. W. Radomski, R. M. J. Palmer, and S. Moncada, *Br. J. Pharmacol.* **92,** 639 (1987).
74. P. S. Macdonald, M. A. Read, and G. J. Dusting, *Thromb. Res.* **49,** 437 (1988).
75. M. W. Radomski, P. Vallance, G. Withley, N. Foxwell, and S. Moncada, *Cardiovasc. Res.* **27,** 1380 (1993).
76. M. W. Radomski, R. M. J. Palmer, and S. Moncada, *Proc. Natl. Acad. Sci. U.S.A.* **87,** 5293 (1990).
77. U. Forstermann, A. Mügge, U. Alheid, A. Haverich, and J. C. Frolich, *Circ. Res.* **62,** 185 (1988).
78. Z. Yang, L. von Segesser, E. Bauer, *et al., Lancet* **337,** 939 (1991).
79. D. S. Houston, J. T. Shepherd, and P. M. Vanhoutte, *Am. J. Physiol.* **248,** H389 (1985).
80. H. Shimokawa, N. A. Flavahan, and P. M. Vanhoutte, *Circ. Res.* **65,** 740 (1989).
81. P. Golino, F. Piscione, J. T. Willerson, M. Cappelli-Bigazzi, A. Focaccio, B. Villari, C. Indolfi, E. Russolillo, M. Condorelli, and M. Chiariello, *N. Engl. J. Med.* **324,** 641 (1991).
82. T. F. Lüscher, D. Diederich, R. Siebenmann, K. Lehmann, P. Stulz, L. Von Segesser, H. Z. Yang, M. Turina, E. Graedel, E. Weber, and F. R. Buehler, *N. Engl. J. Med.* **319,** 462 (1988).
83. Z. Yang, R. P. Siebenmann, L. von Segesser, P. Stultz, M. Turina, and T. F. Lüscher, *Circulation* **89,** 2266 (1994).
84. P. M. W. Bath, D. G. Hassall, A. M. Cladwill, R. M. J. Palmer, and J. F. Martin, *Arterioscler. Thromb.* **11,** 254 (1991).
85. P. Kubes, M. Suzuki, D. N., and Granger, *Proc. Natl. Acad. Sci. U.S.A.* **86,** 4651 (1991).
86. C. L. Jackson and S. M. Schwartz, *Hypertension (Dallas)* **20,** 713 (1992).
87. M. Viswanathan, C. Strömberg, A. Seltzer, and J. M. Saavedra, *J. Clin. Invest.* **90,** 1707 (1992).
88. V. J. Dzau and G. H. Gibbons, *J. Cardiovasc. Pharmacol.* **21**(Suppl. 1), S1 (1993).
89. U. C. Garg and A. Hassid, *J. Clin. Invest.* **83,** 1774 (1989).
90. O. B. Dubey and T. F. Lüscher, *Hypertension (Dallas)* **22,** 412 (1993).
91. J. J. Castellot, M. L. Addonizio, R. D. Rosenberg, and M. J. Karnovsky, *J. Cell Biol.* **90,** 372 (1981).
92. R. K. Dubey, E. K. Jackson, and T. F. Lüscher, *J. Clin. Invest.* **96,** 141 (1995).

93. R. L. Hannan, S. Kourembanas, K. O. Flanders, S. J. Rogeli, A. B. Roberts, D. V. Faller, and M. Klagsbrun, *Growth Factors* **1,** 7 (1988).
94. E. J. Battegay, E. W. Raines, R. A. Seifert, D. F. Bowen-Pope, and R. Ross, *Cell (Cambridge, Mass.)* **63,** 515 (1990).
95. P. E. Dicorleto and P. L. Fox, *in* ''Endothelial Cells'' (U. Ryan, ed.), p. 51. CRC Press, Boca Raton, Florida, 1990.
96. D. Dubin, R. E. Pratt, J. P. Cooke, and V. J. Dzau, *J. Vasc. Med. Biol.* **1,** 13 (1989).
97. T. F. Lüscher, R. K. Dubey, E. Espinosa, and Z. Yang, *Curr. Opin. Cardiol.* **8,** 963 (1993).
98. T. F. Lüscher, *N. Engl. J. Med.* **330,** 1081 (1994).

[20] Function of Nitric Oxide in Neuronal Cell Death

Valina L. Dawson and Ted M. Dawson

Introduction

Primary neuronal cell cultures can be an ideal model system for investigating isolated cellular mechanisms. Cultures retain many of the physiological and biochemical characteristics of *in situ* neurons (1) and are free from the influences of the microvasculature. Furthermore, culture conditions can be optimized to study the mechanism of interest through manipulation of neuronal circuitry, composition of supportive cells, and extracellular buffering. Because the overall phenotype of neuronal cultures is easily altered by culture conditions, it is important to characterize the cultures carefully.

Primary neuronal cultures are typically generated from tissue that is still differentiating and maturing. This maturation process continues in the culture dish, altering the physiology and biochemistry of the culture system over time. This maturation is dependent on the culture media and supplements which will ultimately determine the phenotype of the culture system. Thus, it is particularly important to determine when and under what conditions the proteins or pathways of interest are present in a particular culture system. This is absolutely critical in modeling nitric oxide (NO) signaling in culture. Nitric oxide synthase (NOS) is not expressed by neurons during the first few weeks of culture (2). To study the cellular consequences of NOS activation and NO production, cultures must be examined after NOS is fully expressed at *in vivo* levels, which occurs between 14 and 20 days *in vitro* (DIV) (2, V. L. Dawson and T. M. Dawson, unpublished observations, 1995). Expression of neuronal NOS (nNOS) is also dependent on the culture conditions. Culturing neurons on glial feeder layers or under serum-free conditions can diminish or prevent the expression of nNOS in neurons. Primary neuronal cultures are useful for optimizing pathways of interest and modeling isolated mechanisms. Because they lack many *in situ* components, cultures may not accurately reflect the intact nervous system. Data derived from neuronal cultures must be regarded as a component of a larger system and must ultimately be judged relative to the results obtained *in vivo*.

Nitric oxide mediates many biological processes including vasodilation, cytotoxicity of activated macrophages, and glutamate-stimulated cGMP formation in the nervous system. Nitric oxide synthase converts L-arginine to

Methods in Neurosciences, Volume 31

NO and citrulline. Endothelial (eNOS) and nNOS are activated by an increase in intracellular calcium. Entry of Ca^{2+} through glutamate receptor channels in neurons can activate NOS. In primary neuronal cultures, glutamate and N-methyl-D-aspartate (NMDA) produce neuronal cell death which can be prevented by several classes of NOS inhibitors. In addition, cell death can be prevented by removal of arginine from the incubation medium or by scavenging NO with reduced hemoglobin. Molecular and biochemical studies indicate that NOS utilizes several cofactors and NO formation is regulated at multiple levels (Fig. 1). Thus, a variety of pharmacological interventions can modify enzyme activity and provide novel agents for neuroprotection. Nitric oxide donors can directly induce neurotoxicity in primary cell cultures that is not attenuated by NOS inhibitors but is blocked by reduced hemoglo-

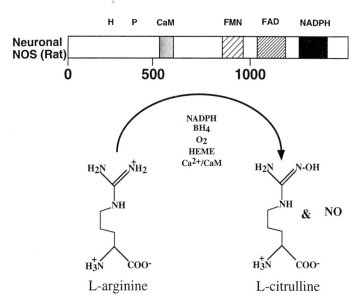

FIG. 1 Arginine analogs compete with arginine (L-ARG) for the catalytic site on NOS and decrease the amount of NO produced. Nitric oxide production can be decreased by the removal of L-arginine from the medium. Calmodulin (CaM) is an essential cofactor, and calmodulin antagonists or agents that bind calmodulin can reduce NOS catalytic activity and decrease NO production. The shuttling of electrons by the flavoprotein moieties is critical for the conversion of L-arginine to citrulline and NO. Nitric oxide syntase catalytic activity can be inhibited by flavoprotein inhibitors. Nitric oxide synthase is phosphorylated by four different kinases on four different amino acids. In the phosphorylated state NOS is inactive. Nitric oxide synthase is dephosphorylated and activated by calcineurin. Calcineurin antagonists keep NOS in the phosphorylated, inactive state and thus decrease NO production.

bin. Both NMDA and NO donor-mediated neurotoxicity is attenuated by superoxide dismutase (SOD) which scavenges the superoxide anion (2). Therefore, the formation of peroxynitrite from the reaction of superoxide anion and NO is likely to be a major mediator of neurotoxicity. One pathway toward neuronal cell death may involve NO induced DNA damage overactivating the nuclear enzyme, poly(ADP-ribose) synthase (PARS; NAD^+ ADP-ribosyltransferase), depleting the cells of NAD followed by depletion of ATP (3).

Culture and Characterization of Primary Neurons from Rodent Forebrain

Preparation of Tissue Culture Plates

Forebrain neurons grow best in 16-mm wells of 4- or 24-well plates and the preference of plastic appears to be NUNC (Naperville, IL) > Costar (Cambridge, MA) > Corning (Corning, NY) > Falcon (Bectin Dikinson Labware, Lincoln Park, NJ). Neuronal cultures grown in 96-well plates have a nonuniform distribution of NOS neurons in each well due to the small number of neurons plated in each well which prevents accurate analysis of NO function. Prior to plating, the culture well must be precoated with an attachment matrix. Stock polyornithine (0.3 mg/ml) (Sigma, St. Louis, MO) is diluted into autoclaved Milli-Q water at a 1 : 100 dilution (1 ml polyornithine/99 ml H_2O) and filtered through a 0.2-μm filter to remove any contaminants or large particles. One milliliter of polyornithine solution is then added to each well. The plates are incubated in an oven for 60 to 90 min at 42°C. Cultured neurons will lie flat and will not aggregate (clump) on a well-coated plate. The wells are then gently rinsed one time with autoclaved Milli-Q water. Let the plates dry overnight in a clean hood under a UV light. Store in a dust-free environment. Plates can be stored for 1 week.

Culture Procedure

Dissect the cortex or striatum from gestational day 15 rat pups under sterile conditions in sterile Brooks–Logan solution (Table I). Begin the tissue dissection by placing the tissue into 1–2 ml of 1× trypsin (GIBCO-BRL, Gaithersburg, MD) and incubating at 37°C for 25 min. Remove the trypsin solution and add 7 ml warm 10 : 10 : 1 MEM (Table II). Finish dissociating the tissue by trituration. This is performed by gently moving the tissue in and out of

TABLE I Brooks–Logan Solution[a]

Reagent	Concentration (mM)
Sucrose	44
D-Glucose	25
NaCl	137
KCl	2.7
Na$_2$HPO$_4$	10
KH$_2$PO$_4$	1.8
1 M HEPES	10

[a] Dissolve all reagents in autoclaved Milli-Q water. Adjust pH to 7.4. Sterile filter through a 0.2-μm filter. This solution can be kept for several weeks at 4°C but should be sterile filtered before each use to prevent contamination of cultures by fungus or mold. The solution should be discarded if there are any visible particles or growth.

sterile 9-inch Pasteur pipettes that have openings of decreasing diameter. Trituration pipettes are made by flaming the tips. The tissue should form a cloudy suspension immediately. Do not triturate more than 10 times. Let the tissue settle for a few minutes. Undissociated tissue will settle to the bottom of the test tube. Determine the number of cells with a hemocytometer. Dilute the cells into the appropriate volume of medium necessary to plate the cells at a particular density by the following equation:

$$[\text{Mean number of cells per grid square} \times 16 \times 10{,}000 \times \text{ml dissociated cells}] \div \text{desired density} = \text{final volume (in ml)}$$

Plate 1 ml of the diluted cells per well of a 24-well plate. Place in 37°C, 7% (v/v) humidified incubator. After 4–5 days change the medium to 5:1:5F2DU MEM (Table II) to inhibit nonneuronal cell growth. Then 3–4 days later the culture medium is changed to 5:1 MEM. The medium is changed twice weekly by aspirating off one-half the volume and replacing it with fresh 5:1 MEM (Table II).

NADPH Diaphorase Stain for Nitric Oxide Synthase in Primary Neuronal Cultures

For a NO component of neuronal signaling to reflect accurately *in vivo* characteristics, it is necessary to have a full complement of nNOS-expressing neurons that is equivalent to *in vivo* expression. In cortical and striatal

TABLE II   Culture Medium Formulations[a]

Medium reagents	Volume (ml)	Source
10:10:1 MEM		
MEM (Eagle's minimal essential medium with Earle's salts, without glutamine)	500	GIBCO-BRL
FBS (Fetal bovine serum,[b] heat inactivated at 56°C for 30 min then sterile filtered)	63.3	GIBCO-BRL
HS (Horse serum,[b] heat inactivated at 56°C for 30 min then sterile filtered)	63.3	GIBCO-BRL
Glutamine[b]	6.3	GIBCO-BRL
5:1:5F2DU MEM		
MEM	500	GIBCO-BRL
HS	26.6	GIBCO-BRL
Glutamine	5.3	GIBCO-BRL
5F2DU (5-Fluoro-2'-deoxyuridine,[c] highly toxic, wear gloves when handling)	3 (10 mM stock)	Sigma
5:1 MEM		
MEM	500	GIBCO-BRL
HS	26.6	GIBCO-BRL
Glutamine	5.3	GIBCO-BRL
MEM + 21 mM Glucose		
MEM	500	GIBCO-BRL
21 mM Glucose	5.25 (2M)	

[a] Media containing glutamine should be used within 4 weeks of being made. Glutamine converts to glutamate with time in aqueous solution and may alter the cell culture conditions. If media cannot be used within 4 weeks, make a smaller volume.

[b] Avoid repeated freeze–thaw cycles with these reagents. For best performance, aliquot once into sterile test tubes the volume needed for each preparation and thaw only the necessary aliquots.

[c] Stock solutions are made with autoclaved Milli-Q water and sterile filtered (0.2 μm) prior to aliquoting and freezing.

cultures, NOS neurons should comprise 1–2% of the neuronal population. Lower expression will prevent an accurate evaluation of the cellular consequences of activation of nNOS (2). Neuronal NOS neurons can be quickly and easily visualized by NADPH diaphorase staining (Table III). NADPH diaphorase staining is performed by fixing the cell cultures in 4% (w/v) freshly depolymerized paraformaldehyde/0.1 M phosphate buffer solution for 30 min at room temperature followed by washing the fixative off the cells with CSS (Table IV). NADPH diaphorase stain mix is then applied (Table III) to the cell cultures. The culture plate is then incubated for 30–90 min at 37°C. The stain is complete when the neurons and projections are a deep purple-black and before the glia begin to stain blue. To stop the reaction, wash the stain

TABLE III   NADPH Diaphorase Stain

Fix: 4% Paraformaldehyde/0.1 M PB
   8% Paraformaldehyde (8 g/100 ml Milli-Q water)
   Heat to 80°C for 30 min (do not boil)
   Clear with 1–2 drops of 10 N NaOH and replace volume of evaporated water
   Add an equal volume of 0.2 M PB (100 ml 0.2 M NaH$_2$PO$_4$:400 ml 0.2 M
     Na$_2$HPO$_4$, pH 7.4) and pass through a 0.45-μm filter
NADPH diaphorase stain solution:
   Buffer:
     0.1 M Tris-HCl (pH 7.2)
     0.2% Triton X-100
     7.5 mg sodium azide/100 ml buffer
   Staining solution (final volume 10 ml):
     1 mM NADPH reduced form (Sigma)   8.33 mg/5 ml
     0.2 mM Nitro blue tetrazolium (NBT) (Sigma)   1.64 mg/5 ml
   Solubilize NADPH and NBT in separate test tubes to prevent precipitation.
     The NBT may need to be sonicated to solubilize. Mix soluble NADPH and
     NBT to form stain mix

off with CSS. The cultures can be kept for a few days at 4°C if 0.05% (w/v) sodium azide is added to the CSS. Greater than 100 NOS neurons are required per well of a 24-well plate to observe an NO component of glutamate neurotoxicity.

TABLE IV   Control Salt Solution[a,b]

Reagent	Final concentration (mM)
NaCl	120
KCl	5.4
CaCl$_2$	1.8
D-Glucose	15
Tris-HCl	25

[a] Dissolve all reagents in Milli-Q water and adjust pH to 7.4 at the temperature the experiment is to be performed (the pH of Tris is temperature sensitive). Sterile filter the solution through a 0.2-μm filter and store at room temperature until use. All solutions should be sterile filtered before use and discarded if there is visible growth.
[b] For glutamate experiments add 0.8 mM MgCl$_2$.

Toxicity Experiments

Exposure (Day 1)

Cultures must have a full complement of nNOS expressing neurons (100–300 neurons per well, DIV 14–21). Prepare exposure solutions which contain the pharmacological agent of interest in sterile CSS (Table IV); CSS is an optimal buffer as it contains few components that can react with NO and either enhance or diminish its cellular activity. Wash the cultures prior to exposure to remove salts and metals that can block the NMDA receptor or react with NO. Expose the cultures to test solutions for 5 min in CSS. Terminate the exposure by aspirating off the exposure solution and replacing with CSS (1–2 ml) and washing two times. Replace the CSS with 1 ml MEM containing 21 mM glucose (Table II). Return the tissue culture plate to the incubator for 18–24 hr. Some neurotoxins require longer exposure times than 5 min. For exposure times greater than 30 min we use an exposure solution containing MEM plus 21 mM glucose and perform the exposures in the incubator. In some culture models, NOS continues to be activated following the initial excitotoxic stimulus and so NOS inhibitors need to be added to the postexposure medium. Note that CSS contains Tris-HCl as the buffering salt which is temperature sensitive. Therefore, it is critical that the pH be adjusted to pH 7.4 at the same temperature that the experiment will be performed.

Assay for Cell Death (Day 2)

There are a variety of methods that can be used to assay for cell death. We have found that the "gold standard" is cell counting and that other methods are not as sensitive in this primary culture system. Stain dead cells with 500 μl of 0.4% trypan blue for 20 min. Trypan blue is a vital stain that stains dead cells dark blue. Live cells exclude the dye and remain phase bright. Very gently remove the trypan blue and wash with CSS until the cells can be visualized under a microscope. Photomicrographs or permanent images of 3–5 random fields of cells should be obtained for counting by an observer blinded to the study design and treatment protocol. Viable versus nonviable cells should be counted with a minimum of 4000–12,000 neurons counted for each data point to reduce potential bias. Alternatively, cultures can be stained with fluorescent dyes based on ethidium homodimer or propridium iodide staining of dead cells (red) and calcien AM or fluorescein diacetate staining live cells (green) (Cytoprobe, Millipore, Bedford, MA; Live/Dead, Molecular Probes, Eugene, OR). Image analysis programs can be used to roughly estimate the proportion of live to dead cells. To determine whether cells are being "washed" away or are destroyed by either the treatment or staining protocol, photomicrographs should be made before and after treatment using a transparent grid etched on the bottom of each culture plate with a diamond tip pen. In our laboratory no appreciable loss of neurons has been identified when comparing before and after photomicrographs from cortical cultures.

Other assays for cytotoxicity are based on biochemical parameters which change in response to cell death. These assays work very well in homogeneous cell cultures. However, in mixed primary neuronal cell cultures which contain glia as well as neurons, the specificity of these methods can be clouded by background noise generated by glia. The lactate dehydrogenase (LDH) assay is widely used for many cytotoxicity paradigms and found to be reliable in some culture models (4, 5). However, LDH activity is affected by the oxidative environment and appears to be a molecular target for transition metal-mediated radical attack in some cultured neuronal systems (6, 7), and thus it may not be suitable for examining NO mediated neurotoxicity. In addition, a relatively large proportion of LDH activity resides in the small astrocytic population present in primary neuronal cultures which become "leaky" under ischemic conditions, contributing nonspecifically and disproportionately to the measurement of LDH release (8). Cytotoxicity assays that measure mitochondrial function such as rhodamine 123 and 3-[4,5-dimethylthiazol-2-y]-2,5-diphenyltetrazolium bromide (MTT) dye assays measure the response of the glia disproportionately to the response of neurons (8): even though glia comprise less than 20% of the total cell population in primary neuronal culture, they contain more than 60% of the mitochondria.

Criteria for Nitric Oxide Synthase Activation

A short application (5 min) of NMDA or glutamate to primary neuronal cultures clearly elicits neuronal cell death when determined 24 hr later. There are both NO-dependent and NO-independent components to this delayed neuronal cell death (2, 9). A number of criteria should be satisfied to determine that the pathway being evaluated is initiated by NO. Since there are few, if any, specific and selective antagonists for any protein, NOS inhibitors with different mechanisms of action should be used (Table V). Potency of inhibition of toxicity should parallel their potencies as NOS inhibitors and stereoisomers should be ineffective. The neuroprotection afforded by arginine analog inhibitors should be reversed by excess substrate L-arginine, while D-arginine should have no effect. Depletion of arginine from the culture medium should be neuroprotective, and hemoglobin which complexes NO should attenuate neurotoxicity. If the toxic insult occurs in part through NO,

TABLE V   Exposure Reagents

Component	Comments
Excitatory amino acids	
NMDA	Readily soluble in water or buffer solutions up to 10 mM. Glycine is necessary for full activation of NMDA receptor; glycine is present in most water supplies but to ensure consistent activation final concentration of 10 μM glycine should be added
Quisqualate	Not readily soluble in water or buffer solutions. Make 1 mM stock solution and heat to 52°C in water bath. Vortex vigorously and heat until fine crystals go into solution
Kainate	Readily soluble in MEM. Add kainate to MEM and expose the cultures for 24 hr in incubator to kainate solution
Glutamate	Readily soluble in water or buffer solutions
NOS inhibitors[a]	
N^G-Nitro-L-arginine (NArg)	Competes with arginine for catalytic site; is moderately soluble in water or buffer solutions up to 1 mM concentration with vigorous vortexing and heat up to 42°C
N-Methylarginine (NMA)	Competes with arginine for catalytic site; is readily soluble in water or buffer solutions
N-Iminoethyl-L-ornithine (NIO)	Competes with arginine for catalytic site; is readily soluble in water or buffer solutions
N^G-Nitro-L-arginine methyl ester (L-NAME)	Competes with arginine for catalytic site; is readily soluble in water or buffer solutions
N^G-Methyl-L-arginine acetate salt (L-NMMA)	Competes with arginine for catalytic site; is readily soluble in water or buffer solutions
Diphenyleneiodonium (DPI)	Inhibits electron shuttling by flavoproteins. Dissolve in dimethyl sulfoxide (DMSO) and dilute in buffer solution. DMSO concentration cannot exceed 0.1% (Kodak)
W-7	Inhibits calmodulin, essential cofactor for NOS activation; soluble in water and buffer solutions
GM$_1$ Ganglioside	Inhibits calmodulin, essential cofactor for NOS activation; barely soluble in buffer solutions; requires a 2-hr preincubation before initiation of experiment
GT$_{1b}$ Ganglioside	Inhibits calmodulin, essential cofactor for NOS activation; barely soluble in buffer solutions; requires 2-hr preincubation before initiation of experiment

(continued)

TABLE V (*continued*)

Component	Comments
FK506 (Immunophilin)	Binds to FKBP-12 and FK506–FKBP12 complex binds to calcineurin, inactivating phosphatase. NOS is activated by calcineurin-mediated dephosphorylation (Fujisuta Pharmaceuticals, Osaka, Japan). FK-506 is readily soluble in DMSO and can then be diluted in biological buffers to a final concentration of DMSO of 0.1%
Arginine depletion	
Arginine	Precursor to NO formation by NOS. Cell cultures can readily be depleted of arginine by two methods: (1) Culture cells for 24 hr in arginine-free MEM (GIBCO, special order) in presence of 2 mM glutamine to inhibit argininosuccinate synthase. Arginine is a semiessential amino acid and is synthesized by cells unless the synthesis pathway is blocked by glutamine. After exposure of cultures to experimental conditions MEM + 21 mM glucose must also be arginine-free. (2) Treat both medium and cell cultures with arginase, 10 units for 2 hr. Following exposure to experimental conditions place cells in arginase-treated MEM + 21 mM glucose
Hemoglobin reduction[b]	
Hemoglobin	Dissolve 2 mM hemoglobin (Sigma) in water; dissolve 20 mM dithionite in water. When both reagents are in solution mix them together (1 : 1, v/v) for final concentration of 1 mM hemoglobin and 10 mM dithionite. The hemoglobin should turn from brown to red. Put reduced hemoglobin in dialysis tubing with pore size less than 60 kDa. Dialyze overnight in 1000-fold excess water at 4°C. The dialysis vessel should be wrapped in foil. Replace water with fresh water and store at 4°C, wrapped in foil until use. Hemoglobin oxidizes very rapidly so prepare cultures for experiment and have all solutions ready prior to diluting the hemoglobin to 500 μM for the experiments. Long exposures to hemoglobin (in hours) should be avoided as hemoglobin can be toxic to neurons
NO donor reagents[c]	
Diethylenetriamine–nitric oxide (DETA–NO)	Specific NO· redox donor (RBI, Natick, MA)

(*continued*)

TABLE V   (*continued*)

Component	Comments
Sodium nitroprusside (SNP)	Readily soluble in water or buffer solutions; releases NO· and $Fe(CN_5)^{2-}$ by photolytic reaction; can also donate NO^+ (Sigma)
S-Nitroso-N-acetylpenicillamine (SNAP)	Soluble in DMSO up to 100 mM then dilute carefully into buffer solutions. Final DMSO concentrations cannot exceed 0.1% as DMSO is a radical scavenger and is neuroprotective at concentrations greater than 0.1%. Additionally, penicillamine is a peroxynitrite ($ONOO^-$) scavenger. In culture, a balance must be obtained between release of NO and scavenging of $ONOO^-$ (RBI, Natick, MA) (Molecular Probes, Eugene, OR)
3-Morpholinosydnonimine (SIN-1)	Soluble in water or buffer solutions; releases both NO and superoxide anion resulting in the generation of $ONOO^-$ (Molecular Probes)

[a] All the arginine analog inhibitors of NOS inhibit all NOS isoforms. If more than one isoform is present in the preparation, care must be taken to assure that inhibition of all isoforms does not confound the results.
[b] Hemoglobin does not exclusively scavenge NO but other oxidants as well, including the superoxide anion.
[c] Nitric oxide donors should be weighed, kept in a foil wrapped test tube, and solubilized immediately before use. In pure deoxygenated water, NO is stable for upward of 30 min, but in buffer solutions the stability of NO is related to the degree of oxygenation and metal ions. For example, the stability of NO in Krebs buffer is less than 5 min. The NO rapidly converts to nitrite in the presence of metal ions. Choice of buffer solutions is therefore very important.

then increased levels of NO should be present during the exposure of the toxic agent. Formation of cGMP can be used as an indirect measure of NO formation in cultures. Kits are commercially available from Amersham (Arlington Heights, IL) or New England Nuclear (DuPont NEN, Boston, MA). The Greiss reaction is usually not sensitive enough in primary neuronal cultures to access NO formation accurately, but under some conditions it can be used. It is based on a colorimetric assay of nitrite (NO_2^-) and nitrate (NO_3^-), reaction by-products of NO (Table VI).

Nitric Oxide Biochemistry

Nitric oxide has emerged as an important neuronal messenger in both the central and peripheral nervous systems. Models of the potential diffusion of NO suggest that even with brief bursts of NO (100 msec) and a short half-life (0.5–5 sec) the physiological sphere of influence could extend as far as 100 μm from the point of origin (encompassing ~2 × 10^6 synapses) (10). Given the potential sphere of influence of NO it is not surprising that NO has been implicated in a variety of neuronal functions.

TABLE VI   Greiss Reaction

Exposure media for greiss reaction
    Dulbecco's modified Eagle's medium, Base (DMEM) (GIBCO) without glucose, glu-
        tamine, phenol red, sodium bicarbonate, sodium pyruvate
    To the base add:
        200 mg $MgSO_4 \cdot 7H_2O$
        3.7 g $NaHCO_3$
        4.5 g D-Glucose
        0.11 g Pyruvic acid (sodium pyruvate)
        0.12 g $CaCl_2 \cdot 2H_2O$
    Bring to 1 liter with distilled, deionized H_2O and filter sterilize through a 0.22-μm
        filter
Greiss reaction reagent
    50 ml H_2O
    2.5 ml 85% Phosphoric acid
    50 mg Naphthylethylenediamine (Sigma)
    500 mg Sulfanilic acid (Sigma)
        Cover beaker with Parafilm and mix on stir plate. It takes approximately 30 min
            to dissolve completely. Reagent begins to turn color 1–1.5 hr after it is made
Greiss reaction
    400 ml sample in exposure medium
    400 ml Greiss reaction reagent
    Mix by inversion
    Wait 5–10 min to allow color to fully develop
    Read absorbance (λ 563), plot against standard curve generated from sodium nitrite

Nitric oxide has been implicated in glutamate neurotoxicity in a variety
of model systems, and prolonged application of NOS inhibitors, after the
initial exposure to excitatory amino acids, leads to enhanced neuroprotection. Others have failed to confirm the role of NO in this phenomenon or have
demonstrated neuroprotection through inhibition of the NMDA receptor
(9, 11). Studies indicate that NO may possess both neurodestructive and
neuroprotective properties depending on the local redox milieu. The NO
radical (NO·) is neurodestructive, whereas NO^+ (i.e., NO complexed to a
carrier molecule) is neuroprotective (11). The chemical composition and
redox environment of the experimental system is therefore critical (12). Care
needs to be taken when employing "NO" donors that the form of "NO"
and the conditions of release are known. For example, the commonly used
NO donor SNP does not spontaneously release NO· in biological buffer
solutions. It exists as NO^+ and donates NO in this form. However, SNP
readily releases NO· by a photochemical reaction. 3-Morpholinosydnonimine
hydrochloride (SIN-1) releases NO· and O_2^-, resulting in generation of
$ONOO^-$. S-Nitrosothiols such as S-nitrosoglutathione are termed NO donors
but generally donate NO as NO^+ to proteins. Adduct compounds such as

diethylenetriamine–nitric oxide (DETA–NO) may provide purer forms of NO·. It is important when designing experiments for evaluating the cellular consequences of NO activity that the reagents and buffers be chosen carefully.

Acknowledgments

V.L.D. is supported by grants from the U.S. Public Health Service (NS22643, NS33142), the American Foundation for AIDS Research, the National Alliance for Research on Schizophrenia and Depression, the American Heart Association, the Alzheimer's Association, and the Muscular Dystrophy Association. T.M.D. is supported by grants from the USPHS (NS01578, NS26643, NS33277) and the American Health Assistance Foundation, Paul Beeson Physician Scholars in Aging Research Program, and the International Life Sciences Institute. The authors own stock in and are entitled to royalty from Guilford Pharmaceuticals, Inc. (Baltimore, MD), which is developing technology related to the research described in this chapter. The stock has been placed in escrow and cannot be sold until a date that has been predetermined by the Johns Hopkins University.

References

1. M. A. Dichter, *Brain Res.* **149,** 279 (1978).
2. V. L. Dawson, T. M. Dawson, D. A. Bartley, G. R. Uhl, and S. H. Snyder, *J. Neurosci.* **13,** 2651 (1993).
3. F. Zhang, J. G. White, and C. Iadecola, *J. Cereb. Blood Flow Metab.* **14,** 217 (1994).
4. D. W. Choi, J. Koh, and S. Peters, *J. Neurosci.* **8,** 185 (1988).
5. R. R. Ratan, T. H. Murphy, and J. M. Baraban, *J. Neurosci.* **14,** 4385 (1994).
6. E. R. Stadtman, *Science* **257,** 1220 (1992).
7. R. R. Ratan, T. H. Murphy, and J. M. Baraban, *J. Neurochem.* **62,** 0376 (1994).
8. B. H. J. Juurlink and L. Hertz, *Dev. Brain Res.* **71,** 239 (1993).
9. T. M. Dawson and S. H. Snyder, *J. Neurosci.* **14,** 5147 (1994).
10. J. Garthwaite and C. L. Boulton, *Annu. Rev. Physiol.* **57,** 683 (1995).
11. S. A. Lipton, Y.-B. Choi, Z.-H. Pan, S. Z. Lei, H.-S. V. Chen, N. J. Sucher, J. Loscalzo, D. J. Singel, and J. S. Stamler, *Nature (London)* **364,** 626 (1993).
12. A. R. Butler, F. W. Flitney, and D. L. Williams, *Trends Pharmacol. Sci.* **16,** 18 (1995).

[21] Nitric Oxide–Cyclic Guanosine Monophosphate Pathway in Central Nervous System: *In Vitro* and *in Vivo* Investigations

Mark Salter, Eric Southam, and John Garthwaite

Introduction

The brain is the richest source of nitric oxide (NO) synthase of any mammalian organ, reflecting the fact that the enzyme is expressed to a greater or lesser extent in almost all brain regions. The target enzyme for NO, soluble guanylyl cyclase, has a similarly widespread distribution, and the anatomical patterns of NO-stimulated cGMP accumulation and NO synthase protein in the brain are broadly correlated, reinforcing the idea that NO signaling is transduced mainly through cGMP. A major stimulus for NO formation is activation of receptors for the excitatory neurotransmitter glutamate. Of special importance is the NMDA (*N*-methyl-D-aspartate) receptor, whose associated ion channel permits substantial influx of Ca^{2+}, but other mechanisms, including neuronal depolarization (leading to the opening of voltage-sensitive Ca^{2+} channels) and the activation of receptors that are coupled to mobilization of Ca^{2+} from internal stores (e.g., G-protein-coupled glutamate and acetylcholine receptors), can also be effective. In all cases it is the resultant rise in cytosolic Ca^{2+} that, through calmodulin, stimulates NO synthase.

Nitric oxide has been implicated in numerous physiological phenomena in the brain and spinal cord, including synaptic plasticity in several of its guises, sensory and motor control, and the regulation of local cerebral blood flow. Cyclic GMP is the likely mediator of NO signal transduction in these situations but NO may also contribute to neurodegeneration in conditions such as cerebral ischemia, through cGMP-independent mechanisms. We describe below the methods we use to investigate the NO–cGMP pathway, both using *in vitro* brain preparations and *in vivo*.

Methods in Neurosciences, Volume 31

In Vitro Systems

Tissue Preparation

Although seemingly obvious, the quality of results obtained and their interpretation depend critically on the quality of the tissue from which they are derived. Incubated brain slices can vary enormously in their health, as judged by direct histological, biochemical, and functional criteria. At one extreme, they appear to be good representations of *in vivo* counterparts; at the other, they are little better than a homogenate. Unfortunately, many authors (usually of neurochemical works, including those on the NO–cGMP pathway) seem unaware of this. A critical determinant of the ultimate viability of a brain slice is the manner in which it is cut and incubated. The methods used in our laboratory are detailed below.

Cerebellar and Hippocampal Slices

For obtaining slices from immature (up to 14 days postnatal) rat cerebellum, the animal is decapitated and the cerebellum quickly removed, placed onto the chopping surface of a McIlwain tissue chopper (Mickle Laboratory Engineering, Gomshall, U.K.), and oriented so that cuts are made in the sagittal plane. The optimal thickness for a slice is about 400 μm: any thicker and an anoxic core develops; any thinner and the region of damaged tissue at the edge of the slice, caused by the act of cutting, is disproportionately large. Chopped tissue is floated off in chilled (10°–12°C) artificial cerebrospinal fluid [ACSF, of composition, in mM: NaCl, 120; KCl, 2; $MgSO_4$, 1.19; KH_3PO_4, 1.18; $NaHCO_3$, 26; glucose, 11; gassed with 5% (v/v) CO_2 in O_2; pH 7.4], contained in a petri dish, and slices of the vermis, usually four, are carefully separated with the aid of blunt tweezers. The immature hippocampus is dissected from the brain, chopped at 400-μm intervals in the transverse plane, floated off into chilled (6°C) ACSF, and the slices (about four) teased apart.

Slices of more mature cerebellum should be prepared using a vibrating tissue slicer, such as the Vibroslice (Camden Instruments, Loughborough, U.K.), because fewer than 10% of neurons in adult rat cerebellar slices, prepared by mechanical chopping, remain healthy (1). The cerebellum is placed onto filter paper moistened with ACSF and a razor blade used to remove both cerebellar hemispheres. One of the cut ends of the vermis is glued (with cyanoacrylate) onto the Vibratome stage, and the tissue is encased in liquid agar [3% (w/v) agar dissolved in 0.9% (w/v) saline, kept stirred at 37°C], which quickly solidifies to provide support for the tissue during slicing. The stage is then locked into position in the Vibratome bath and is immersed in chilled ACSF (10°C). Four or five 400-μm parasagittal slices of vermis can be obtained from each cerebellum.

Adult hippocampal tissue is less sensitive to chopping but in our experience the Vibratome produces better quality slices, as judged by histological criteria. The brain is removed and is blocked by razor blade cuts made coronally, to remove the forebrain and the brain stem/cerebellum, and sagittally, to bisect the brain. The medial surface is glued (with cyanoacrylate) onto the Vibratome stage which is then locked into position in the Vibratome bath and immersed in chilled ACSF (6°C). Improved preservation of mature hippocampal tissue is achieved if slicing is performed in ACSF modified to contain 4 mM MgSO$_4$ and 1 mM CaCl$_2$. An initial Vibratome cut is made to trim the block so that the hippocampus is visible in the transverse plane. Thereafter cuts are made, beginning at the dorsal surface, at 400-μm intervals, and the resulting slices (usually four) are excised from the surrounding tissue by cutting through the subiculum, fimbria, and any adhering connective tissue with microscissors.

Once prepared, slices are transferred to incubation vessels containing fresh gassed ACSF (usually 20 ml in a 50-ml conical flask) using a spatula, maintained at 37°C in a shaking water bath, and are allowed to recover for about 30 min. They are then redistributed, so that slices, usually four, obtained from different animals, are grouped together for a further hour of preincubation before the experiment begins.

Cerebellar Cell Suspensions

When glutamate is the receptor agonist of choice, highly efficient uptake mechanisms ensure that the concentration of this transmitter falls rapidly with increasing distance from the surface of a slice (2). An alternative is to dissociate tissue into a suspension of cells which will have equal access to applied glutamate. These preparations are also useful if direct sampling of substances released from cells, such as NO, is required. Cell suspensions are prepared from cerebella from 6- to 14-day-old rats. If the differentiating cells are the ones of interest (as is the case for the NO–cGMP pathway), the animals can be treated 2 days before needed with the mitotic inhibitor hydroxyurea (2 mg/kg, intraperitoneally), which depletes the tissue of the numerous proliferating cells present at these ages.

Prepare the following solutions:

1. Calcium-free Kreb's solution containing (in mM): NaCl, 120; KCl, 2; NaHCO$_3$, 26; MgSO$_4$, 1.19; KH$_2$PO$_4$, 1.18; glucose, 11; which is well gassed with 95% O$_2$, 5% CO$_2$ (v/v)
2. 200 ml solution 1 plus 0.6 g bovine serum albumin (BSA) (fraction V)
3. 50 ml solution 2 plus 25 mg trypsin
4. 30 ml solution 2 plus 2.4 mg DNase, 15.6 mg soybean trypsin inhibitor, and 46.5 μl of 1 M MgSO$_4$
5. 8 ml solution 4 plus 42 ml solution 2

6. 25 ml solution 2 plus 2.5 μl of 1 M CaCl$_2$ and 31 μl of 1 M MgSO$_4$
7. Ungassed incubation medium containing (in mM): NaCl, 130; KCl, 3; MgSO$_4$, 1.2; Na$_2$HPO$_4$, 1.2; Tris-HCl, 15; CaCl$_2$, 2; adjusted to pH 7.4 at 37°C; glucose, 11; add 1% BSA (essentially fatty acid free) and filter (0.45-μm pore size) before use

Rats are decapitated and the cerebella removed, cleaned of choroid plexus, and chopped (two to four at a time) at 400-μm intervals in both the parasagittal and coronal planes with a McIlwain tissue chopper. The resulting tissue blocks are washed into conical tubes containing 10 ml solution 2 (at room temperature) and dispersed by gentle trituration. Following centrifugation (50 g, 5 sec) and aspiration of the supernatant, the tissue blocks are resuspended in 10 ml solution 3, transferred to 50-ml conical flasks, and maintained for 15 min at 37°C in a shaking water bath under a constant stream of 95% O$_2$, 5% CO$_2$ (v/v). Dilute with 10 ml solution 5 (at room temperature), centrifuge (50 g, 5 sec), and aspirate the supernatant. Add about 2 ml solution 4, resuspend, and then shear tissue by trituration through a siliconized Pasteur pipette (15 times per cycle). Allow clumps to settle out and, using a clean pipette, transfer the clump-free milky suspension to 3 ml solution 6 in a conical tube. The cycle is repeated until the remaining tissue is disrupted. Remove any contaminating clumps of tissue which settle out before centrifuging (50 g, 3 min), aspirating the supernatant, and resuspending the cells in a volume of incubation medium 7 (at room temperature) sufficient to achieve the desired cell density (20×10^6 cells/ml is optimal if NO-mediated cGMP accumulation in response to excitatory amino acid receptor agonists is being measured). Cell counts are made with a Coulter counter or hemocytometer.

Citrulline Assay of Nitric Oxide Synthase Activity

Nitric oxide synthase activity may be assessed by measuring the rate of conversion of L-[^{14}C]arginine to L-[^{14}C]citrulline. The citrulline assay is most frequently used to determine the NOS activity of brain homogenates and may be modified to permit the measurement of the NOS activity in intact rat brain slices.

Homogenate

Tissue is homogenized in 50 mM Tris (pH 7.0), 250 mM sucrose, 1 mM EDTA, 1 mM dithiothreitol (DTT), 10 μg/ml leupeptin, and 10 μg/ml soybean trypsin inhibitor (4°C) using an Ystral homogenizer and centrifuged (12,000 g for 10 min at 4°C). Supernatant (18 μl) is added to 100 μl prewarmed (37°C) 50 mM potassium phosphate (pH 7.2), 240 μM CaCl$_2$, 120 μM NADPH, 1.2

mM L-citrulline, 24 μM L-arginine, and L-[U-^{14}C]arginine [15,000 disintegrations per minute (dpm)] and incubated at 37°C for up to 10 min. The reaction is stopped by mixing with 1.5 ml of 50% Dowex 50W ion-exchange resin (200–400, 8% cross-linked, converted to the sodium form by washing four times in 1 M NaOH and then repeatedly rinsing in distilled water until the pH drops to 8.0). Add 5 ml of H$_2$O and leave to settle before removing 4 ml of supernatant to scintillant for determination of ^{14}C. The blank is determined by including 100 μM L-N^ω-nitro-L-arginine (L-NA) in the assay. Calculate the NOS activity as follows:

100 μl assay medium contains 2400 pmol L-arginine, radioactivity added to assay = 150,000 dpm, therefore, specific activity of arginine = 62.5 dpm/pmol.

Assuming 1 mol of citrulline is formed from 1 mol of arginine, citrulline

$$\text{formed} = \frac{\text{dpm in sample} - \text{dpm in blank (L-NA)}}{\text{specific activity of L-arginine}}$$

Nitric oxide synthase activity is citrulline formed/time of assay/amount of tissue added.

(A correction factor should be included because only 4.0 ml of resin/sample mix is added to scintillant.)

Slices

By conserving the integrity of the tissue, links between glutamate receptor activation and NOS activity may be investigated. For example, a 10-min incubation of adult rat cerebellum with the glutamate receptor agonist α-amino-3-hydroxy-5-methoxy-4-isoxazole propionate (AMPA, 30 μM) causes an increase in NOS activity of 13-fold over basal (Fig. 1). Slices are transferred into medium (about 1 ml per slice of adult rat cerebellum) composed of ACSF plus 157 nM L-[^{14}C]arginine (specific activity 11.8 GBq/mmol), 1 mM L-citrulline, and 20 μM L-arginine (37°C in a shaking water bath). To eliminate non-NOS-induced citrulline formation (the blank in the following calculation), control slices are incubated in the presence of 100 μM L-NA. After 30 min basal citrulline formation, test compounds are added for predetermined times before the slices are inactivated in 1 ml of boiling 50 mM Tris and 4 mM EDTA buffer for 3 min. They are then homogenized by sonication, and 50 μl of homogenate is removed for protein determination. Following centrifugation (10,000 g, 5 min), radiolabeled arginine is separated out of the supernatant (about 900 μl) by adding 4 ml 50% ion-exchange resin (see above). When the resin has settled out, 2 ml of the supernatant is carefully removed and added to 12 ml of scintillant prior to counting.

Under conditions where the tissue concentration is high enough to produce

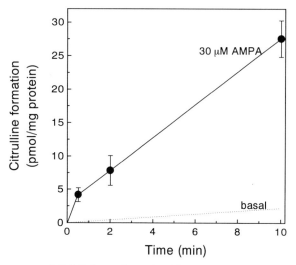

FIG. 1   Time course of NOS-dependent citrulline production in slices of adult rat cerebellum incubated with AMPA (30 μM). Each data point represents the mean ($\pm$ standard error) of four slices. Basal citrulline production (0.21 $\pm$ 0.05 pmol/mg/min) was measured over 30 min ($n = 4$).

significant increases in extracellular citrulline ($\sim$100 mg tissue/ml for rat cerebral cortex and 10 mg/ml for cerebellum), radiolabeled citrulline production can be measured by sampling the incubation medium. For these experiments the reaction is stopped by adding the resin to the medium (including the intact slices) and processing as described above.

Measurement of Nitric Oxide, Nitrite, and Nitrate

The instability of NO in oxygenated buffer means that it is rapidly oxidized to NO_2^- and, under some conditions, NO_3^-. Therefore, measurement of these ions provides an indirect assay of NO production. There are a number of chemical methods that may be used for the measurement of NO or its oxygenation products. Unfortunately, they are generally insufficiently sensitive to measure transient NO production by tissue maintained *in vitro*. However, they are useful when incubations are long-term (>3 hr), particulary when the use of culture medium containing relatively high concentrations of arginine rules out the use of the citrulline assay to monitor NOS activity due to the difficulty in obtaining a high enough specific activity of the radiolabeled arginine. Because of the high amounts of contaminating NO_3^- in normal distilled water and even in clean glassware and plasticware it is important to use and rinse all materials with Milli-Q water.

The Griess Reaction

The NO_2^- content of incubation medium may be measured by mixing the medium, in equal parts, with Griess reagent (0.1% (w/v) naphthylethylenediamine dihydrochloride, 1% (w/v) sulfanilamide in 5% (v/v) H_3PO_4, stable at 4°C for up to 12 hr before use), allowing the mixture to stand for 10 min, and then centrifuging at 12,000 *g* for 2 min at room temperature. Dilutions of $NaNO_3$ (0–100 μM) are used as standards. The absorbance of the 150-μl aliquots of supernatant, transferred to 96-well plates, is read at two wavelengths (550 and 620 nm) using a multiwell plate reader. The NO_3^- may be assayed by the Griess reaction by first reducing it to NO_2^-. To do this, add 100 μl of sample to 20 μl of a 1 : 1 : 1 : 1 mixture of 1 *M* potassium phosphate (pH 7.5), 14.4 m*M* NADPH, 120 m*M* FAD, and 3 mg/0.5 ml nitrate reductase from *Aspergillus* (prepared on ice immediately before use), and incubate at 37°C for 1 hr. After NO_3^- reduction is complete the NADPH must be oxidized to avoid interference with the Griess reaction (this is not required for the chemiluminescence method; see below) by incubating samples with 10 units/ml lactate dehydrogenase (from rabbit muscle) and 10 m*M* pyruvate for 5 min at 37°C.

Chemiluminescence

Nitric oxide may be detected by a chemiluminescence method that depends on the reaction of NO with O_3 which generates NO_2 in an excited state. On decay to the ground state, a photon is emitted that may be detected using a photomultiplier tube. NO_2^- is first reduced back to NO by injecting the sample through a 1–3% solution of KI in glacial acetic acid under reflux. The NO generated is carried on a stream of N_2 gas and mixed with O_3, and the resulting chemiluminescence compared with that produced by $NaNO_3$ standards. The technique is rapid and is moderately sensitive to NO (submicromolar range). Its major limitation is that NO must be evaporated into the gas phase for analysis.

Nitric oxide may also be measured by monitoring the NO-dependent conversion of oxyhemoglobin to methemoglobin spectrophotometrically (see section on *ex vivo* NOS Assay).

Electrochemical Probe for Nitric Oxide

Electrochemical probes designed to detect NO are now commercially available (e.g., World Precision Instruments, Sarasota, FL; Intracel, Royston, U.K.). The principle of such probes is that NO will become oxidized at the surface of a platinum wire cathode held at a voltage more positive to the anode:

$$NO + 4OH^- \rightarrow NO_3^- + 2H_2O + 3e^-$$

The current generated is proportional to the NO oxidized. A thin membrane over the pore of the probe restricts access to the cathode by gaseous substances, while the voltage maintained between the cathode and anode facilitates the detection of oxidizable, rather than reducible (e.g., O_2) molecules. The probe is sensitive to changes of NO concentration in the submicromolar range and is capable of responding within a few seconds. However, although the probe has successfully been used to detect the generation of NO in the molecular layer of rat cerebellar slices following climbing fiber stimulation (3), until development of the probe progresses to the point where the diameter of the sensor is reduced from the present 30 μm, its use is largely restricted to solutions and cell suspensions.

Measurement of cGMP

Measurement of cGMP has long been used to study indirectly the pharmacology of glutamate receptors. After incubating slices with test compounds for predetermined periods, the tissues are inactivated by plunging them into 200 μl of boiling 50 mM Tris, 4 mM EDTA (pH 7.6) buffer for 3 min. Tissue is homogenized by sonication and 50 μl removed for protein estimation. Following centrifugation (10,000 g, 5 min), the cGMP content of a 100-μl aliquot is measured by radioimmunoassay (kits are commercially available from Amersham Radiochemical Centre, Amersham, U.K.). The quantity of cGMP, particularly in stimulated cerebellar slices, may necessitate the dilution of the supernatant, perhaps hundredfold, in order to obtain values within the limits of the standard curve (0.5–8 pmol per 100-μl aliquot).

Cyclic nucleotides are continually metabolized by phosphodiesterases. Therefore, when studying *in vitro* preparations of brain tissue with high cGMP-phosphodiesterase activity, for example, hippocampal slices, it may be necessary to include the broad spectrum phosphodiesterase inhibitor 3-isobutyl-1-methylxanthine (IBMX, 1 mM) in the incubation medium for the preincubation and test periods in order to limit cGMP catabolism.

Soluble Guanylyl Cyclase Activity Assay

A crude homogenate of cerebellar soluble guanylyl cyclase is prepared by removing the cerebellum of an adult rat and quickly immersing it in about 5 ml ice-cold 10 mM Tris containing 1 mM DTT (pH 7.4). The tissue is homogenized (Ystral motorized homogenizer), taking care not to allow the temperature of the homogenate to rise above 4°C, diluted (about 1 in 5 with homogenization buffer), and 50 μl removed for protein determination (ideally 2–10 mg/ml). After centrifugation (10,000 g, 30 min, 4°C), the supernatant is removed and kept on ice until used. The enzyme activity of 50-μl aliquots of homogenate is assayed by mixing with 150 μl prewarmed (37°C) reaction buffer (consisting of 50 mM Tris (pH 7.6), 4 mM IBMX, 5 mM creatine phosphate, 0.05% (w/v) BSA, 150 μg/ml rabbit muscle creatine kinase, 1 mM GTP, and 3 mM MnCl$_2$), together with the test compounds, in 1.5-ml Eppendorf tubes, and incubating for 10 min at 37°C. Enzyme activity is terminated by adding 40 μl NaEDTA buffer and heating to 100°C for 3 min. The cGMP content of the reaction mixture is determined by radioimmunoassay (see above) and enzyme activity expressed as pmol cGMP formed/min/mg protein.

In Vivo Systems

By correlating the activity of the NO–cGMP pathway with changes in *in vivo* physiology and/or behavior, we may gain insights into the role(s) of the pathway in various physiological and pathophysiological states. We describe below how the pathway can be monitored *in vivo* by both microdialysis and *ex vivo* techniques.

Microdialysis

The implantation of a microdialysis probe allows the spatial and temporal monitoring of the glutamate–NO–cGMP pathway in the brains of both conscious and unconscious animals and thereby facilitates the study of the *in vivo* effects of either stimulating or downregulating the pathway by the infusion or systemic administration of substances. Although the following parameters are usually monitored only after infusion or administration of ionotropic glutamate receptor agonists to activate the pathway some studies have examined function under basal conditions.

Nitric oxide may be monitored directly by infusing 1 μM oxyhemoglobin in ACSF through the microdialysis probe and the product, methaemoglobin, detected by spectrophotometry at 401 nm (4). Nitric oxide may also be monitored indirectly by measuring the NO$_2^-$ content of the microdialyzate using the Griess reaction as described above. There is substantial conversion of NO$_2^-$ to NO$_3^-$ *in vivo* and because the Griess reaction does not measure NO$_3^-$ samples must first be reduced to NO$_2^-$ (see Griess reaction, *in vitro* section). Alternatively, *in vivo* NOS activity may be monitored by high-performance liquid chromatography (HPLC) to measure the level of L-citrulline in the dialyzate. Thirty microliters of the microdialyzate is derivatized with *o*-phthalaldehyde (10 μl) and injected onto a 4.6 mm × 150 mm Eicompak MA-5ODS column (Eicom, Kyoto, Japan), running a mobile phase consisting of 100 mM sodium phosphate (pH 6.0), 54 mg/liter EDTA and 30% (v/v) methanol (flow rate 1.0 ml/min) and detecting electrochemically (ECD-100, Eicom) using a graphite working electrode at +650 mV versus an Ag|AgCl reference electrode. Finally, the implantation of a custom made NO porphyrinic microsensor has been used to monitor NO directly (5).

The *in vivo* activity of the glutamate–NO–cGMP pathway can also be monitored by determining the level of cGMP in brain microdialyzates. Stimulation or inhibition of the pathway is reflected by the modulation of cGMP levels within the extracellular space and, therefore, within microdialyzates. The cGMP content of the microdialyzate is measured by radioimmunoassay as described above with greater sensitivity achieved by acetylating the sample (mix 1 ml acetic anhydride with 2 ml triethylamine and add 5 μl to 100 μl of sample).

Ex Vivo Systems

The *in vivo* activity of the glutamate–NO–cGMP pathway may be assessed by assaying NOS or by measuring tissue cGMP levels *ex vivo*.

Nitric Oxide Synthase Assay

Ex vivo assays may be used to assess the efficacy of NOS inhibitors *in vivo*, providing there is no significant dissociation of inhibitor from enzyme during the assay. L-NA, frequently perceived to be an irreversible inhibitor of NOS, does, in fact, slowly dissociate from the enzyme. It is, therefore, important to determine the initial rate (<30 sec) of NOS activity in order to minimize any underestimation of inhibitor efficacy. The inhibitory effects of L-NA on brain NOS activity and cGMP accumulation *in vivo* (see method below), as assessed by *ex vivo* assay, are closely correlated (Fig. 2), suggesting that there is little dissociation of L-NA from NOS either during tissue preparation or rapid assay. Of the other arginine analogs frequently used to inhibit brain NOS, N^ω-monomethyl-L-arginine (L-NMMA) dissociates too rapidly for *ex vivo* assay while N^ω-nitro-L-arginine methyl ester (L-NAME), although not a significant inhibitor of brain NOS itself, is converted to L-NA in both *in vitro* and *in vivo* systems.

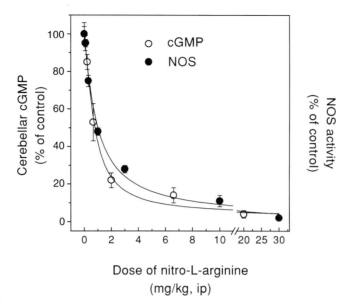

FIG. 2   The effect of L-NA on harmaline-induced increases in cerebellar cGMP and on NOS activity measured as an initial rate *ex vivo*. Each data point represents the mean (± standard error) of three rats. The cGMP content of cerebella from harmaline-alone-treated animals was 232 ± 12 pmol/g tissue and the NOS activity of cerebella from control animals was 52 ± 2.5 nmol/min/g tissue.

In order to assay NOS activity *ex vivo,* brain tissue is rapidly removed and freeze-clamped in liquid nitrogen (tissue can be safely stored at $-70°C$ for several months without loss of NOS activity). It is homogenized (Ystral motorized homogenizer) at 4°C in 10 volumes of 50 mM Tris (pH 7.0), 250 mM sucrose, 1 mM EDTA, 1 mM dithiothreitol, 10 μg/ml leupeptin, and 10 μg/ml soybean trypsin inhibitor, centrifuged (12,000 g, 10 min, 4°C), and the supernatant kept on ice until used. Nitric oxide synthase activity is determined by monitoring the conversion of oxyhemoglobin to methemoglobin by NO spectrophometrically. Fifty microliters of tissue homogenate supernatant is mixed with 450 μl prewarmed (37°C) 50 mM potassium phosphate (pH 7.2), 278 μM CaCl$_2$, 1.6 μM oxyhemoglobin, 22.2 μM L-arginine, and 111 μM NADPH, and the absorption difference between 421 and 401 nm ($\varepsilon = 77,200$ M^{-1} cm^{-1}) is immediately and continuously monitored at 37°C using a dual-wavelength spectrophotometer (Schimadzu UV-3000, Kyoto, Japan) with a slit width of 5 nm [a single wavelength spectrophotometer can be used, with reduced sensitivity, by monitoring at 401 nm only ($\varepsilon = 38\ 600$)]. The sensitivity of this spectrophotometric assay of NOS activity is comparable to the citrulline assay. However, the presence of substantial amounts of hemoglobin in some tissues can lead to a underestimation of NOS activity levels or even to the apparent absence of NOS activity.

Cyclic GMP Determination

The NO–cGMP pathway may be stimulated in the rodent cerebellum *in vivo* by the systemic administration of harmaline. This β-carboline increases the firing of olivocerebellar afferents (climbing fibers), resulting in the calcium-dependent activation of NOS within the climbing fibers themselves and/or within other cerebellar neurons depolarized by this excitatory activity. This increased activity of the pathway, and its downregulation by enzyme inhibitors, may be monitored by the *ex vivo* measurement of cerebellar cGMP levels.

Harmaline (100 mg/kg) is administered subcutaneously. After 10–15 min, animals are killed: rats by cervical dislocation and, following decapitation (decapitation alone leads to a large variation in cGMP values), the cerebellum is rapidly removed and freeze-clamped; mice by focused microwave irradiation. The cGMP content of freeze-clamped or irradiated cerebellar tissue is determined as described above. The temporal effects of NOS inhibitors can be studied by their administration for predetermined periods prior to the harmaline-induced increase in cerebellar cGMP.

References

1. J. Garthwaite, P. L. Woodhams, M. J. Collins, and R. Balazs, *Brain Res.* **173,** 373 (1979).
2. J. Garthwaite, *Br. J. Pharmacol.* **85,** 297 (1985).
3. K. Shibuki and D. Okada, *Nature (London)* **349,** 326 (1991).
4. A. Balcioglu and T. J. Maher, *J. Neurochem.* **61,** 2311 (1993).
5. T. Malinski, F. Bailey, Z. G. Zhang, and M. Chopp, *J. Cereb. Blood Flow Metab.* **13,** 355 (1993).

[22] Nitric Oxide in Regulation of Microvascular Tone: Methods of Detection

Ulrich Pohl

Introduction

It is the functional state of the microcirculation that determines the supply of oxygen and substrates to tissues. It is therefore not surprising that microcirculatory functions like arteriolar and capillary perfusion and the regulation of capillary pressure are under the control of various and partly redundant mechanisms. The role of endothelium-derived vasoactive factors, especially nitric oxide (NO), in these processes has gained much interest in recent years. However, in the microcirculatory bed it is perhaps most difficult to analyze whether the endothelium is involved. Mechanical removal of the endothelium, which is a standard procedure to prove that it plays a role in vasomotor responses, is nearly impossible in the microcirculation. Mechanical endothelial removal is likely to destroy the thin muscle layer as well. The injection of air can destroy the endothelium (1), but it obstructs perfusion in small vessels. Chemical removal, for example, by perfusion of saponin (2), inevitably activates thrombocytes and leukocytes and increases vascular permeability. Thus, all methods alter the experimental conditions so decisively that the interpretation of the results is uncertain. Moreover, it is difficult to quantify the role of NO. On the one hand, methods to detect NO in the microcirculation directly are not readily available or applicable. Indirect methods, on the other hand, are subject to experimental errors. In view of the substantial effects of endothelial factors on the tone of isolated vessels, it is nevertheless indispensible to analyze their role in the intact microcirculation. This chapter briefly reviews the present knowledge of the role of NO in the control of microvessels *in vivo* and describes direct or indirect methods to analyze it.

Physiological Role of Nitric Oxide in Microcirculation

There is ample indirect evidence in various tissues and species that endothelium-derived nitric oxide contributes to vascular control. Many compounds, among them transmitters, hormones, and blood-borne substances, exert their dilator action on blood vessels by stimulating the release of NO (3). Some, like platelet-derived nucleotides, induce vasoconstriction at sites of destroyed endothelium but dilate at sites with intact endothelium. Functionally, these platelet-derived compounds help to constrict vessels at the site of injury but keep the vessel patent at sites distal to these areas and prevent further activation of platelets (4). Local inflammatory mediators like bradykinin or histamine also elicit dilation in an NO-dependent manner (3). Moreover, there is a considerable continuous basal release of NO as inhibitors of NO synthase (see below) induce vasoconstriction in most vascular beds (5). It is not clear to what extent the basal release is constitutive because it is well known that the flowing blood, by exerting a shear stress on the endothelial surface, stimulates the production of NO (6) and the expression of NO synthase (7). This shear-induced dilation is crucial for a coordinated response of blood vessels to elevated flow to increase vascular conductivity adequately (8). Furthermore, the shear-induced dilation opposes the myogenic constriction, which occurs especially in small vessels following an increase of transmural pressure (6). The latter is an important component of the autoregulation of blood flow. This mechanism allows the body to keep capillary pressure constant but tends to reduce blood flow by acting as a positive feed-back loop. It is therefore not surprising that an inhibition of NO synthesis results in reduced tissue blood flow, tissue hypoxia, and a reduced adaptation of blood flow to altered tissue demands (9).

Nitric oxide can affect brain vessels in several ways; NO has been detected in endothelial cells, perivascular nerve endings, and in nerve cells (10). Depending on the experimental conditions NO is a mediator or modulator of CO_2-induced dilation (11, 12). It also participates in autoregulation of cerebral blood flow (13) and contributes to the control of cerebrovascular resistance (14).

Detection of Nitric Oxide or Nitric Oxide-Mediated Dilation

Stimulation of Nitric Oxide Release

Nitric oxide synthase catalyzes the oxidation of a guanidino nitrogen of L-arginine to NO with concomitant formation of L-citrulline. There are three isoforms of this enzyme. In endothelial cells the endothelial nitric oxide synthase (NOS III), a mainly membrane-bound enzyme is found (15). The activity of this enzyme is Ca^{2+}–calmodulin-dependent, and hence many stimuli that increase intracellular free calcium (particularly by eliciting Ca^{2+} influx) are stimulators of NO production. The classic stimulus is acetylcholine (ACh) (16), whose physiological role is not yet clear because it does not circulate. Many other biological stimuli have been identified, among them bradykinin (BK), ATP, and substance P (3), compounds that are released either from nerve endings or cells in the close vicinity of endothelial cells. Only a few endothelial stimulators have been used so far in *in vivo* experiments on brain microvessels of mice, cats, rats, and rabbits (14, 17–19). Mostly, ACh is used as endothelial stimulator. Other NO-stimulating compounds include histamine and bradykinin (dilations by the latter compound are not mediated by NO but rather by prostaglandins or oxygen radicals in rabbit and cat vessels) (19, 20). L-Arginine, the substrate of all nitric oxide synthase isoforms, also induces vasodilation in brain vessels (21), but it is not clear whether the dilating NO is derived from the endothelium of the brain vessels or from the neuronal tissue.

It should be emphasized that most stimuli elicit the release not only of NO but also of other endothelial factors such as prostaglandins and endothelium-derived hyperpolarizing factor (EDHF). Any dilation induced by these compounds can therefore not be taken automatically as a measure of nitric oxide release. It has been shown further that these factors partly interact with one another at the level of smooth muscle or thrombocytes (22, 23), thereby augmenting the effects of one another. This is important to know for making analyses of the role of NO, solely based on changes of vascular diameter.

Use of Nitric Oxide Synthase Inhibitors

A number of compounds that inactivate NO, like methylene blue (50 μM), gossypol (50 μM), and hemoglobin (5 μM), were used in the past to analyze the role of NO in the microcirculation of intact animals or isolated organs. However, these compounds act in a nonspecific manner and, at least in part, are cytotoxic. At present, specific inhibitors of the synthesis of NO, the L-arginine analogs N^G-nitro-L-arginine (L-NNA). N^G-monomethyl-L-arginine (L-NMMA), and N^G-nitroarginine methyl ester (L-NAME), which are substituted at a guanidino nitrogen, are widely used. L-NAME has the best solubility in water, which is important for intravenous application. The interpretation of results requires caution when acetylcholine is used as NO stimulator because L-NAME was found to block muscarinergic receptors as well (24). L-NMMA can be degraded to citrulline and recycled to L-arginine such that the inhibition by this compound is reversible, and its application can even deliver the substrate of NOS (25). It should also be kept in mind that all these inhibitors need to be transported into the cell. Different transporters are involved, but, in general, the transport is dependent on cell membrane potential, the state of cellular activation, and on time. To achieve a complete blockade of NOS in the vascular endothelium, the inhibitors have to be applied at concentrations between 10 and 100 m*M* over periods of at least 30 min. Higher concentrations, in our laboratory, did not reduce further the remaining ACh-induced dilation *in vivo* but inhibited the effect of other dilators in a nonspecific manner.

It should be emphasized that NO is not only produced in endothelial cells but also in nerve tissue as well as in macrophages and vascular smooth muscle under conditions of inflammation. Different types of NO synthase (NOS) are involved (15) as discussed elsewhere in this volume. In smooth muscle and macrophages NOS (iNOS) is induced on exposure to certain cytokines. Perfusates and media may contain lipopolysaccharides or cytokines in amounts sufficient to induce iNOS within 4–6 hr. To avoid this, the use of dexamethasone is recommended by some authors (26).

Production of NO by iNOS can affect vascular tone decisively. With the NOS inhibitors presently used, it is difficult to inhibit a particular NOS selectively and thereby to separate the different NO sources and their role in certain vascular reactions. In the brain the problem is even more pronounced. Nitric oxide can be potentially derived from the vascular endothelium, perivascular nerve endings (mainly in larger vessels), and from neuronal tissue (10). At present it is difficult to decide whether all dilations of brain vessels that can be inhibited by L-arginine analogs are endothelial in nature. It has been shown that NMDA-dependent dilation is mediated by neuron-derived rather than by endothelium-derived NO (27). Some NOS inhibitors with preferential selectivity for one of the three subtypes of NOS have been described. The selectivity is usually only 10- to 20-fold, in certain cases up to 100-fold. Anyhow, some compounds such as 7-nitroindazole reduce the neuronal NOS (nNOS) activity by up to 60% *in vivo* without affecting blood pressure, which suggests that endothelial NOS (eNOS) is only minimally affected. Table I lists NOS inhibitors with a certain selectivity for nNOS or iNOS.

Selective Local Endothelial Damage

A method to damage the endothelium locally has been described by Rosenblum and colleagues (17). Briefly, a laser spot [10–40 μm; HeNe (helium–neon), 6 mW] is directed toward an arteriole through the microscope lens. This beam is innocuous unless a 0.5% solution of Evans Blue (25 mg/kg) in normal saline is injected intravenously. Most likely, the dye acts as an energy-absorbing target at the plasma endothelium interface. Short exposures to the laser light (up to 20 sec) had no demonstrable morphological effects but eliminated vasodilator responses to ACh and bradykinin in rat pial arterioles (17). Longer exposure times produced visible injuries and platelet adhesion.

TABLE I   Inhibitors of Nitric Oxide Synthase with Predominant Effects on Certain Isoforms

Substance	Selectivity against	Inhibitory doses	System tested	Ref.[a]
Preferential inhibition of nNOS				
7-Nitroindazole	eNOS	20 mg/kg i.p.	Rats *in vivo* } nNOS ↓ /no RR ↑	(1)
7-Nitroindazole	eNOS	50 mg/kg i.p.	Rats *in vivo*	(2)
S-Methyl-L-thiocitrulline	eNOS (10×)	k_i 1.2 vs. 11 μM	Human NOS	(3)
	eNOS (17×)	IC_{50} 0.3 vs. 5.4 μM	Rat: brain slices/aorta	(3)
S-ethyl-L-thiocitrulline	eNOS (50×)	k_i 0.5 vs. 24 μM	Human NOS	(3)
Preferential inhibition of iNOS				
Aminoguanidine	eNOS (10–100×)	IC_{50} 5 vs. 160 μM	Rat aorta	(4)
	eNOS		Rat aorta (no effect on endothelial NO)	(5)
L-N^6-Iminolysine	nNOS (28×)	IC_{50} 3 vs. 92 μM	Mouse iNOS/rat nNOS	(6)
Isothioureas	eNOS (2–190×)	k_i 0.05 vs. 9 μM	Human NOS preparation	(7)

[a] *Key to References:* (1) B. P. Connop, N. G. Rolfe, R. J. Boegman, K. Jhamandas, and R. J. Beninger, *Neuropharmacology* **33**, 1445 (1994); (2) T. Yoshida, V. Limmroth, K. Irikura, and M. A. Moskowitz, *J. Cereb. Blood Flow Metab.* **14**, 929 (1994); (3) E. S. Furfine, M. F. Harmon, J. E. Paith, R. G. Knowles, M. Salter, R. J. Kiff, C. Duffy, R. Hazelwood, J. A. Oplinger, and E. P. Garvey, *J. Biol. Chem.* **269**, (1994); (4) T. P. Misko, W. M. Moore, T. P. Kasten, G. A. Nickols, J. A. Corbett, R. G. Tilton, M. L. McDaniel, J. R. Williamson, and M. G. Currie, *Eur. J. Pharmacol.* **233**, 119 (1993); (5) G. A. Joly, M. Ayres, F. Chelly, and R. G. Kilbourn, *Biochem. Biophys. Res. Commun.* **199**, 147 (1994); (6) W. M. Moore, R. K. Webber, G. M. Jerome, F. S. Tjoeng, T. P. Misko, and M. G. Currie, *J. Med. Chem.* **37**, 3888 (1994); (7) E. P. Garvey, J. A. Oplinger, G. J. Tanoury, P. A. Sherman, M. Fowler, S. Marshall, M. F. Harmon, J. E. Paith, and E. S. Furfine, *J. Biol. Chem.* **269**, 26669 (1994).

Although the exact mechanism of laser-induced endothelial impairment is not clear this method seems to be useful as an additional test of the endothelial nature of vasomotor responses, independent of the use of NOS inhibitors or inhibitors of cyclooxygenase.

Measurements of Cyclic GMP

Nitric oxide stimulates the soluble guanylate cyclase in endothelial and smooth muscle cells, thus increasing intracellular cGMP, which is partly released from these cells. One can make use of the cGMP increasing effect of NO in order to obtain indirect (quantitative) information on endothelial NO release. In isolated saline-perfused rabbit hearts it is possible to measure a 100% increase of cGMP in the coronary effluent under endothelial stimulation with acetylcholine (1 μM), which can be blocked by pretreatment with the NOS inhibitor L-NNA (30 μM) (28). However, an acceptable signal-to-noise ratio is obtained only when superoxide dismutase (SOD) (16 U/ml) is added to the perfusate. This enzyme catalyzes the conversion of superoxide anions to H_2O_2 and thus prolongs the half-life of NO (29). Even then it is not possible to assess "basal" NO-dependent cGMP production in unstimulated hearts. This can be achieved, however, using platelets as NO detectors (30). Platelets have a high intrinsic activity of soluble guanylate cyclase and represent a natural target of luminally released NO during their passage through the microcirculation. Using washed human platelets injected into the coronary circulation of isolated rabbit hearts, it is possible to show NO-dependent increases of cGMP (radioimmunoassay) in these platelets not only on endothelial stimulation with ACh but also after their passage through the pharmacologically nonstimulated microcirculation (30). The platelets are prepared as follows: Blood from healthy human donors is collected into 3.8% sodium citrate (20% of final volume). Platelet rich plasma (PRP) is obtained by centrifugation at 200 g for 20 min. After administration of acid citrate–dextrose (20 vol%) and prostacyclin (300 ng/ml), the PRP is centrifuged again and the pellet finally suspended in a modified Tyrode's solution containing indomethacin (1 μM) to obtain a final platelet count of 2–3 $\times$ 10^6 platelets/ml. Experiments should finish within 180 min after the end of platelet preparation.

We also tested whether platelet cGMP levels could be used as an NO detection system in the blood-perfused microcirculation of rabbit hindlimbs, taking venous blood samples during local intraarterial infusion of ACh (0.3 mM, 0.2 ml/min). We found significant increases of cGMP (from 0.7 to 0.9 pmol/mg protein) under maximal stimulation of the endothelium with acetylcholine, but there was a considerable "noise," produced by nonspecific stimulation of the platelets during their isolation from the venous blood samples (900 μl). An isolation is necessary because the plasma contains considerable amounts of cGMP (6 pmol/ml plasma in rabbits) that is released from endothelial and possibly other cells on stimulation with ANP or CNP and therefore is not related to NO activity. For isolation, the blood was spun at 200 g for 10 min at 3°C in the presence of sodium citrate and the PDE inhibitor zaprinast. The supernatant containing platelet-rich plasma was then spun at 800 g for 15 min after dilution with the ACD stabilizer (4/1, v/v).

To avoid the isolation procedure and hence the noise, cGMP-dependent protein phosphorylation in platelets could be shown rather than cGMP levels themselves, thus avoiding time-consuming isolation of platelets from the plasma. The platelet protein VASP is phosphorylated in a cAMP- and cGMP-dependent manner during passage through the saline-perfused coronary vascular bed (28). Because the phosphorylation produces an apparent shift of the protein band from 46 to 50 kDa in the Western blot, it is possible to detect this phosphorylation without radioactive labeling of the platelets (31). Again, in the whole blood this method cannot be used without isolation of platelets because plasma protein bands mask the band of phosphorylated VASP. Taken together, the measurement of cGMP release from the endothelium as well as of cGMP changes in platelets is a useful way to detect the NO production of the endothelium of saline-perfused organs. Even a quantitative approach is possible. However, due to the necessary isolation procedures which nonspecifically increase platelet cGMP and due to the need for relatively large blood sample volumes (about 1 ml) this method is not feasible in the microcirculation of small animals *in vivo*.

Photometric Detection: Hemoglobin Assay

Another technique to detect NO released from the whole vasculature can also only be used in saline-perfused organs. Kelm and Schrader made use of the previously described fact that oxyhemoglobin is rapidly oxidized to methemoglobin by NO. By measurement of the extinction difference between the absorption maximum of methemoglobin (401 nm) and the isosbestic point (411 nm) in a flow-through cell of a double-beam spectrophotometer, NO concentrations can be calculated after determination of the extinction coefficient. This method allows kinetic measurements by use of the flow-through cell. Using this method it was possible to detect NO concentrations as low as 20 nM in the coronary effluent of isolated guinea pig hearts, which were due to the "basal" NO release from the coronary endothelium. The threshold for detection is around 1 nM NO (32). To obtain an acceptable signal-to-noise ratio, relatively low hemoglobin (Hb) concentrations have to be used (~4 μM). Commercially available hemoglobin is at least 50% methemoglobin and must be reduced with a 10-fold molar excess of dithionite. Dithionite is later removed by dialysis against N_2-saturated distilled water at 4°C. It should be noted that Hb in millimolar concentration even reduces the vasomotor responses to ACh and Bk in isolated organs or vessels. In contrast to *in vivo* conditions, Hb penetrates the wall of saline-perfused vessels and locally scavenges NO before it reaches the smooth muscle cell.

Bioassay

A relatively simple and sensitive method to detect NO is the bioassay technique. Effluent from a perfused donor tissue (vascular ring or intact organ) is superfused over a detector, usually an endothelium-denuded vascular ring. By adjustment of the superfusion rate and selection of a short transit time between donor and detector, even the low amount of NO released from a vascular ring [concentration range of NO in the effluent is 5–20 nM at a 4 ml/min superfusion rate as measured simultaneously by chemiluminescence (33)] can be detected by a dilation of the detector ring. The best sensitivity can be obtained by very short transit times (<10 sec) and by addition of SOD to the perfusate (see above) in order to stabilize NO. Unfortunately, the effluent of an autoperfused microcirculatory preparation is normally limited and the perfusion rate cannot be varied, rendering this technique inpracticable for *in vivo* conditions.

Kontos and colleagues described a bioassay technique for the detection of NO in cerebral microvessels (18). Anesthetized cats were equipped with two cranial windows. In the donor window NO production was elicited by superfusion with ACh (10^{-7} M) at a constant superfusion rate of 1 ml/min. The effluent from the donor window potentially containing NO that had diffused out of the vessels to the surface was directed through the assay window via a thin tubing allowing a transit time of 6 sec. The vessels of the assay windows were pretreated with topical atropine (2 mg/ml) to exclude direct effects of ACh on them. Under these conditions, the effluent from the donor window induced a dilation which was not affected by indomethacin. Since it was attenuated or abolished at longer transit times (indicating an unstable factor) and by addition of hemoglobin there was indirect evidence that the transferable dilating factor was NO. However, this technique does not allow to identify the origin of NO which may not only be derived from the endothelium of brain vessels but also from neuronal tissue.

Nitric Oxide Sensors

Electrochemical detection is a standard technique for the measurement of oxygen in a biological specimen. Because this technique should be applicable to the measurements of other gases as well, Clark-type electrodes were modified to measure nitric oxide instead of oxygen. Several of these electrodes have been described (34–36). Modified versions are commercially available. In principle, NO is oxidized at the anode at a positive voltage of 0.5 to 0.9 V. The resulting current (in the range of pA) is proportional to the amount of oxidized NO molecules. To exclude other oxidizable compounds such as catecholamines, the electrode must be coated with a gas-permeable membrane. However, NO_2^- can then still be detected, especially at voltages above 0.9 V. Therefore the measurements have to be interpreted with caution. The detection limit of the electrodes is around 10 pmol NO, which in a usual sample corresponds to a concentration of 0.5 μM. This is not sufficient to measure basal NO release in isolated vessels or cultured cells. In most publications to date, stimulated NO release could also only be detected at supramaximal doses of endothelial stimulators. This does not allow study of the role of NO in the microcirculation under physiological conditions.

An NO microsensor has been described whose detection limit is about three orders of magnitude better than that of a modified Clark electrode (32). A carbon fiber is coated with polymeric metalloporphyrin, which catalyzes the oxidation of NO to NO^+. Measurements are performed at a voltage of 0.62 V. To obtain selectivity against NO_2^-, the electrode is coated further with a thin layer of a negatively charged cation-exchange membrane (Nafion). Reportedly, this microsensor is able to measure NO release from a single cell (37). Nitric oxide has been measured in brain tissue (38), in blood (32), as well as in the effluent of saline-perfused rat lungs (39). When this electrode becomes available to more laboratories, its practicability in physiological measurements in the microcirculation can be tested. At present, the amperometric detection of NO with such a microsensor appears to be the most promising method to obtain quantitative kinetic NO signals from the microcirculation.

Other Methods

Quantitative detection of nitric oxide is possible by measuring the chemiluminescence generated by the interaction of NO with ozone ($NO + O_3 \rightarrow NO_2^* + O_2$; $NO_2^* \rightarrow NO_2 + 2$ hr). The chemiluminescence is measured by a sensible photomultiplier tube and is directly proportional to the NO level of the sample. This method is designed to measure NO gas. However, NO-containing fluids can also be used. Nitric oxide is driven into the gas phase by bubbling the solution with an inert gas which, however, gives rise to experimental errors if one wants to determine the NO release quantitatively. Moreover, as part of the NO to be detected may have already been oxidized in the biological sample, detection can be improved by a reflux technique, that is, exposure of the sample to acid and KI (potassium iodide) which induces the reduction of nitrite and nitrate to NO. Because it is not known to what extent nitrate and nitrite in biological samples are derived from endothelial NO or from other sources, a considerable error for the determination of authentic NO is possible. This is one of the reasons why chemiluminescence is not used in studies in whole blood samples but rather in saline buffers or culture medium. Chemiluminescence is an established and widely used method to study NO release from cultured cells and isolated vessels or tissue pieces. The detection limit is in the subnanomolar range (32). However, for the reasons discussed it has not been used for *in situ* determination of the NO release from vessels of the microcirculation.

Because NO has an impaired electron it is subject to electron paramagnetic resonance (EPR) assay. To stabilize NO for detection, spin traps such as nitroxide compounds or hemoglobin are necessary. Electron paramagnetic resonance is the least sensible detection method, with a detection limit of only 0.1 mM. It was, however, successfully used to measure nitrate levels in whole blood (40) of humans and in rats (41) under pathophysiological conditions with enhanced NO production.

Summary

Nitric oxide is an important signal molecule in the endothelial control of vascular tone. Therefore, it is of great interest to determine its production and release under physiological and pathophysiological conditions. There are a number of established direct and indirect methods to determine NO release. Whereas many of these methods can be easily used in saline-perfused organs the methods to determine NO in the true microcirculation are limited. At present, the most useful tools are the use of selective inhibitors of nitric oxide synthase. An adequate selectivity for one of the subtypes of nitric oxide synthase seems especially important for functional studies in the brain because nitric oxide is not only released from endothelial cells but also from neuronal tissue. A very promising method to determine nitric oxide in the microcirculation is the amperometric detection of NO by means of special microsensors. However, at present the method is not yet widely used and further experiments have to show its feasibility with respect to NO measurements in the microcirculation.

References

1. J. R. Guyton, D. T. Dao, and K. L. Lindsay, *Exp. Mol. Pathol.* **40**, 340 (1984).
2. E. Wiest, V. Trach, and J. Dammgen, *Basic Res. Cardiol.* **84**, 469 (1989).
3. S. Moncada, R. M. J. Palmer, and E. A. Higgs, *Pharmacol. Rev.* **43**, 109 (1991).
4. H. Nishimura, W. I. Rosenblum, G. H. Nelson, and S. Boynton, *Am. J. Physiol.* **261**, H15 (1991).
5. S. M. Gardiner, A. M. Compton, T. Bennett, R. M. J. Palmer, and S. Moncada, *Hypertension (Dallas)* **15**, 486 (1990).
6. U. Pohl, K. Herlan, A. Huang, and E. Bassenge, *Am. J. Physiol.* **261**, H2016 (1991).
7. A. M. Malek and S. Izumo, *J. Hypertens.* **12**, 989 (1994).
8. T. M. Griffith, D. H. Edwards, R. L. I. Davies, T. J. Harrison, and K. T. Evans, *Nature (London)* **329**, 442 (1987).
9. U. Pohl, K. Wagner, and C. de Wit, *Eur. Heart J.* **14**(Suppl. 1), 93 (1993).
10. H. Tomimoto, M. Nishimura, T. Suenaga, S. Nakamura, I. Akiguchi, H. Wakita, J. Kimura, and B. Mayer, *J. Cereb. Blood Flow Metab.* **14**, 930 (1994).
11. G. Bonvento, J. Seylaz, and P. Lacombe, *J. Cereb. Blood Flow Metab.* **14**, 699 (1994).

12. Q. Wang, D. A. Pelligrino, H. M. Koenig, and R. F. Albrecht, *J. Cereb. Blood Flow Metab.* **14,** 944 (1994).
13. P. A. Kelly, C. L. Thomas, I. M. Ritchie, and G. W. Arbuthnott, *Neuroscience (Oxford)* **59,** 13 (1994).
14. F. M. Faraci, *Am. J. Physiol.* **261,** H1038 (1991).
15. R. G. Knowles and S. Moncada, *Biochem. J.* **298,** 249 (1994).
16. R. F. Furchgott and J. V. Zawadzki, *Nature (London)* **288,** 373 (1980).
17. W. I. Rosenblum, *Stroke* **17,** 494 (1986).
18. H. A. Kontos, E. P. Wie, and J. J. Marshall, *Am. J. Physiol.* **255,** 1259 (1988).
19. J. R. Copeland, K. A. Willoughby, T. M. Tynan, S. F. Moore, and E. F. Ellis, *Am. J. Physiol.* **268,** H458 (1995).
20. H. A. Kontos, E. P. Wei, J. T. Povlishock, and C. W. Christman, *Circ. Res.* **55,** 295 (1984).
21. W. I. Rosenblum, H. Nishimura, and G. H. Nelson, *Am. J. Physiol.* **259,** 1396 (1990).
22. C. de Wit, P. von Bismarck, and U. Pohl, *Cardiovasc. Res.* **28,** 1513 (1994).
23. D. H. Maurice and R. J. Haslam, *Mol. Pharmacol.* **37,** 671 (1990).
24. I. L. O. Buxton, D. J. Cheek, D. Eckman, D. P. Westfall, K. M. Sanders, and K. D. Keef, *Circ. Res.* **72,** 387 (1993).
25. M. Hecker, J. A. Mitchell, H. J. Harris, M. Katsura, C. Thiemermann, and J. R. Vane, *Biochem. Biophys. Res. Commun.* **167,** 1037 (1990).
26. R. G. Knowles, M. Salter, S. L. Brooks, and S. Moncada, *Biochem. Biophys. Res. Commun.* **172,** 1042 (1990).
27. F. M. Faraci and K. R. Breese, *Circ. Res.* **72,** 476 (1993).
28. U. Pohl, C. Nolte, A. Bunse, M. Eigenthaler, and U. Walter, *Am. J. Physiol.* **266,** H606 (1994).
29. G. M. Rubanyi and P. M. Vanhoutte, *Am. J. Physiol.* **250,** H822 (1986).
30. U. Pohl and R. Busse, *Circ. Res.* **65,** 1798 (1989).
31. M. Hallbrügge, C. Friedrich, M. Eigenthaler, P. Schanzenbacher, and U. Walter, *J. Biol. Chem.* **265,** 3088 (1990).
32. F. L. Kiechle and T. Malinski, *Am. J. Clin. Pathol.* **100,** 567 (1993).
33. P. R. Myers, Q. Zhong, J. J. Jones, M. A. Tanner, H. R. Adams, and J. L. Parker, *Am. J. Physiol.* **268,** H955 (1995).
34. K. Shibuki, *Neurosci. Res.* **9,** 69 (1990).
35. H. Tsukahara, D. V. Gordienko, and M. S. Goligorsky, *Biochem. Biophys. Res. Commun.* **193,** 722 (1993).
36. K. Ichimori, H. Ishida, M. Fukahori, H. Nakazawa, and E. Murakami, *Rev. Sci. Instrum.* **65,** 1 (1994).
37. T. Malinski and Z. Taha, *Nature (London)* **358,** 676 (1992).
38. T. Malinski, F. Bailey, Z. G. Zhang, and M. Chopp, *J. Cereb. Blood Flow Metab.* **13,** 355 (1993).
39. D. Wang, K. Hsu, C. P. Hwang, and H. I. Chen, *Biochem. Biophys. Res. Commun.* **208,** 1016 (1995).
40. A. Wennmalm, B. Lanne, and A. Petersson, *Anal. Biochem.* **4,** 128 (1990).
41. U. Westenberger, S. Thanner, H. Ruf, K. Gersonde, G. Sutter, and O. Trentz, *Free Radical Res. Commun.* **11,** 167 (1990).

[23] Regulation of Cerebral Circulation by Nitric Oxide

Frank M. Faraci

Introduction

Nitric oxide (NO) is a highly diffusible messenger molecule that plays an important role in a diverse group of biological systems (1, 2). Nitric oxide is produced from L-arginine by the NO synthase family of enzymes (1). Three separate genes control the expression of endothelial, neuronal, and inducible (or immunologic) isoforms of NO synthase (3, 4).

Nitric oxide plays an important role in the regulation of the cerebral circulation (5, 6). All three genes for NO synthase can be expressed in the wall of cerebral vessels. Endothelial and neuronal NO synthases are expressed constitutively in endothelium and in NO synthase-containing neurons in the adventitia, respectively. Inducible NO synthase can be expressed in cerebral endothelium and presumably in vascular muscle in response to stimuli such as lipopolysaccharide and some cytokines. In addition, NO synthase in parenchymal neurons, and perhaps glial cells, can affect the cerebral circulation.

The goal of this chapter is to summarize the role of NO in the regulation of cerebral vascular tone. Emphasis will be placed on functions of NO under physiological conditions and, when data are available, on molecular biology of NO synthase in cerebral vessels.

Sources of Nitric Oxide

Cerebral blood vessels are normally exposed to NO from both endothelial and neuronal sources. The presence of endothelial NO synthase messenger RNA (mRNA) and protein has been demonstrated using *in situ* hybridization (7) and immunocytochemistry (8). A series of studies performed both *in vitro* and *in vivo* have provided functional evidence for the presence of endothelial NO synthase in cerebral vessels (5, 6). Nitric oxide synthase is also present in perivascular fibers in the adventitia of large cerebral vessels (8, 9) and in parenchymal neurons, some of which are in close association with parenchymal microvessels (5). In addition to neurons and endothelium, there may be low levels of constitutive NO synthase in glial cells (10). In some disease states or in response to brain injury, the gene for inducible NO synthase can

Methods in Neurosciences, Volume 31

be expressed in glia including astrocytes and microglia (10–12). Induction of NO synthase in glia is a major source of NO that potentially can influence cerebral vascular tone.

Methods to Study Role of Nitric Oxide

The role of NO in regulation of the cerebral circulation has been examined by measuring contraction or relaxation of cerebral arteries or arterioles, changes in diameter of vessels on the surface of the brain using a cranial window *in vivo*, and changes in cerebral blood flow.

Most studies have examined responses to either application of NO (or more commonly a nitrovasodilator) to cerebral vessels or treatment with pharmacological inhibitors of NO synthase, including N^G-monomethyl-L-arginine (L-NMMA), N^G-nitro-L-arginine (L-NNA), or its methyl ester (L-NAME) (5, 6). In addition, a few studies have determined activity of NO synthase (by measurement of conversion of L-arginine to NO and L-citrulline) in cerebral vessels (13, 14) or extracellular accumulation of nitrite (a chemical breakdown product of NO) (15, 16). Nitric oxide produces relaxation of vascular muscle by activation of soluble guanylate cyclase which produces cyclic GMP. Thus, an additional approach to obtain an index of activity of NO synthase has been to measure NO-dependent accumulation of cyclic GMP intracellularly or in extracellular fluid (5, 17, 18).

Basal Production of Nitric Oxide

Nitric oxide is a potent dilator of cerebral blood vessels (5, 19). Under control conditions, most studies indicate that basal production of nitric oxide exerts a tonic dilator influence in the cerebral circulation. Inhibition of production of nitric oxide produces contraction of cerebral arteries and arterioles (5) and reduces cerebral blood flow in a variety of species (5, 20). Although more than one cell type could potentially produce nitric oxide that influences basal cerebral vascular tone, some evidence suggests that endothelium is the major source (5).

Evidence suggests that the endothelium in cerebral arterioles produces nitric oxide in response to increases in shear stress (21). Thus, acute changes in shear stress may be a key determinant of basal production of nitric oxide in cerebral vessels (Fig. 1). In addition, the 5′ flanking region of the endothelial NO synthase gene contains a shear stress-responsive element (5′-GAGACC-3′) (22) and sustained exposure of endothelium in culture to shear stress increases both mRNA and protein expression for endothelial NO synthase

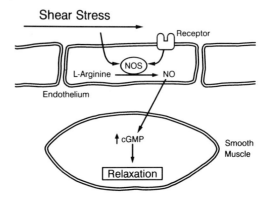

FIG. 1   Synthesis of nitric oxide (NO) or endothelium-derived relaxing factor in endothelium of cerebral blood vessels. Nitric oxide synthase (NOS) may be activated in response to shear stress and receptor-mediated agonists. Once produced, NO diffuses to vascular muscle and activates guanylate cyclase, which increases levels of cyclic GMP (cGMP) causing relaxation.

(23). These findings suggest that shear stress may have a chronic influence on vascular tone by regulating transcription of the endothelial NO synthase gene.

Ablumenal release of NO influences vascular tone. In addition, bioassay studies demonstrate that endothelium also releases NO into the bloodstream (5, 24). This lumenal release of NO probably protects cerebral endothelium by inhibiting aggregation of platelets and adhesion of leukocytes (25, 26).

Endothelium-Dependent Relaxation

By releasing vasoactive factors, endothelium is an important regulator of vascular tone (5). Nitric oxide, or possibly a closely related substance (27), is a major mediator of endothelium-dependent relaxation in cerebral blood vessels (5). Production of NO by endothelium is increased above basal levels in response to a diverse group of agonists (Fig. 1). Activity of endothelial NO synthase is calcium-dependent and activation of several receptors increases intracellular levels of calcium which stimulates enzyme activity. Receptor-mediated agonists which produce endothelium-dependent (NO-dependent) relaxation of cerebral arteries include acetylcholine, substance P, arginine vasopressin, bradykinin, oxytocin, endothelin (via activation of endothelin-B receptors), histamine, and ATP (5, 28). The calcium ionophore, A23187, also produces NO-dependent relaxation of large cerebral arteries (5). A similar mechanism of endothelium-dependent relaxation has been described in human cerebral arteries (29).

   In addition to exogenous inhibitors of NO synthase such as L-NNA and L-NAME, enzymatic formation of NO in cerebral vessels can also be inhibited by N^G,N^G-dimethyl-L-arginine (asymmetric dimethylarginine, ADMA) (30). ADMA is an endogenous substance which is normally distributed throughout the brain (31). Physiologically relevant concentrations of ADMA inhibit basal production of nitric oxide and relaxation of cerebral blood vessels in response to acetylcholine (Fig. 2). ADMA is also a potent inhibitor of activity of brain NO synthase (30). Thus, ADMA may be an important endogenous modulator of activity of brain NO synthase and cerebral vascular tone.

Hypercapnia

   Hypercapnia produces marked increases in cerebral blood flow which appears to be dependent on formation of nitric oxide in several species (5, 6, 20, 32). The precise mechanism which accounts for this dependency is not clear, however. One possibility is that NO is the mediator of relaxation of vascular muscle during hypercapnia. This possibility is supported by the observation that acidosis (which occurs during hypercapnia) increases activity of brain NO synthase (33) and increases cyclic GMP in brain (34). A second possibility is that normal basal levels of NO (or cyclic GMP) are required for vasodilatation during hypercapnia to occur (35). Although vaso-

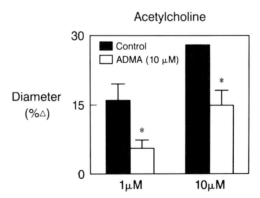

FIG. 2   Change in diameter of the basilar artery in the rat in response to acetylcholine in the absence and presence of N^G,N^G-dimethyl-L-arginine (asymmetric dimethylarginine, ADMA), an endogenous inhibitor of NO synthase. Values are means ± SE. *, $p < 0.05$ versus control response. Data are redrawn from Faraci *et al.* (30).

dilatation during hypercapnia is at least partially dependent on the production of NO, the cellular source of NO is not known (5).

Neuronal Activation

Studies suggest that NO mediates increases in cerebral blood flow in response to increases in neuronal activity and metabolism. Several experimental models support this concept. First, activation of receptors for glutamate, the predominant excitatory amino acid in the mammalian brain, activates NO synthase and causes neurons to release NO extracellularly (36). Dilatation of cerebral arterioles and increases in cerebral blood flow in response to glutamate, and the glutamate analogs N-methyl-D-aspartate and kainate, are dependent on formation of NO (37–40). Studies using 7-nitroindazole, a relatively selective inhibitor of neuronal NO synthase, suggest that NO that produces vasodilatation in response to activation of glutamate receptors is neuronally derived (41, 42). Second, local increases in cerebral blood flow in response to activation of glutamatergic pathways are also mediated by NO (43, 44). Third, cerebral vasodilatation during seizures, which involves endogenous release of excitatory amino acids, is also dependent on activity of NO synthase (45, 46). Fourth, cerebral vasodilatation during cortical spreading depression, which appears to involve activation of glutamate receptors, is attenuated by inhibitors of NO synthase (47, 48). Thus, neuronally derived NO appears to mediate increases in local blood flow during increases in neuronal activity in response to several stimuli including excitatory amino acids and seizures. Although NO may mediate increases in blood flow in response to increases in metabolism, NO itself does not affect oxygen consumption or glucose utilization in brain (6, 49).

Inducible Nitric Oxide Synthase

Inducible NO synthase is a gene product which, in contrast to endothelial and neuronal isoforms, is not expressed under basal conditions in most cells (4). Expression of inducible NO synthase may occur in response to proinflammatory factors such as lipopolysaccharide and certain cytokines (3, 4). Expression of mRNA for inducible NO synthase has been detected in response to proinflammatory factors in cerebral endothelium (50–52), neurons (53), and glia (10–12) using either Northern analysis or reverse transcription coupled with the polymerase chain reaction (RT-PCR).

The gene for inducible NO synthase is also expressed in brain following ischemia (54–56), during meningitis (57), rabies (58), and encephalitis (59).

Inducible NO synthase has also been detected in cerebral microvessels from patients with Alzheimer's disease (13).

Expression of inducible NO synthase in brain may affect cerebral vascular tone and permeability. Local application of lipopolysaccharide causes marked, progressive dilatation of cerebral arterioles which is reduced substantially by inhibitors of NO synthase including aminoguanidine (17, 60). Cerebral vasodilatation that occurs during experimental meningitis is also dependent on production of NO (15). Increases in arteriolar diameter in response to lipopolysaccharide are attenuated by pretreatment with dexamethasone (60), which inhibits expression of the inducible NO synthase gene (3). Expression of inducible NO synthase may contribute to increased permeability of the blood–brain barrier that occurs in response to lipopolysaccharide and during meningitis (14).

Molecular Biology

Some studies have begun using molecular approaches which provide insight into the role of NO in cerebral vascular biology. This includes detection of changes in levels of mRNA for NO synthases using *in situ* hybridization (57), Northern analysis or RT-PCR (50, 51, 55, 61), and change in protein levels using Western blotting (62).

A potentially valuable approach to study the role of NO is the use of models which either overexpress or do not express one of the NO synthase genes. Genetic models have been developed which are deficient in expression of the genes for neuronal (63) or inducible (64) NO synthase (NO synthase knockouts). The role of neuronal NO synthase during focal ischemia has been examined in mice that are deficient in neuronal NO synthase (63).

It may be of value to overexpress NO synthase in transgenic animals or by using gene transfer techniques. The relatively recently developed method of gene transfer to cerebral vessels using adenoviral vectors (65) may prove useful in future studies of the role of NO synthase in the cerebral circulation.

Acknowledgments

The authors thanks Dr. Donald Heistad for critical review of the manuscript. Original studies cited in this review were supported by National Institutes of Health Grants HL-38901 and NS-24621 and by a Grant-in-Aid from the American Heart Association (95014510). F. M. Faraci is an Established Investigator of the American Heart Association.

References

1. R. G. Knowles and S. Moncada, *Biochem. J.* **298,** 249 (1994).
2. S. Moncada and A. Higgs, *N. Engl. J. Med.* **329,** 2002 (1993).
3. S. M. Morris and T. R. Billiar, *Am. J. Physiol.* **266,** E829 (1994).
4. C. Nathan and Q. Xie, *Cell (Cambridge, Mass.)* **78,** 915 (1994).
5. F. M. Faraci and J. E. Brian, *Stroke* **25,** 692 (1994).
6. C. Iadecola, D. A. Pelligrino, M. A. Moskowitx, and N. A. Lassen, *J. Cereb. Blood Flow Metab.* **14,** 175 (1994).
7. P. A. Marsden, K.T. Schappert, H. S. Chen, M. Flowers, C. L. Sundell, J. N. Wilcox, S. Lamas, and T. Michel, *FEBS Lett.* **307,** 287 (1992).
8. D. S. Bredt, P. M. Hwang, and S. H. Snyder, *Nature (London)* **347,** 768 (1990).
9. K. Nozaki, M. A. Moskowitz, K. I. Koketsu, T. M. Dawson, D. S. Bredt, and S. H. Snyder, *J. Cereb. Blood Flow Metab.* **13,** 70 (1993).
10. S. Murphy, M. L. Simmons, L. Agullo, A. Garcia, D. L. Feinstein, E. Galea, D. J. Reis, D. Minc-Golomb, and J. P. Schwartz, *Trends Neurosci.* **16,** 323 (1993).
11. E. Galea, D. J. Reis, and D. L. Feinstein, *J. Neurosci. Res.* **37,** 406 (1994).
12. H. Fujisawa, T. Ogura, A. Hokari, A. Weisz, J. Yamashita, and H. Esumi, *J. Neurochem.* **64,** 85 (1995).
13. M. A. Dorheim, W. R. Tracey, J. S. Pollock, and P. Grammas, *Biochem. Biophys. Res. Commun.* **205,** 659 (1994).
14. K. M. K. Boje, *Eur. J. Pharmacol.* **272,** 297 (1995).
15. U. Koedel, Bernatowicz, R. Paul, K. Frei, A. Fontana, and H.-W. Pfister, *Ann. Neurol.* **37,** 313 (1995).
16. R. C. Kukreja, E. P. Wei, H. A. Kontos, and J. N. Bates, *Stroke* **24,** 2010 (1993).
17. Brian, Jr., J. E., D. D. Heistad, and F. M. Faraci, *Am. J. Physiol.* **269,** H783 (1995).
18. W. M. Armstead, *Am. J. Physiol.* **268,** H1436 (1995).
19. J. E. Brian, D. D. Heistad, and F. M. Faraci, *Stroke* **25,** 639 (1994).
20. R. W. McPherson, J. R. Kirsch, R. F. Ghaly, and R. J. Traytsman, *Stroke* **26,** 682 (1995).
21. A. C. Ngai and H. R. Winn, *Circ. Res.* **77,** 832 (1995).
22. K. Miyahara, T. Kawamoto, K. Sase, Y. Yui, K. Toda, L.-X. Yang, R. Hattori, T. Aoyama, Y. Yamamoto, Y. Doi, S. Ogoshi, K. Hashimoto, C. Kawai, S. Sasayama, and Y. Shizuta, *Eur. J. Biochem.* **223,** 719 (1994).
23. K. Nishida, D. G. Harrison, J. P. Navas, A. A. Fisher, S. P. Dockery, M. Uematsu, R. M. Nerem, R. W. Alexander, and T. J. Murphy, *J. Clin. Invest.* **90,** 2092 (1992).
24. R. Paterno, F. M. Faraci, and D. D. Heistad, *Stroke* **25,** 2459 (1994).
25. H. Nishimura, W. I. Rosenblum, G. H. Nelson, and S. Boynton, *Am. J. Physiol.* **261,** *H15* (1991).
26. W. I. Rosenblum, H. Nishimura, E. F. Ellis, and G. H. Nelson, *J. Cereb. Blood Flow Metab.* **12,** 703 (1992).
27. H. A. Kontos, *Stroke* **24**(Suppl. 1), I-155 (1993).

28. T. Kitazono, D. D. Heistad, and F. M. Faraci, *J. Pharmacol. Exp. Ther.* **273,** 1 (1995).
29. H. Onoue, N. Kaito, M. Tomii, S. Tokudome, M. Nakajima, and T. Abe, *Am. J. Physiol.* **267,** H880 (1994).
30. F. M. Faraci, J. E. Brian, and D. D. Heistad, *Am. J. Physiol.* in press.
31. S. Ueno, A. Sano, K. Kotani, K. Kondoh, and Y. Kakımoto, *J. Neurochem.* **59,** 2012 (1992).
32. F. M. Faraci, K. R. Breese, and D. D. Heistad, *Stroke* **25,** 1679 (1994).
33. M. Hecker, A. Mulsch, and R. Bussi, *J. Neurochem.* **62,** 1524 (1994).
34. K. Irikura, K. I. Maynard, W. S. Lee, and M. A. Moskowitz, *Am. J. Physiol.* **267,** H837 (1994).
35. C. Iadecola, F. Zhang, and X. Xu, *Am. J. Physiol.* **267,** R228 (1994).
36. J. Garthwaite and C. L. Boulton, *Annu. Rev. Physiol.* **57,** 683 (1995).
37. F. M. Faraci and K. R. Breese, *Circ. Res.* **72,** 476 (1993).
38. W. Meng, J. R. Tobin, and D. W. Busija, *Stroke* **26,** 857 (1995).
39. F. J. Northington, J. R. Tobin, R. C. Koehler, and R. J. Traystman, *Am. J. Physiol.* **269,** H215.
40. F. M. Faraci, K. R. Breese, and D. D. Heistad, *Stroke* **25,** 2080 (1994).
41. F. M. Faraci and J. E. Brian, *Stroke* **26,** 2172 (1995).
42. F. Zhang, S. Xu, and C. Iadecola, *Neuroscience (Oxford)* **69,** 1195 (1995).
43. N. Akgoren, M. Fabricius, and M. Lauritzen, *Proc. Natl. Acad. Sci. U.S.A.* **91,** 5903 (1994).
44. J. Li and C. Iadecola, *Neuropharmacology* **33,** 1453 (1994).
45. F. M. Faraci, K. R. Breese, and D. D. Heistad, *Am. J. Physiol.* **265,** H2209 (1993).
46. A. P. De Vasconcelos, R. A. Baldwin, and C. G. Wasterlain, *Proc. Natl. Acad. Sci. U.S.A.* **92,** 3175 (1995).
47. D. M. Colonna, W. Meng, D. D. Deal, and D. W. Busija, *Stroke* **25,** 2463 (1994).
48. M. Wahl, L. Schilling, A. A. Parsons, and A. Kaumann, *Brain Res.* **637,** 204 (1994).
49. S. Takahashi, M. Cook, J. Jehle, C. Kennedy, and L. Sokoloff, *J. Neurochem.* **65,** 414 (1995).
50. I. P. Oswald, I. Eltoum, T. A. Wynn, B. Schwartz, P. Caspar, D. Paulin, A. Sher, and S. L. James, *Proc. Natl. Acad. Sci. U.S.A.* **91,** 999 (1994).
51. M. Bereta, J. Bereta, I. Georgoff, F. D. Coffman, S. Cohen, and M. C. Cohen, *Exp. Cell Res.* **212,** 230 (1994).
52. J. Murata, S. B. Corradin, R. C. Janzer, and L. Juillerat-Jeanneret, *Int. J. Cancer* **59,** 699 (1994).
53. D. Minc-Golomb, I. Tsarfaty, and J. P. Schwartz, *Br. J. Pharmacol.* **112,** 720 (1994).
54. C. Iadecola, F. Zhang, and X. Xu, *Am. J. Physiol.* **268,** R286 (1995).
55. C. Iadecola, F. Zhang, S. Xu, R. Casey, and M. E. Ross, *J. Cereb. Blood Flow Metab.* **15,** 378 (1995).
56. C. Iadecola, X. Xu, F. Zhang, E. E. El-Fakahany, and M. E. Ross, *J. Cereb. Blood Flow Metab.* **15,** 52 (1995).
57. I. L. Campbell, A. Samimi, and C. S. Chiang, *J. Immunol.* **153,** 3622 (1994).
58. A. M. Van Dam, J. Bauer, W. K. M. Man-A-Hing, C. Marquette, F. J. H. Tilders, and F. Berkenbosch, *J. Neurosci. Res.* **40,** 251 (1995).

59. M. I. Bukrinsky, H. S. L. M. Nottet, and H. Schmidtmayerova, *J. Exp. Med.* **181,** 735 (1995).
60. J. E. Brian, Jr., D. D. Heistad, and F. M. Faraci, *Stroke* **26,** 277 (1995).
61. Z. G. Zhang, M. Chopp, and S. Gautam, *Brain Res.* **654,** 85 (1994).
62. H. Kasuya, B. K. A. Weir, M. Nakane, J. S. Pollock, J. Johns, L. S. Marton, and K. Stefansson, *J. Neurosurg.* **82,** 250 (1995).
63. Z. Huang, P. L. Huang, N. Panahian, T. Dalkara, M. C. Fishman, and M. A. Moskowitz, *Science* **265,** 1883 (1994).
64. J. D. MacMicking, C. Nathan, G. Hom, N. Chartrain, D. S. Fletcher, M. Trumbruer, K. Stevens, Q.-W. Xie, K. Sokol, N. Hutchinson, H. Chen, and J. S. Mudgett, *Cell (Cambridge, Mass.)* **81,** 641 (1995).
65. H. Ooboshi, M. J. Welsh, C. D. Rios, B. L. Davidson, and D. D. Heistad, *Circ. Res.* **77,** 7 (1995).

[24] Nitric Oxide in NAD/NADH-Dependent Protein Modification

Bernhard Brüne and Eduardo G. Lapetina

Introduction

The role of nitric oxide (NO) as a biological messenger in both physiological and pathophysiological pathways has only recently been established. Nitric oxide is unique among biological messengers because it is a free radical and a gas with known activities such as an endothelium-derived relaxing factor (EDRF)-like function, neuronal signaling, and immune-mediated cytotoxicity. In mammalian cells, NO is produced by the enzyme NO synthase (NOS) in response to a variety of signals. Various constitutive and inducible isoenzymes are responsible for the conversion of L-arginine to NO and stoichiometric amounts of citrulline. Competitive inhibitors of NOS such as N^G-monomethyl-L-arginine (L-NMA), have been used in pharmacological interventions involving NO-mediated signaling pathways (1, 2).

Intracellular signaling pathways can be categorized as either cGMP-dependent or cGMP-independent. Interactions of NO with transition metals, oxygen, or superoxide give rise to the formation of various metal–NO adducts, NO_x, and peroxynitrite ($ONOO^-$), respectively. Additional nitrosative reactions at nucleophilic centers, involving thiol groups, explain the propensity of S-nitrosothiol formation. Target interactions via redox chemistry and additive chemistry consist of both covalent modification of proteins and oxidation events that do not involve attachment of the NO group. Accordingly, metal- and thiol-containing proteins serve as major physiological and pathophysiological targets for NO (1, 2).

Nitric oxide signaling unrelated to soluble guanylyl cyclase activation and cGMP formation involves S-nitrosylation of proteins and an ADP-ribosylation (like) reaction, both of which rely on the covalent modification of the glycolytic enzyme glyceraldehyde-3-phosphate dehydrogenase (GAPDH), a likely function of the intracellular prevalence of the latter (3).

Both exogenously derived NO [application of NO donors like sodium nitroprusside (SNP), and 3-morpholinosydnonimine (SIN-1)] or endogenously generated NO (stimulation of the cytokine inducible NOS) are known to mediate NAD^+-dependent posttranslational modification of a 39-kDa cytosolic protein. This was evident in several studies in which NO was used in the presence of [^{32}P]NAD^+ in human platelet cytosol (4), in various rat tissues, and in rat brain (5, 6). Isolation of the radiolabeled 39-kDa protein and amino-terminal sequence analysis revealed its identity with GAPDH (7, 8).

Evidence suggests that S-nitrosylation of the active-site thiol of the protein is a prerequisite for subsequent modification with NAD^+. Nitric oxide-group transfer chemistry is an important aspect of the initial step of this mechanism (9). Furthermore, the reaction with nucleotide is more easily rationalized if NADH, rather than NAD^+ is the cosubstrate. Reduction of nicotinamide makes it susceptible to nitrosative attack. Therefore, probing GAPDH modification in the presence of [^{32}P]NADH gives rise to more pronounced incorporation of radioactivity and enzyme inhibition. It is likely that transnitrosation from active-site RS-NO (protein S-nitrosothiol) to NADH would facilitate protein thiolate attack on the nucleotide, leading to efficient protein modification (10). Although the reaction was originally addressed as an ADP-ribosylation mechanism, some evidence argues for the binding of a whole molecule of [^{32}P]NAD^+ during the NO-stimulated, NAD^+-dependent modification of GAPDH (see Ref. 11 for references).

In this chapter we briefly describe the experimental conditions we have used to investigate [^{32}P]NAD(H)-dependent labeling of cytosolic proteins, including GAPDH. Experimental conditions for establishing thiol-mediated modifications of proteins are described as well. We review experimental evidence for S-nitrosylation during covalent enzyme modification and the identification of the amino acid that is used as a substrate.

[^{32}P]NAD(H)-Dependent Modification of Cytosolic Proteins Including Glyceraldehyde-3-Phosphate Dehydrogenase

Cytosolic proteins (up to 75 μg protein/assay) from individual cells or various tissues can be used to investigate [^{32}P]NAD^+-dependent, NO-induced protein modification. Cultured cells are scraped off the culture dishes, collected by centrifugation, and resuspended in 100 mM HEPES (pH 7.5) and 2 mM dithiothreitol (DTT). Cell lysis is accomplished via hypotonic buffer shock treatment followed by several rounds of freeze–thawing, nitrogen cavitation, or ultrasonication [cells are gassed with nitrogen for 5–10 min and then sonicated 10 sec each with 1-min cooling intervals in an ice-water bath using a Branson sonifier (Branson Ultrasonic, Danbury, CT)]. The cytosolic fraction from different tissues also can be used (12). Perfusion of the organ or extensive washing it recommended to remove erythrocytes. Homogenization in 5 volumes of 100 mM HEPES buffer (pH 7.5), 2 mM dithiothreitol, and 100 μg/ml phenylmethylsulfonyl fluoride (PMSF) followed by centrifugation (15 min at 10,000 g at 4°C, followed by 30 min at 100,000 g at 4°C) enables recovery of the proteins of interest.

Labeling of cytosolic proteins (up to 75 μg protein/assay) is carried out in 100 mM HEPES buffer (pH 7.4) containing 1 to 10 μM NAD$^+$ and 0.5 to 1 μCi [^{32}P]NAD$^+$ in a total volume of 100 μl. The application of NO-donors and thiols is described below. Because protein labeling is virtually independent of the buffer system used, no particular system is required. However, incorporation of radioactivity increases at more alkaline pH values (see Ref. 3 for references). Assays are run for 30 to 90 min at 37°C, after which 20% of ice-cold trichloroacetic acid is added. Samples are left on ice for 30 min. Then protein is collected by centrifugation (10,000 g, 10 min at 4°C). Pellets are washed twice with 1 ml cold, water-saturated ether, each time followed by a centrifugation step (10,000 g, 10 min at 4°C). Residual traces of ether are removed after the last centrifugation by heating samples at 37°C for 15 min. Protein is suspended in dithiothreitol-reduced SDS–PAGE sample buffer (20 to 60 μl), denatured at 95°C for 5 min, and subjected to electrophoretic separation in 10% SDS–polyacrylamide gels along with molecular weight markers, preferentially employing radioactive and colored ones (Amersham, Arlington Heights, IL). Gels are fixed, stained, and destained, which simultaneously removes unbound radioactivity moving in or slightly behind the running front of the gel. When the time is limited, gels are dried directly without fixation. As unbound radioactivity does not interfere with the specific labeling of proteins like GAPDH, both protocols are reasonable.

Protein modification using the reduced form of the nucleotide basically follows the same protocol. Under these conditions we used [^{32}P]NADH (120000 cpm/assay) in the presence of 10 μM cold NADH. Because GAPDH-labeling with [^{32}P]NADH is much faster than labeling with [^{32}P]NAD$^+$, incubation times can be reduced to less than 30 min.

[^{32}P]NAD$^+$/[^{32}P]NADH-labeled proteins are detected using the phosphor imager system (Molecular Dynamics, Krefeld, Germany) (13). In this case, 0.5 to 1 μCi of radioactivity results in significant labeling, when gels are subjected to the exposure plates for 12–24 hr. When traditional autoradiography is used, we recommend 1 μCi of [^{32}P]NAD$^+$ and longer exposure times, namely 24–48 hr. For the detection of proteins other than GAPDH, exposure times have to be prolonged, sometimes up to 4 days. Instead of [^{32}P]NAD$^+$, [*adenine*-^{14}C]NAD$^+$ may be used, in which case exposure times must be increased to 1 week.

For mechanistic considerations or as a positive control experiment, commercially available GAPDH (Boehringer Mannheim, Mannheim, Germany; supplied as a 10 mg/ml suspension in 3.2 M ammonium sulfate) may be used. The purified enzyme should be directly diluted to 2 mg/ml with 100 mM HEPES buffer (pH 7.4), and then further diluted in the final assay. Alternatively, the original protein–sulfate suspension should be centrifuged (15,000 g, 10 min at 4°C), the supernatant removed, and the pellet suspended at 2 mg/ml in 100 mM HEPES buffer (pH 7.4). GAPDH concentrations of 10 to 30 μg protein per assay are routinely used in a fashion similar to that already described. In general, incubating GAPDH or cytosolic fractions with [^{32}P]NAD$^+$ or [^{32}P]NADH result in some residual protein modification unrelated to the action of NO. Labeling increases with time and should be monitored. This is not a consideration with incubation times of less than 30 min.

Preparation of [^{32}P]NADH from [^{32}P]NAD$^+$

To prepare [^{32}P]NADH, 10 μl (50 μCi) of [^{32}P]NAD$^+$ (800 Ci/mmol) is incubated with 2 mM MnCl$_2$, 10 mM isocitrate (pH 7.5), 1 μM cold NAD$^+$, 22 U isocitrate dehydrogenase (NAD$^+$ specific, 31 U/mg), and 100 mM HEPES (pH 7.5) in a total volume of 100 μl at 37°C for 45 min. Separation is achieved on a Nucleosil C$_{18}$ reversed-phase column (Macherey and Nagel, Düren, Germany) equilibrated with buffer A (200 mM potassium phosphate buffer, pH 6.0) at a flow rate of 1 ml/min. Elution is done with buffer B (200 mM potassium phosphate buffer, pH 6.0/5% methanol) using the following gradient elution: 0–10 min, 70% B; 11–26 min 100% B; 27–35 min 70% B at flow rate of 0.7 ml/min. UV-Detection is accomplished at 254 nm. In our experiments, NAD$^+$ eluted at 15 min whereas NADH came off the column at 20 min. Fractions are collected, and radioactivity is determined using a liquid scintillation counter (PW 4760 from RayTest, Straubenhardt, Germany). In assays performed to label GAPDH, 120,000–150,000 cpm of [^{32}P]NADH/assay were used.

Nitric Oxide-Stimulated, [^{32}P]NAD$^+$- and [^{32}P]NADH-Dependent Protein Modification

The easiest way to achieve NO-stimulated protein modification is to use commercially available NO donors, such as SNP, SIN-1, or NONOates like spermine-NO or diethylamine–NO (DEA–NO). Depending on the assay conditions, the stimulatory potency of an NO donor is detectable at concentrations as low as 10 μM. However, for starting conditions we recommend concentrations of roughly 500 μM SNP and 200 to 400 μM of SIN-1. As a general rule, all NO-releasing compounds should be stored in a cool dry place and protected from light. Aqueous, pH-buffered solutions should be made up fresh each time, just prior to their use. NO donors are introduced by adding a small amount (1–5 μl) of a stock solution, which is immediately followed by the addition of radioactivity.

In the case of some NO donors special care must be taken, because hydrolysis of agents like BF_4NO (nitrosonium tetrafluoroborate; supplied by Aldrich, Milwaukee, WI) or BF_4NO_2 (nitronium tetraflouroborate; distributed by Fluka, Ronkonkoma, NY) becomes rate limiting for labeling experiments. With BF_4NO, which hydrolyzes to nitrosonium ion (NO^+) and tetrafluoroborate (BF_4^-), we add radioactivity first and followed by the NO donor. Freshly prepared stock solutions (pH values below 2; 0.2 M HCl) promote sufficient NO^+ stability to add 1 or 2 μl of the stock solution to our basic labeling assay (pH control is necessary). The same applies to the use of BF_4NO_2, which hydrolyzes to nitronium ion (NO_2^+) and BF_4^-. When decomposition of NO donors becomes rate limiting, maximal labeling is difficult to achieve.

Similar precautions must be taken when $ONOO^-$ (peroxynitrite) is used to induce covalent GAPDH modification. Peroxynitrite is rapidly diluted into HEPES buffer, maintaining an alkaline pH. For GAPDH modification, 1–10 μl is then transferred to the complete labeling assay (pH 7.4).

Alternatively, NOS produces sufficient NO to cause enhanced radioactive labeling of cytosolic GAPDH. This has been demonstrated using rat cerebellum cytosol (14) fortified with all NO synthase cofactors and $[^{32}P]NAD^+$ and with rat brain containing added purified brain NOS (8). In this case, NOS inhibitors can be used to trace back protein modification to an active NO output system.

Role of Small Molecular Weight Thiol Cofactors

With the exception of BF_4NO and BF_4NO_2, the NO donor-driven modification of cytosolic proteins, including GAPDH, is enhanced by thiols like DTT. Therefore, we recommend that DTT, at a concentration of 1 to 2 mM, be added for routine assays. Although they are less active, glutathione or cysteine can replace DTT. Care should be taken at DTT concentrations of 5 mM or higher, as the stimulatory effect is reversed to an inhibitory one at high DTT concentrations (15). In contrast to the thiol enhanced NO-stimulated GAPDH modification, thiols suppress the basal, NO-independent $[^{32}P]NAD^+$-dependent protein modification.

It is conceivable that thiols accelerate the release of NO from NO-donors like SNP, reduce oxidized protein sulfhydryl groups to a reduced state, or stabilize NO^+ by S-nitrosothiol formation. Using thiol-blocking agents like dinitrobenzene or N-ethylmaleimide (16), we substantiated the requirement of free sulfhydryl groups on GAPDH as a prerequisite for posttranslational modification. These considerations also apply for the activity of peroxynitrite. Even 50 to 100 μM $ONOO^-$ did not produce substantial protein modification unless 2.5 mM DTT was included in the assay. Notably, with thiols in relative excess over peroxynitrite, conditions are conclusive to the formation of nitrosothiol (9). In contrast, when present in relative excess to thiols, $ONOO^-$ oxidizes thiols, thereby inhibiting subsequent posttranslational protein modification. Generally, oxidizing conditions, including exposure to $ONOO^-$ in the absence of a sufficiently reducing environment, attenuates incorporation of radioactivity due to subsequent stimulation with NO donors. The exception for this rule is the action of NO^+- or NO_2^+-donating agents. Thus, the competing reaction with thiol, present in relative excess over enzyme, prevents the reaction of NO^+ or NO_2^+ with protein active-site thiol.

Subtractive Protein Modification; Determination of Endogenously Modified GAPDH

Cellular studies are limited by the cell impermeable character of pyridine nucleotides. To investigate protein modification under cellular conditions, a subtractive, differential detection method must be employed. Following endogenous protein modification in response to NO-donor addition, or as a result of inducible NOS induction, the cytosolic fraction of individual cells is prepared as outlined. Cytosolic fractions of NO-treated and control cells are subjected to maximal NO-stimulated protein/GAPDH modification, that is, around 100 μg protein, 2 mM DTT, 10 μM NAD^+, 1 μCi [^{32}P]NAD^+, 400 μM SIN-1, and 100 mM HEPES buffer (pH 7.4) during a 60-min incubation period. This is followed by the standard procedure to locate radioactively modified proteins on the SDS gels. A decreased incorporation of radioactivity into GAPDH in the cytosol of treated cells compared to that from control cells suggest the occurrence of endogenous protein/GAPDH modification. Endogenous protein modification in response to NO actions can be further substantiated using NOS inhibitors (17). Although the differential method detects variations in the amount of radioactivity associated with individual proteins that can be traced back to endogenous NO action, this does not necessarily reflect endogenous NAD^+/NADH-dependent protein modification. Reduced incorporation of radioactivity also may arise from some other kind of NO-mediated protein thiol group modification, such as thiol oxidation or protein S-thiolation. In brain tissue, this method has been successfully employed to detect alterations of endogenous NO-mediated protein modifications due to long-term potentiation (18) or kindling progression (19).

Protein Sulfhydryl Group Modification; Hg^{2+} Cleavage Experiments

The modification of GAPDH sulfhydryl group by $[^{32}P]NAD^+/[^{32}P]NADH$ is deduced from $HgCl_2$ cleavage experiments (15, 20). To assess the pattern of chemical stability of modified GAPDH, we follow the method employed for ADP-ribosylcysteine modified in the heterotrimeric G protein G_i, formed by pertussis toxin (21). The pellet of NO-stimulated, $[^{32}P]NAD^+$-modified GAPDH was taken up in 10 μl 100 mM HEPES buffer (pH 7.5) and incubated for 30 min at room temperature. Thereafter, 10 μl 1 M HEPES buffer (pH 7.5) and water is added to a final volume of 0.1 ml. Extensive vortexing and a 10-min exposure to sonication resulted in resuspension of the protein pellet. Next, a $HgCl_2$ solution was added to achieve a final concentration of 1 to 10 mM Hg^{2+}, followed by intensive vortexing. Cleavage was carried out during a 1- to 2-hr incubation period at 37°C. For the control, $HgCl_2$ was replaced by NaCl. Samples were processed further by adding 0.5 ml 20% trichloroacetic acid, incubating on ice for 30 min, centrifuging at 15,000 g for 15 min at 4°C, and extracting with ether twice (see above). Radioactivity associated with the proteins after cleavage was detected as described, following SDS gel electrophoresis.

Concluding Remarks

Nitric oxide-stimulated $NAD^+/NADH$-dependent posttranslational modification of the glycolytic enzyme GAPDH is a cGMP-independent NO-signaling pathway and involves the intermediate formation of protein S-nitrosothiol. After the S-nitrosothiol is formed, covalent binding of NAD^+ or, preferentially NADH probably occurs. Experimental evidence for this is the marked incorporation of radioactivity into the protein (GAPDH) in the presence of an active NOS or in response to NO-releasing compounds. Figure 1 summarizes the overall reaction sequence and exemplifies the reaction of purified GAPDH with $[^{32}P]NAD^+$ or $[^{32}P]NADH$ in the presence of SNP.

The successful labeling of proteins like GAPDH with $[^{32}P]NAD^+$ suggests an ADP-ribosylation (like) mechanism. However, the covalent binding of the whole molecule of NAD^+ had been reported (15), as has the ability of GAPDH to transfer the ADP-ribose moiety of NAD^+ to free cysteine, resulting in a true thioglycosidic linkage (22). The observed greater efficiency of labeling with NADH suggests a mechanism apart from ADP-ribosylation; the exact nature of the binding remains unknown. Nitric oxide-stimulated protein modification by $NAD^+/NADH$ in general, and of GAPDH in particular, has been achieved *in vitro* and in intact cells, substantiating the role of NO-group transfer chemistry in NO signaling. The protein modification

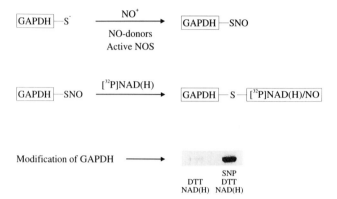

FIG. 1   Proposed scheme for the reaction of GAPDH with NAD$^+$/NADH in the presence of NO. *Top*: Reaction of GAPDH with NO$^+$ forming *S*-nitroso-GAPDH. The S-nitrosylated protein covalently binds NADH preferentially over NAD$^+$, with the exact nature of the modification still unknown (for details, see text). *Bottom*: Autoradiography showing incorporated radioactivity in the presence of the NO donor sodium nitroprusside (SNP), dithiothreitol (DTT), and either [^{32}P]NAD$^+$ or [^{32}P]NADH. For details, see text.

mechanism may prove useful as an indirect measurement of intracellular protein S-nitrosylation and, therefore, reveal aspects of the mechanism of NO signaling under pathophysiological conditions related to massive NO release. Several functions of GAPDH, in addition to a role in glycolysis, have been addressed (3). It will be intriguing to determine some of these activities in relation to release of the unique messenger nitric oxide.

References

1. J. S. Stamler, *Cell (Cambridge, Mass.)* **78,** 931 (1994).
2. H. H. H. W. Schmidt and U. Walter, *Cell (Cambridge, Mass.)* **78,** 919 (1994).
3. B. Brüne, S. Dimmeler, L. Molina y Vedia, and E. G. Lapetina, *Life Sci.* **54,** 61 (1994).
4. B. Brüne and E. G. Lapetina, *J. Biol. Chem.* **264,** 8455 (1989).
5. R. S. Duman, R. Z. Terwilliger, and E. J. Nestler, *J. Neurochem.* **57,** 2124 (1991).
6. M. B. Williams, X. Li, X. Gu, and R. S. Jope, *Brain Res.* **592,** 49 (1992).
7. S. Dimmeler, F. Lottspeich, and B. Brüne, *J. Biol. Chem.* **267,** 16771 (1992).
8. J. Zhang and S. H. Snyder, *Proc. Natl. Acad. Sci. U.S.A.* **89,** 9382 (1992).
9. S. Mohr, J. S. Stamler, and B. Brüne, *FEBS Lett.* **348,** 223 (1994).
10. S. Mohr, J. S. Stamler, and B. Brüne, *J. Biol. Chem.* **271,** 4209 (1996).
11. L. J. McDonald and J. Moss, *Mol. Cell. Biochem.* **138,** 201 (1994).

12. S. Dimmeler, U. K. Messmer, G. Tiegs, and B. Brüne, *Eur. J. Pharmacol.* **267,** 105 (1994).
13. R. F. Johnston, S. C. Pickett, and D. L. Barker, *Electrophoresis* (*Weinheim*) **11,** 355 (1990).
14. S. Dimmeler and B. Brüne, *Biochem. Biophys. Res. Commun.* **178,** 848 (1991).
15. L. J. McDonald and J. Moss, *Proc. Natl. Acad. Sci. U.S.A.* **90,** 6238 (1993).
16. B. Brüne and E. G. Lapetina, *Arch. Biochem. Biophys.* **279,** 286 (1990).
17. S. Dimmeler, M. Ankarcrona, P. Nicotera, and B. Brüne, *J. Immunol.* **150,** 2964 (1993).
18. R. S. Duman, R. Z. Terwilliger, and E. J. Nestler, *J. Neurochem.* **61,** 1542 (1993).
19. A. Vezzani, S. Sparvoli, M. Rizzi, M. Zinetti, and M. Fratelli, *NeuroReport* **5,** 1217 (1994).
20. S. Dimmeler and B. Brüne, *Eur. J. Biochem.* **210,** 305 (1992).
21. T. Meyer, R. Koch, W. Fanick, and H. Hilz, *Biol. Chem. Hoppe-Seyler* **369,** 579 (1988).
22. V. Pancholi and V. A. Fischetti, *Proc. Natl. Acad. Sci. U.S.A.* **90,** 8154 (1993).

[25] Assessment of Synaptic Effects of Nitric Oxide in Hippocampal Neurons

Charles F. Zorumski, Steven Mennerick, and Yukitoshi Izumi

Introduction

Nitric oxide (NO) appears to be an important modulator of synaptic transmission in the mammalian central nervous system (CNS) (1, 2) and may play roles in neural development, synaptic plasticity, and neurodegeneration (1–3). Although still controversial, the evidence for NO participating in synaptic plasticity is best for N-methyl-D-aspartate (NMDA) receptor-dependent long-term potentiation (LTP) in the CA1 hippocampal region (1, 2). Here, the induction of lasting synaptic enhancement requires at least three postsynaptic events: membrane depolarization, NMDA receptor activation, and CA^{2+} influx. Despite the requirement for these postsynaptic events, it appears that presynaptic changes in the release of glutamate contribute to the synaptic enhancement (1, 2). If presynaptic changes are important in CA1 LTP, yet the process critically depends on postsynaptic events, then a messenger is needed to communicate between the postsynaptic neuron and the presynaptic terminal. Several candidate "retrograde messengers" have been described, including NO, arachidonic acid, carbon monoxide, and platelet-activating factor, with the best support being for NO (1, 2).

In addition to a possible role in LTP, NO has other effects that could alter synaptic transmission in the hippocampus including blockade of NMDA responses by actions at a redox site on NMDA ion channels (4). Furthermore, untimely release of NO by NMDA receptor activation has been shown to block LTP in the CA1 region (5) and release of NO during certain patterns of synaptic stimulation contributes to homosynaptic long-term depression (LTD) (6). These diverse, and at times, conflicting effects of NO, coupled with the proposed role of NO in certain forms of neurodegeneration, make it important to understand the cellular and synaptic mechanisms of this diffusible messenger. Because the hippocampus has been a favored preparation for synaptic studies, particularly studies of synaptic plasticity, this chapter will describe *in vitro* hippocampal preparations that have been useful for studying the effects of NO.

Methods in Neurosciences, Volume 31

Hippocampal Cell Cultures

Dissociated cell cultures offer a great deal of control over experimental variables for studies of synaptic transmission. In particular, pre- and postsynaptic neurons can be visually identified and monosynaptic connections can be studied in isolation from polysynaptic modulation. Additionally, because cells are grown as a monolayer, it is relatively easy to achieve rapid and complete exchange of experimental solutions. Unfortunately, dissociated cultures must usually be prepared from embryonic or young postnatal animals. Thus, cells are at immature stages of development. Additionally, normal synaptic connections are disrupted during culture preparation, and it is an assumption that the reformed connections in culture are similar to those found *in vivo*. To the extent that this has been studied, it appears that synapses in culture mirror very well synaptic connections studied in more intact preparations.

General Culture Methods

In our laboratory, hippocampal cells are harvested from 1- to 3-day-old albino rat pups using procedures modified from Huettner and Baughman (7) and Nakajima *et al.* (8). Prior to dissection, surgical instruments are sterilized in 70% (v/v) methanol for 30 min or longer, and the dissection is carried out in a tissue culture flow hood.

Rat pups are anesthetized with halothane and oxygen, and are decapitated using a pair of sharp scissors. The scissors are used to open the skull along the midline and the brain is rapidly dissected into 1–2 ml Leibowitz L-15 medium containing 0.2 mg/ml essentially fatty acid-free bovine serum albumin (BSA). The brain is bisected along the midline and the hippocampi are dissected from the medial aspects of the cerebral hemispheres. Hippocampi are sliced transversely into about 500 μm thick slices using a sharp scalpel. The hippocampal slices are placed in 1–2 ml of L-15 medium containing 1 mg/ml papain plus 0.2 mg/ml BSA at 33°–35°C and are gently gassed via a sterile pipette with 95% O_2–5% CO_2 (v/v). The enzymatic solution and slices are swirled intermittently using a microstirring bar on a magnetic stir plate.

The enzymatic digestion is stopped after 20–30 min by exchanging the enzyme-containing Leibowitz solution with culture medium (1–2 ml) composed of Eagle's minimal essential medium (MEM) supplemented with 5% (v/v) horse serum, 5% (v/v) fetal calf serum, 17 mM D-glucose, 400 μM glutamine, 50 units/ml penicillin, and 50 μg/ml streptomycin. In this culture medium, the hippocampal slices are dissociated into single cells by gentle trituration for several minutes using a fire-polished Pasteur pipette. The supernatant is transferred to a sterile 15-ml tube, fresh culture medium is added to the slices, and the trituration is repeated with a smaller bore fire-polished Pasteur pipette. This cycle is repeated until the chunks of tissue have been dissociated.

The resulting suspensions of isolated cells are pooled and counted using a hemocytometer. The cells are diluted in the supplemented MEM to a final density of 2–3×10^5 cells/ml and are plated on 35-mm Falcon tissue culture dishes (Becton Dickinson Labware, Lincoln Park, NJ) that have been previously coated with a thin layer of rat tail collagen [Type I: 10 mg collagen/ 20 ml of $1:1000$ (v/v) acetic acid]. The culture dishes are prepared prior to beginning the dissection and are allowed to dry under an ultraviolet light during the dissection and dissociation of cells. After drying, the culture dishes are rinsed once with sterile MEM. For some experiments we also plate cells on 35-mm Falcon dishes that have been modified using a Mechanex BB-Form 2 tissue culture dish imprinter (Medical Systems, Greenvale, NY). This tool creates a central well in the culture dish into which cells are plated. Cells can also be plated onto glial monolayers. The glial cultures are obtained using the same methods, either from hippocampus or neocortex. After the glial background grows to confluence, neurons are killed by feeding with ice-cold medium.

Following plating, cultures are kept in a 95% air–5% CO_2 humidified incubator (NAPCO Scientific Co., Tualatin, OR). After 3 days *in vitro*, cultures are treated with 10 μM cytosine arabinoside (ARA-C) to retard glial growth. We do not routinely feed cells after plating and use them for experiments between 3 and 16 days in culture.

We have obtained best results using newborn pups and plating onto collagen-coated dishes. Other substrates can be used including poly(D-lysine), poly(L-lysine), or Matrigel. Because a glial underlayer forms even with a single plating, we have not routinely used a double-plating technique to provide a glial feeding layer. All culture media, sera, antibiotics, and glutamine are purchased in solution from GIBCO (Grand Island, NY).

Hippocampal Microisland Cultures

There are several problems in using conventional hippocampal cultures to study synaptic transmission. First, cultured neurons send out extensive processes that course over relatively long distances. This makes it difficult to study the time course of synaptic events that arise in distal dendritic compartments. Second, the frequency of encountering monosynaptically connected pairs of neurons, although higher than in slice preparations, is only about 40% (9). To overcome these problems, we have used a preparation of cultured hippocampal neurons grown in a microisland culture environment in which one or more hippocampal neurons are isolated on an astrocytic background that is restricted in size. Table I lists some of the advantages of working with microcultures for the study of CNS synaptic transmission. Of particular importance are the increased probability of encountering monosynaptically connected pairs of neurons (>80 versus 40%), the ability to study autaptic (self) synapses on single neuron microislands, and the simpler passive electrical properties of the neurons (9).

Our microisland procedure has been adapted from methods originally described by Segal and Furshpan (10, 11). The dissection and harvesting of hippocampal neurons is done using the methods described for hippocampal mass cultures. The major differences in preparation of microisland cultures are the density of cells plated and the method of substrate application. Following dissociation, cells are diluted with the supplemented Eagle's MEM to a density of 20,000 cells/ml. This prevents the formation of microislands that are overcrowded with neurons. The cells are plated onto 35-mm Falcon tissue cultures dishes that have been previously coated with 0.15% (w/v) agarose. The agarose is dried overnight and, just prior to hippocampal dissection, sprayed with small droplets of rat tail collagen using a microatomizer (Thomas Scientific, Swedesboro, NJ). With practice, the size of the collagen droplets range from 20 μm to 1000 μm. The dishes containing the

TABLE I   Advantages of Hippocampal Microcultures for Study of Central Nervous System Synaptic Transmission

Characteristics	Comments
Strong synaptic connections	Autapses: Presynaptic and postsynaptic functions can be studied with single recording pipette (single-neuron microcultures)
	High probability of obtaining monosynaptically connected pairs of neurons (two-neuron microcultures)
Isolation of synaptic networks	Can study monosynaptic transmission free from polysynaptic modulatory influences
	Agents that increase excitability can be studied without increase in polysynaptic activity
	Can study miniature and evoked synaptic responses from same presynaptic source
Drug delivery	Ability to make rapid (millisecond) applications of drugs
	Spatial restriction of microcultures allows solution exchange across soma and processes
Favorable passive membrane properties	High input resistance, simpler capacitance charging curves than mass culture neurons

collagen microdroplets are then placed under an ultraviolet light until cells are plated (usually about 1 hr).

Individual microislands consist of one or more nonneuronal cells that stain with antibodies against glial fibrillary acidic protein (GFAP) and are presumably type I astrocytes. These glial cells adhere to the collagen droplets and, over time in culture, come to occupy nearly the entire extent of the droplet. Neurons appear to adhere to the glial cell in small colonies of one or more cells. Importantly, the agarose substrate does not permit either cell adhesion or process growth, limiting both the glial cells and the neurons to a restricted physical space. To prevent glial overgrowth, the microisland cultures are treated with 10 μM ARA-C after 3 days in culture. Cultures are routinely used for synaptic studies between 8 and 15 days *in vitro* and are not fed with fresh medium following plating.

There can be a great deal of variability in the microisland cultures between and within platings. We have tried a variety of substrates other than rat tail collagen to minimize this variability, most without success. Other substrates we have used include collagen/poly(D-lysine) mixture, human placental collagen, and Matrigel (Collaborative Research, Bedford, MA). We have had some success with each of these substrates, particularly Matrigel, but in no case were the results more consistent than with rat tail collagen.

We believe that the adhesive substrate application is critical for producing small microcultures (50–100 μm diameter) that give best results in synaptic studies. It is extremely difficult to control the size of the collagen droplets precisely. However, we now use a technique of holding both the tissue culture dish and the microatomizer in the air and spraying collagen at the dish horizontally, rather than from above, with a distance of 18–20 inches between them. Two to four sprays from the atomizer are used per tissue culture dish. Using these methods we routinely find about 8–10 single neuron microislands per 35-mm culture dish.

Electrophysiological Studies in Hippocampal Cultures

Extracellular Solutions

At the time of study, the culture medium is replaced with a standard extracellular solution containing (in mM): 140 NaCl, 4 KCl, 10 D-glucose, and 10 HEPES, adjusted to pH 7.3 using NaOH. Then CaCl$_2$ and MgCl$_2$ are added in varying concentrations depending on the experiment. The osmolarity of the extracellular solution is measured using a vapor pressure osmometer (Wescor, Logan, UT) and is adjusted to 310 mOsM with sucrose. For study of spontaneous miniature synaptic currents (mini's), 1 μM tetrodotoxin (TTX) is added to block action potential-mediated transmitter release. Tetrodotoxin is also used for all experiments in which currents are evoked by exogenous applications of excitatory amino acids (EAA) or γ-aminobutyric acid (GABA). When NMDA receptor-mediated events are studied, 1–20 μM glycine is added. Depending on the experiment and whether glutamate- or GABA-mediated synaptic events are being studied, specific receptor antagonists and ion channel blockers are added to the extracellular solution.

Patch Electrodes and Intracellular Solutions

For synaptic studies in cultured hippocampal neurons, we routinely use whole-cell patch clamp recording techniques. Patch electrodes are pulled on a Model P-87 horizontal Brown–Flaming electrode puller (Sutter Instrument, Novato, CA) using 1.2 mm o.d. borosilicate glass (World Precision Instruments, Sarasota, FL) and a four-stage pull. Patch pipettes are fire-polished using a microforge (Narishige Instruments, Tokyo, Japan), and have resistances of 2–6 MΩ. To diminish problems with series resistance, pipettes of lower resistance are preferred.

Depending on the experiment, we have used a variety of intracellular solutions. For presynaptic neurons and for studies of excitatory autaptic currents on single neuron microislands (11, 12), we use a solution containing (in mM): 140 potassium gluconate, 5 NaCl, 5 EGTA, 0.5 CaCl$_2$, 10 HEPES, 2 Mg^{2+}-ATP, and 0.5 Na$^+$-GTP with pH adjusted to 7.3 using KOH. The pipette solution used for postsynaptic neurons and for recording miniature synaptic currents contains (in mM): 140 cesium methanesulfonate (Aldrich, Milwaukee, WI), 5 NaCl, 5 1,2-bis-(o-aminophenoxy)ethane-N,N,N',N'-tetraacetic acid (BAPTA), 2 magnesium ATP, and 10 HEPES with pH adjusted to 7.3 using CsOH. A variety of other salts including KCl, potassium methyl sulfate, potassium acetate, CsCl, and cesium acetate have been used successfully.

For study of GABA-mediated inhibitory autaptic current, intracellular and extracellular solutions with symmetric chloride concentrations have given best results. This is because the chloride reversal potential is near 0 mV and at −70 mV (the potential at which most autaptic studies are conducted) there is a large driving force for chloride flux. Since autaptic studies are conducted in a single neuron, it is not possible to depolarize the cell much beyond −40 mV because ionic currents needed to evoke transmitter release are inactivated.

One problem in recording synaptic currents with whole-cell techniques is that responses tend to run down over the course of an experiment. This rundown varies considerably from cell-to-cell and is difficult to predict. Preliminary experiments suggest that rundown primarily results from presynaptic factors. For this reason, we have sometimes used a nystatin perforated patch technique to record from presynaptic cells (12). Amphotericin has also been tried with similar results. The perforated patch pipette solution contains (in mM): 125 potassium gluconate, 10 KCl, 5 NaCl, 3 MgCl$_2$, and 10 HEPES at pH 7.3. Nystatin (Sigma, St. Louis, MO) and Pluronic F127 (Molecular Probes, Eugene, OR) are dissolved in dimethyl sulfoxide (DMSO) and added to the solution at 300 μg/ml and 250 μg/ml, respectively. The DMSO is present in the solution at a final concentration of 0.35%. The Pluronic F127 may speed intracellular access, which is usually achieved within 5–10 min. While reliable recordings can be achieved using this method, run down of synaptic responses sometimes still occurs. Additionally, the series resistance using the perforated patch technique (15–20 MΩ) is not as low as we routinely achieve with conventional whole-cell recordings (<10 MΩ). For these reasons, we prefer conventional whole-cell recordings for synaptic studies.

Neuronal Recordings

Physiological experiments are conducted on the stage of an inverted microscope equipped with phase-contrast optics (Nikon Diaphot or TMS, Tokyo, Japan). Most experiments are conducted at room temperature (22°–24°C). In hippocampal mass cultures, neurons with bipolar or pyramidal cell bodies and smooth cell membranes are typically selected for study. On microisland cultures, phase bright neurons are selected. Typically, larger round cells on microislands are excitatory while smaller, somewhat darker neurons tend to be inhibitory. About half the cells in these cultures are inhibitory (9, 11). Excitatory and inhibitory cells may be easier to discriminate in older (>10 day) cultures in which sizes and shapes have reached a more mature stage. With two cell microislands, we routinely select the larger of the two cells to be the presynaptic cell when we are examining excitatory synaptic events.

Glial Cell Recordings

An advantage of the microisland preparation for certain studies is that it allows simultaneous recording from neurons and glial cells during synaptic transmission. The techniques for recording from hippocampal glial cells ion a microisland are the same as those described above, though the flat morphology of type I astrocytes in culture creates some extra difficulties. Microislands often contain more than one glial cell, but it is possible to voltage clamp the entire extent of glia on microislands because the cells are electrotonically coupled (13). Spatial clamp considerations preclude similar studies of glial cells in confluent mass cultures. Three criteria appear to helpful in selecting glial cells for study. First, it is helpful to select smaller islands that have more favorable membrane characteristics (higher input resistances and smaller membrane capacitances). Second, "clean" appearing glia in which the glial nucleus is clearly visible under phase contrast optics give best results in terms of membrane seals and intracellular access. Finally, islands that appear thicker, usually with multiple nuclei, give better results than islands that are thinner and have apparent gaps in the glial cytoplasm. Electrode contact with the glial cell can be monitored either visually, observing for a slight deflection in the glial cell membrane, or electrically, monitoring an oscilloscope trace for the increase in electrical resistance that accompanies cell contact. Either method gives acceptable results with practice.

One problem encountered with glia is a propensity for access resistance to increase during recording. Although this is also a problem with neurons, changes in access resistance are more common with glial recordings. To help maintain intracellular access, slight suction is often applied to the patch pipette during whole-cell recording and is maintained during an experiment.

Synaptic Studies in Cultured Neurons

Two different types of amplifiers are used for synaptic studies. Recordings from postsynaptic neurons and glial cells are obtained using a patch clamp amplifier (either an Axopatch 1D, Axon Instruments, or List EPC-7 amplifier, Adams and List, Inc., Westbury, NY). These amplifiers are useful for experiments where low noise recordings are needed, particularly in the case of miniature synaptic events. During recordings of evoked postsynaptic currents, series resistance compensation is typically set at 70–90%. This is generally necessary because microculture synaptic currents can often exceed 5 nA, causing voltage errors of over 40 mV with a typical 8 MΩ access resistance.

For recordings from presynaptic neurons and for recordings of large autaptic currents on single neuron microislands, an Axoclamp 2A amplifier (Axon Instruments) is often used. This amplifier has good current clamp capabilities and is useful when presynaptic action potentials are being monitored. Additionally, the Axoclamp amplifier is used in the discontinuous single-electrode voltage clamp mode (8–11 kHz switching frequency) to stimulate and record large (>2 nA) autaptic currents. The discontinuous voltage clamp time shares the functions of voltage sampling and current injection. Because current is not injected during the period of membrane voltage sampling, the voltage drop across the pipette access resistance does not contribute to the sampled voltage. Thus, the injected current is proportional to the difference between the true membrane potential and the command potential, with no contribution of the potential drop across the access resistance. The circuitry of a discontinuous voltage clamp adds some electrical noise to the recordings and requires the use of an oscilloscope to monitor the duty cycle of the amplifier.

For stimulation of either presynaptic neurons or autaptic currents we use a 1.5 msec voltage step to +20 mV from a holding potential between −50 and −70 mV. The tetrodotoxin sensitivity of synaptic currents elicited with this protocol suggests that a break away action potential in unclamped regions of the axon is necessary for evoked transmitter release (14). Alternatively, in current clamp mode a 1.5 msec depolarizing current step of 1–2 nA is used. In dual cell recordings, we have found it particularly advantageous to use voltage clamp methods for recording from presynaptic neurons. Under these conditions, it is possible to monitor two monosynaptic connections of a single cell (the postsynaptic response in the second neuron and the autaptic response) (9, 15). Evoked synaptic responses are obtained every 15–30 sec to avoid problems with frequency dependent synaptic changes (paired pulse depression/potentiation) that can occur during more frequent stimulation.

Data Collection and Analysis

Currents are typically filtered at 2 kHz using an 8-pole Bessel filter and digitized at 3–5 kHz using either commercially available software (pCLAMP, Axon Instruments) or programs written in the Axobasic language (Axon Instruments). During experiments, evoked responses are collected to computer disk while miniature synaptic currents are collected to a video cassette recorder (VCR) tape using a PCM-2 A/D Recorder Adapater (Medical Systems, Greenvale, NY) and a commercial VCR at a resolution of 14 bits, a sampling rate of 44 kHz and a bandwidth of DC to 16 kHz. Data are analyzed off-line using pCLAMP software or routines written in Axobasic (Axon Instruments).

Drug Delivery

During synaptic studies, extracellular solution exchanges and drug applications are achieved using a multibarrel, gravity-fed local perfusion system. The system consists of 6–7 independent lines of PE-10 (polyethylene) tubing emptying into a common port that is tapered to minimize dead space and to provide directed flow onto the cell(s) being studied. The common port, fashioned from SMI (American Hospital Supply Corp., Miami, FL) micropipette glass (size E), is pulled on a vertical pipette puller, broken back to an inner diameter of 100–200 μm, and fire polished. During experiments, the drug delivery pipette is positioned 100–200 μm away from the recorded microisland. Control experiments using junction potential changes and responses to the nondesensitizing AMPA receptor agonist, kainate, indicate that complete solution exchange over a microisland can be achieved in about 200 msec with careful construction of the pipette tip. This system provides gentle solution exchange that does not disrupt the dual cell recordings often used in synaptic studies. During an experiment the cells of interest are continuously perfused with either control or experimental solutions. Suction is applied intermittently to keep the bath level low, thus minimizing pipette capacitance.

Hippocampal Slices

Brain slice preparations were first introduced by McIlwain and colleagues (16) for use in biochemical studies. Subsequently, brain slices have proven useful for physiological (17, 18) and histological studies (19, 20). Major advantages of brain slice preparations include control over experimental variables and drug applications. Furthermore, slices can be prepared from animals of various ages and from a variety of CNS regions. In contrast to primary cell cultures, brain slices maintain relatively intact local structure and synaptic connections. In this section, we focus on the preparation of hippocampal slices and the use of this preparation to study NO effects on synaptic transmission and synaptic plasticity.

Slice Preparation

In our laboratory, hippocampal slices are prepared from albino rats of various postnatal ages. Animals are deeply anesthetized with halothane [1–2% (v/v) with O_2 in an approved fume hood] and are rapidly decapitated using a guillotine. The scalp is cut along the midline and the skull is opened longitudinally along the sagittal suture using sharp scissors. As soon as the dura is exposed, the brain is cooled by superfusion with ice-cold artificial cerebrospinal fluid (ACSF) containing (in mM): 124 NaCl, 5 KCl, 2 $CaCl_2$, 2 Mg_2SO_4, 22 $NaHCO_3$, 1.25 NaH_2PO_4, and 10 D-glucose. To diminish trauma to the slices, it is sometimes helpful to prepare slices in ACSF containing 0.1 mM $CaCl_2$, especially when using adult rats (>60 days of age). The ACSF is continuously bubbled with 95% O_2–5% CO_2 (v/v) to maintain pH at 7.4. The brain is removed from the cranial cavity by blunt dissection and is kept cold with ACSF. The cerebral hemispheres are separated along the midline and the hippocampus is dissected from each hemisphere and placed in cold ACSF.

Transverse slices (300–600 μm thick) are cut using a McIlwain tissue chopper, a Vibratome (World Precision Instruments Vibroslicer), or a rotary tissue slicer (21). Tissue choppers have been the instruments most commonly used for preparing hippocampal slices, and have the advantages of allowing good control over the thickness of the slices, producing slices with smooth cut surfaces, and being easy to use. Some compression of the slice generally occurs with a tissue chopper, making this method less useful for histological studies. Problems can also occur because slices stick to the chopper blade and may not be completely separated in the cutting process.

When using a vibrating blade, the tissue is help on an agar base and slices are cut in a chamber that is submerged in ACSF (22, 23). In our hippocampal slice preparation, we suspend the dissected hippocampus with fine pins along a trench cut into a 3.3% agar base in a 5.5-cm culture dish (Falcon, Bectin Dickinson Labware, Lincoln Park, NJ) that fits our submerged cutting chamber. Slices are cut from bottom to top under mirror guidance. Some raggedness occurs at the cut edge of the tissue, but there is little slice compression. At a depth of 200 μm in the slice there is good histological preservation. Disadvantages of Vibratome slice preparation include slow cutting time and the need to develop manipulative skills that are not required with tissue choppers.

Rotary tissue slicers (21) combine aspects of tissue choppers and Vibratomes. Tissue is cut in a submerged chamber and it is relatively easy to separate individual slices. There is also little compression of the sliced tissue. However, some skill is necessary to avoid damage to the cut edge of the slice. Precise vertical positioning of the rotating blade and slow advancement through the tissue produce the most favorable preparations.

After cutting, slices are transferred to a holding chamber where they are kept in a stainless steel mesh holder (we use a stainless steel German tea strainer) and submerged in gassed ACSF. Slices are kept in the holding chamber for at least 1 hour prior to experimentation to allow recovery from dissection.

At the time of study, slices are carefully transferred via a blunt Pasteur pipette to a recording chamber where they are held below the surface of the ACSF by nylon mesh (24). Recording chambers in which slices are suspended on nylon at the interface between ACSF and gas can also be used. These interface chambers are useful when recording synaptic events extracellularly, but tend to be less stable for intracellular recordings. In the recording chamber, submerged slices are continuously perfused with ACSF at a rate of 2 ml/min. The ACSF is perfused using either gravity or an electrical pump (Masterflex, Cole-Parmer Instruments, Chicago, IL). Drugs and experimental solutions are applied by bath perfusion in the ACSF. Alternatively drugs can be applied by local perfusion using the multibarrel gravity-fed perfusion system described in the hippocampal culture experiments. The temperature in the recording chamber is kept at 30°C using a Peltier cell. Slices are visualized using a dissecting microscope and indirect lighting.

Electrophysiological Studies in Hippocampal Slices

Extracellular Recordings

Under visual control, glass recording pipettes are placed in stratum radiatum for purposes of monitoring field excitatory postsynaptic potentials (EPSPs) and in stratum pyramidale for monitoring population spikes. Field EPSPs are the most useful extracellular measurement of synaptic input to the CA1 hippocampal region whereas population spikes reflect the summed firing of a large group of pyramidal neurons in response to excitatory synaptic drive. Glass electrodes of 5–10 MΩ when filled with 2 M NaCl are used for extracellular recordings. In some experiments, we have used pipette solutions that are the same as the ACSF, but have generally preferred a concentrated NaCl solution for recording the low amplitude extracellular potentials (typically only several millivolts in amplitude). Signals are amplified using conventional amplifiers (AM Systems, Everett, WA, and Warner Instruments, Hamden, CT). After amplification, signals are digitized using commercially available software (pCLAMP) and programs written in Axobasic.

Conventional Intracellular Recordings

An advantage of slice preparations is the ability to perform stable intracellular recordings using sharp microelectrodes (resistances of 30–100 MΩ). Best results have been achieved using an approach along the CA1 cell body layer at an angle of about 30° above the slice surface. Although sharp electrode recordings can be useful for either current clamp or single electrode voltage clamp studies, the considerably better signal-to-noise ratio provided by whole-cell patch clamp techniques has caused this technique to largely replace sharp electrode recordings for voltage clamp studies.

Whole-Cell Patch Clamp Recordings

The application of patch clamp recording techniques to the study of neurons in tissue slices was a major advance in physiology. Two general approaches to patch recordings from slices have been used. The first uses thin slices (150–250 μm thick) in which the surface of the slice is cleaned with extracellular solution from a local perfusion pipette and neurons are identified visually using an upright microscope (25). In the second method, neurons are approached "blindly" by advancing the patch pipette through the slice until a neuron is contacted. Cell contact is identified by monitoring the change in current evoked by small voltage steps applied through the patch pipette (26). In our laboratory we have exclusively used the blind patch technique for recording from hippocampal slices.

The general methods for patch clamp recording from slices are similar to those used in cultured neurons but several factors seem to help in obtaining good membrane-electrode seals. It is important that extracellular and intracellular solutions are clean and as free from debris as possible at the time of use. Positive pressure is applied to the patch pipette as it crosses the air–fluid interface in the slice chamber. This positive pressure is applied using a locking syringe (AM Systems) and is maintained until electrical contact with a cell is identified. Following cell contact, the syringe pressure is reversed to allow membrane seal formation. This sometimes occurs instantly, but can take up to several minutes. After seal formation, additional suction is applied until intracellular contact (as judged by changes in the transients produced by voltage steps) is achieved. At that point, the syringe is removed because additional pressure decreases the stability of recordings.

For patch recordings from slices, we have typically used pipettes of 3–6

$M\Omega$ that have been lightly fire-polished at the tip. Lower resistance pipettes are preferred to diminish problems with series resistance. During whole-cell recordings some run down of synaptic responses occurs and after about 20–30 min it is rare to induce LTP (27, 28). Run down of baseline responses seems to be diminished by using low concentrations of calcium chelators in the patch solution [e.g., 0.2 mM ethyleneglycol-bis(β-aminoethyl-ether)N,N,N',N'-tetraacetic acid (EGTA) or BAPTA]. We have also used nystatin perforated patch recordings in slices but the access resistance is rarely as good as that achieved with conventional whole-cell recordings.

Electrical Stimulation of Slices

Synaptic responses are evoked by means of a bipolar electrode placed in the Schaffer collateral–commissural pathway. Typically a concentric stimulating electrode with a 25 μm tip (Rhodes Medical Instruments, Tujunga, CA) is used, although twin bipolar electrodes can also be used. Constant current pulses of 0.1–0.2 msec are applied using a Grass S-88 stimulator and a constant current generator. Single test responses are typically evoked at a frequency of every 30–60 sec to prevent changes in synaptic responses that can accompany more frequent electrical stimulation. More frequent stimuli and paired stimuli can also be applied depending on the goals of the experiment.

In some experiments, particularly studies of synaptic plasticity, two stimulating electrodes are placed in the Schaffer collateral pathway at different distances along the dendritic region to activate independent inputs to CA1 neurons. This provides a control for the effects of electrical stimulation since only one pathway is stimulated at a time and LTP or LTD can be induced selectively in one or the other pathway. The independence of the pathways is verified by demonstrating a lack of paired pulse plasticity when the two pathways are independently activated at brief (~50 msec) intervals.

The intensity of the stimulating current is determined for each slice and synaptic input by performing an input–output (IO) curve at the beginning of an experiment. The IO curve is generated by applying single stimuli to the Schaffer collateral pathway at a low intensity and increasing the intensity in a stepwise fashion until maximal responses are obtained. Based on the IO curve for each slice, we typically use a stimulus intensity that evokes a 50% maximal population spike to monitor synaptic responses and for tetanization to produce LTP. This stimulation produces reliable LTP in more than 95% of slices from 30-day-old albino rats when used in a stimulus train consisting of 100 pulses at 100 Hz. The disadvantage of this stimulus is that at the 50% point on the IO curve for population spikes, the field EPSPs are about 70% maximal. Thus following LTP induction, we typically observe a 20–30% increase in field EPSPs and a 40–50% increase in population spikes. Homosynaptic LTD in the CA1 region is induced by applying 1 Hz stimulation for

10–15 min. In experiments examining LTP or LTD, we repeat complete IO curves at 20 and 60 min after the delivery of the tetanus.

Hippocampal Slice Cultures

The pioneering work of Gähwiler (29, 30) has provided a third *in vitro* alternative for studying hippocampal synaptic transmission. Hippocampal slice cultures combine advantages of cell culture with those of brain slices. Two general methods are typically used for preparing hippocampal slice cultures. In the Gähwiler roller-tube method (29, 30), slices are prepared in a chicken plasma clot on glass coverslips. These coverslips are submerged in culture medium in culture tubes and are gently rotated during incubation in a roller drum. In roller tube cultures, the hippocampal slices gradually spread over the coverslip and become a cellular monolayer in which synaptic connections between the various hippocampal regions remain intact.

A second method for preparing hippocampal slice cultures was described by Stoppini and colleagues (31). In this method, hippocampal slices are placed on a porous membrane (Millicell-CM, Millipore Corp., Marlborough, MA) and are incubated at the interface between gas and culture medium. These cultured slices maintain intact synaptic connections and remain two to four cell layers thick. The culture medium consists of Eagle's MEM with 25% (v/v) horse serum, 25% (v/v) Hanks' solution, and 6.5 mg/ml glucose at pH 7.2.

Using either method, slices remain viable for 1 month or longer. Slice cultures have the advantage over dissociated cultures that they can be prepared from animals up to 20–30 days of age.

Nitric Oxide and Synaptic Transmission

Nitric Oxide Inhibitors

Both hippocampal cultures and hippocampal slices have been used to study synaptic effects of NO, with emphasis on the role of NO in synaptic plasticity. Several studies have examined effects of NO inhibitors on the development and maintenance of LTP in hippocampal slices. Competitive NO synthase inhibitors, including L-N^G-monomethylarginine (L-NMMA), L-N^G-nitroarginine (L-NOArg) and others have been applied by preincubation for up to several hours in the slice holding chamber, bath application in the recording chamber, or both. Because many drugs have nonspecific effects, it has been important in studies using competitive NO synthase inhibitors to show that the effects of the inhibitor are reversed by L-arginine, the natural substrate for NO synthase. In contrast, D-arginine does not reverse the effects of the competitive NO synthase inhibitors. Similarly, D-isomers of several competitive NO synthase inhibitors are available and do not inhibit the production of NO. These D-isomers are useful controls for the active L-isomers. Although many NO synthase inhibitors inhibit several isoforms of NO synthase, more selective agents are available. For example, 7 nitroindazole (7-NI) has relatively good selectivity for neuronal NO synthase (32), whereas aminoguanidine (33) and L-N^G-(1-iminoethyl)lysine (34) are more selective for inducible forms of NO synthase.

An alternative strategy for inhibiting presumed NO effects is to use hemoglobin, an agent that binds NO with fairly high affinity. Hemoglobin is a large protein that does not readily cross cell membranes. Thus, when bath applied, an inhibitory effect of hemoglobin suggests that an extracellular event is being modified. Unfortunately, hemoglobin is not specific and binds several molecules including NO, carbon monoxide, and O_2. It can also be difficult to work with hemoglobin at high concentrations because of bubbling in the solution.

Application of Nitric Oxide

There are several difficulties in examining the direct effects of NO on synaptic transmission. Perhaps most important is the fact that NO is a highly reactive molecule with a short half-life (measured in seconds). Thus delivery of NO to appropriate sites for defined periods of time has been difficult to achieve with confidence. Two primary strategies have been used to apply NO. The most direct method is to bubble NO gas into recording solutions at known concentrations (35). For these studies, NO stock solutions are prepared by bubbling NO gas into deionized water that has been saturated with helium. The NO stock solution is diluted to final concentration in helium-saturated ACSF and is injected into the recording chamber using gas-tight syringes. To prevent breakdown of NO by reaction with superoxide anions, superoxide dismutase is included in the NO solutions (35). Studies using this method have found that NO increases the frequency but not the amplitude or duration of spontaneous miniature synaptic events in cultured hippocampal neurons (36), strongly suggesting a presynaptic locus of action. Additionally, Zhou *et al.* (35) found that NO, in conjunction with weak presynaptic stimulation allows the generation of robust LTP in the CA1 region of hippocampal slices.

A second strategy for delivering NO has been to use agents that spontaneously release NO. Because of variability in release and in tissue penetration of the parent compound, it is difficult to know the exact concentration of NO without some independent measure. Agents that have been used as NO releasers include sodium nitroprusside (SNP), *S*-nitrosocysteine (SNC), 3-

morphilinosydnonimine (SIN-1), and nitroglycerin (NTG), among others (4). Appropriate controls for these agents include inactivation prior to application (e.g., light treatment of SNP). However, the inactivated species may have unexpected and unwanted properties. For example, light-inactivated SNP inhibits NMDA synaptic responses and excitotoxicity in hippocampal slices (37).

Nitric oxide can also be released in tissues by photolysis of certain compounds. In particular, potassium ruthenium nitrosyl pentachloride releases NO (and other products) when activated by a light flash at 320 nm. Murphy and colleagues (38) used a xenon arc flashlamp with a UG11 filter to provide a 90-nm bandpass and a peak of 320 nm. The filter was coated with a reflector to prevent wavelengths beyond 700 nm. Problems with this system include damage to the experimental preparation if the intensity of the light flash is not carefully adjusted. An advantage of the release of "caged NO" is that it produces NO release in defined regions depending on the precision of the light source. Additionally, this method has the advantage that actual concentrations of NO may be easier to control. The final concentration of NO achieved in tissues can be monitored using an NO-sensitive electrode (World Precision Instruments) (38).

An important consideration in determining the effects of NO concerns the redox status of the molecule. Lipton and colleagues (4) demonstrated that the toxic and neuroprotective effects attributed to NO depend greatly on redox state. The reduced form of the molecule ($NO \cdot$) has neurotoxic properties, presumably from reaction with superoxide anions to form peroxynitrite ($OONO^-$). In contrast oxidized forms of NO (e.g., nitrosonium ions, NO^+) exert neuroprotective actions by inhibiting NMDA receptor function via nitrosylation of sulfhydryl groups on the NMDA receptor. Certain agents that release NO tend to favor formation of specific redox products. SIN-1 and SNC tend to produce $NO \cdot$ and have been found to produce dose-dependent toxicity in cortical cell cultures (4). SNP and TNG do not spontaneously liberate $NO \cdot$ but can be metabolized to $NO \cdot$ via NO^+. The effects of various redox states of NO on synaptic transmission are not as well understood as the toxic effects. However, it is clear that NO (via NO^+) can depress NMDA mediated synaptic responses and, under certain circumstances, alter non-NMDA receptor mediated responses as well (37, 38).

Conclusions

Hippocampal cell cultures, hippocampal slices, and hippocampal slices cultures have proven useful for studying synaptic transmission and synaptic plasticity *in vitro*. Each of these preparations has clear strengths and limitations. Dissociated cultures have the major advantage of allowing precise control over many experimental variables, but are usually derived from immature animals and are far removed from the living animal. Hippocampal slices and slice cultures have the advantage that certain synaptic pathways remain relatively intact and there is good control over experimental variables. Although the process of slice preparation adds some uncertainty as to the relationship of observed synaptic responses to those that occur *in vivo*, the advantages of greater experimental control make these preparations attractive for studying cellular mechanisms involved in synaptic transmission and synaptic plasticity.

Acknowledgments

The authors thank A. M. Benz and J. Que for technical assistance, and L. L. Thio, G. D. Clark, K. Kato, R. Rader, and D. B. Clifford for many helpful discussions. Work in the our laboratory is supported by NIMH Research Scientist Development Award MH00964, Grants MH45493 and AG11355, and a fellowship from the Bantly Foundation.

References

1. C. F. Zorumski and Y. Izumi, *Biochem. Pharmacol.* **46,** 777 (1993).
2. R. D. Hawkins, M. Zhou, and O. Arancio, *J. Neurobiol.* **25,** 652 (1994).
3. T. M. Dawson and V. L. Dawson, *Neuroscientist* **1,** 7 (1995).
4. S. A. Lipton, Y. B. Choi, Z. H. Pan, S. Z. Lei, H. S. V. Chen, N. J. Sucher, J. Loscalzo, D. J. Singel, and J. S. Stamler, *Nature (London)* **364,** 626 (1993).
5. Y. Izumi, D. B. Clifford, and C. F. Zorumski, *Science* **257,** 1273 (1992).
6. Y. Izumi and C. F. Zorumski, *NeuroReport* **6,** 1131 (1993).
7. J. E. Huettner and R. W. Baughman, *J. Neurosci.* **8,** 160 (1988).
8. Y. Nakajima, S. Nakajima, R. J. Leonard, and K. Yamaguchi, *Proc. Natl. Acad. Sci. U.S.A.* **83,** 3022 (1986).
9. S. Mennerick, J. Que, A. M. Benz, and C. F. Zorumski, *J. Neurophysiol.* **73,** 320 (1995).
10. M. M. Segal and E. J. Furshpan, *J. Neurophysiol.* **64,** 1390 (1990).
11. M. M. Segal, *J. Neurophysiol.* **65,** 761 (1991).
12. M. T. Lucero and P. A. Pappone, *J. Gen. Physiol.* **95,** 523 (1990).
13. S. Mennerick and C. F. Zorumski, *Nature (London)* **368,** 59 (1994).
14. J. M. Bekkers and C. F. Stevens, *Proc. Natl. Acad. Sci. U.S.A.* **88,** 7834 (1991).
15. S. Mennerick and C. F. Zorumski, *J. Neurosci.* **15,** 3178 (1995).
16. H. McIlwain, L. Buchel, and J. D. Chesire, *Biochem. J.* **48,** 12 (1951).
17. C. L. Li and H. McIlwain, *J. Physiol. (London)* **139,** 178 (1957).
18. C. Yamamoto and H. McIlwain, *J. Neurochem.* **13,** 1333 (1966).
19. K. K. Skrede and R. H. Westgaard, *Brain Res.* **35,** 589 (1971).

20. Y. Izumi, A. M. Benz, D. B. Clifford, and C. F. Zorumski, *Neurosci. Lett.* **135,** 227 (1992).
21. M. Kaneda, Y. Higashitani, R. Ohtani, S. Fujii, and H. Kato, *Yamagata Med. J.* **4,** 81 (1986).
22. J. G. R. Jeffreys, *J. Physiol. (London)* **324,** 2P (1982).
23. B. E. Alger, S. S. Dhanjal, R. Dingledine, J. Garthwaite, G. Henderson, G. L. King, P. Lipton, A. North, P. A. Schwartzkroin, T. A. Sears, M. Segal, T. S. Whittingham, and J. Williams, *in* "Brain Slices" (R. Dingledine, ed.), p. 381. Plenum, New York, 1984.
24. R. A. Nicoll and B. E. Alger, *J. Neurosci. Methods* **4,** 153 (1981).
25. F. A. Edwards, A. Konnerth, B. Sakmann, and T. Takahashi, *Pfluegers Arch.* **414,** 600 (1989).
26. M. Blanton, J. LoTurco, and A. Kriegstein, *J. Neurosci. Methods* **30,** 203 (1989).
27. K. Kato, D. B. Clifford, and C. F. Zorumski, *Neuroscience (Oxford)* **53,** 39 (1993).
28. K. Kato and C. F. Zorumski, *J. Neurophysiol.* **70,** 1260 (1993).
29. B. H. Gähwiler, *J. Neurosci. Methods* **4,** 329 (1981).
30. B. H. Gähwiler, *Trends Neurosci.* **11,** 484 (1988).
31. L. Stoppini, P. A. Buchs, and D. Muller, *J. Neurosci. Methods* **37,** 173 (1991).
32. P. K. Moore, R. C. Babbedge, P. Wallace, Z. A. Gaffen, and S. L. Hart, *Br. J. Pharmacol.* **108,** 296 (1993).
33. T. P. Misko, W. M. Moore, T. P. Kasten, G. A. Nickols, J. A. Corbett, T. G. Tilton, M. L. McDaniel, J. R. Williamson, and M. G. Currie, *Eur. J. Pharmacol.* **233,** 119 (1993).
34. W. M. Moore, R. K. Webber, G. M. Jerome, F. S. Thoeng, T. P. Misko, and M. G. Currie, *J. Med. Chem.* **37,** 38867 (1994).
35. M. Zhou, S. A. Small, E. R. Kandel, and R. D. Hawkins, *Science* **260,** 1946 (1993).
36. T. J. O'Dell, R. D. Hawkins, E. R. Kandel, and O. Arancio, *Proc. Natl. Acad. Sci. U.S.A.* **88,** 11285 (1991).
37. Y. Izumi, A. M. Benz, D. B. Clifford, and C. F. Zorumski, *Exp. Neurol.* **121,** 14 (1993).
38. K. P. S. J. Murphy, J. H. Williams, N. Bettache, and T. V. P. Bliss, *Neuropharmacology* **33,** 1375 (1994).

[26] Synaptic Transmission in Hippocampal Slice

Erin M. Schuman

Introduction

The hippocampal slice is a preparation used by many scientists to study the mechanisms of synaptic transmission and plasticity. There are many features of this preparation that make it a particularly attractive experimental system. The most noteworthy is perhaps, the cortical anatomy of the hippocampus (1, 2). Because the cell and dendritic layers have an extremely organized cortical layout, extracellular field potentials can be recorded almost effortlessly in many of the subfields of the hippocampus. In addition, "blind" intracellular and whole-cell recordings can be easily obtained from pyramidal cell somata owing to the compact distribution of the cell bodies in both the dentate gyrus and the stratum pyramidale of the CA fields. An additional advantage, applicable to all brain slice preparations, is the ability to manipulate both the extra- and intracellular milieu of the neurons pharmacologically. Agonists and antagonists can be added directly to the Ringer's solution bathing the slice. In addition, drugs such as enzyme inhibitors can be injected directly into neurons through micropipettes or patch pipettes. Based on the relative efficacy of extracellular versus intracellular introduction of enzyme inhibitors, one can sometimes infer the synaptic locus of the drug effect of the drug; for example, a postsynaptic effect or a presynaptic effect.

Hippocampal Anatomy

The hippocampal formation is comprised of several distinct regions including the dentate gyrus, the hippocampus proper (which can be divided into three CA fields: CA1, CA2, and CA3), the subicular complex (which can also be divided into the subiculum, the presubiculum, and the parasubiculum), and the entorhinal cortex (which is divided into cell layers 1–5 and medial and lateral divisions) (3, 4) (see Fig. 1). There is a reasonably linear flow of information through the hippocampal formation beginning with the axons of pyramidal neurons in entorhinal cortex which form a fiber tract known as the perforant path. Most of the perforant path makes synapses onto the granule cells of the dentate gyrus. The axons of the granule cells, called the mossy fibers, in turn make connections with the CA3 pyramidal neurons.

Methods in Neurosciences, Volume 31

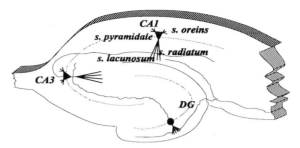

FIG. 1   Schematic diagram of the hippocampal slice showing principal cell types and some major excitatory synaptic connections. Labeled areas include the dentate gyrus (DG), the CA3 and CA1 pyramidal neuron fields, and some of the layers of the hippocampus including strum oriens, stratum pyramidale, stratum radiatum, and stratum lacunosum-moleculare.

The Schaffer collaterals (axons of CA3 cells) contact the CA1 neurons whose axons ultimately connect to neurons in the subiculum. The above description is, in fact, deceivingly simple. One of the major criticisms of hippocampal slice physiologists is that we fail to take into account many of the other relevant synaptic connections in the hippocampal formation (3). This offense comes into several forms: first, we sometimes ignore the importance of interneurons present in the slice and their potency in both producing and shaping synaptic response (5). Second, there are other synaptic connections present in some slices which are often overlooked. For example, the axons of CA3 neurons also contact other CA3 neurons (6) and a significant fraction of fibers in the perforant path project directly into stratum radiatum of the CA1 field to contact directly the CA1 pyramidal neurons (1, 2, 4). Last, the notion that the mechanisms of synaptic transmission and hippocampal function can be elucidated by the deconstructionist slice approach is probably incorrect. There are many transverse connections which run along the long axis of the hippocampus that are not spared in the slice preparation (4). These connections are likely just as important as those preserved in the slice preparation in determining the functional output of the hippocampus.

Hippocampal Slice Preparation and Storage

As with other brain slice preparations, hippocampal slices are relatively easy to prepare and maintain. The most important considerations in preparing healthy slices is the speed and the gentility of the dissection. Ideally, the whole process, from decapitation of the animal to storage of the slices, should take no more than 5 min. A short description of the procedure is as follows: the rat is decapitated and the brain is removed and quickly placed in chilled and oxygenated (95% O_2, 5% CO_2) artificial cerebrospinal fluid (ACSF). The brain is then placed on filter paper moistened with ACSF, bisected, and the hippocampus is removed from each hemisphere and placed in chilled, oxygenated ACSF. Slices can be prepared by using a Vibratome or by using a device which utilizes a razor blade and gravity. In the latter case, the hippocampi are placed on a stage driven by a micrometer. The arm with the mounted razor blade is dropped from a height approximately 3 inches above the hippocampi. The arm is then lifted slowly and carefully so that the newly cut slices stick to the side of the razor blade. Slices are removed from the razor blade with a soft artist type brush. Once cut, slices can be stored submerged in oxygenated ACSF or at the interface of ACSF and oxygen. In the interface environment, slices can be placed on filter paper moistened by ACSF and placed in an oxygenated and humidified holding chamber. Slices maintained in the interface environment will remain healthy for up to 10 hr following the dissection, but it is usually best if experiments are conducted within 4–6 hr.

Field Potential Recording: Theory and Practice

One of the most attractive features of the hippocampus is the ability to obtain extracellular field potential recordings, allowing one to monitor the synaptic strength of an entire population of synapses. The regular and ordered anatomical arrangement of the hippocampus make it an ideal preparation for field recordings. Typically, researchers record extracellular population excitatory postsynaptic potentials (fEPSP) or action potentials (population spikes) in the dendritic (stratum radiatum) or cell body (stratum pyrimidale) region of the hippocampus, respectively. For measurements of synaptic transmission that are relatively uncontaminated by changes in membrane excitability the fEPSP is the recording of choice. A representative example of an fEPSP is shown below in Fig. 2. At strong stimulus intensities one may record an fEPSP that is contaminated at its peak by a population spike. In some cases, there may be contaminating population spikes which are not always visible in the fEPSP traces. Thus, measurements of peak potential are not always accurate and most researchers prefer to measure the initial slope of the fEPSP. The true peak of an fEPSP is equivalent to the slope multiplied by the time-to-peak. If the time-to-peak is constant then the slope of the fEPSP is roughly equivalent to the true peak potential. Measuring the initial slope of the fEPSP has the additional advantage of circumventing the contribution of inhibitory synaptic transmission to the population response.

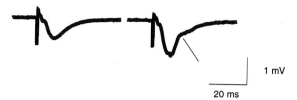

1 mV

20 ms

FIG. 2 Representative field excitatory postsynaptic potentials (*f*EPSPs) obtained by recording extracellularly in the stratum radiatum in response to electrical stimulation of the Schaffer collaterals. The first downward deflection is the stimulus artifact, the small downward hump that follows is the presynaptic fiber volley, and the largest negativity is the actual synaptic response. The arrow on the trace on the right indicates the presence of a small population spike.

Obtaining Two Independent Afferent Pathways in Single Slice

Synaptic transmission and plasticity can be extremely variable from day to day and from slice to slice. Therefore, it is often desirable to conduct both control and experimental conditions in the same slice. This can be achieved by stimulating two independent sets of afferent fibers in the same slice and recording responses with a single electrode (Fig. 3). Because one cannot actually see the stimulated fibers, a method for ascertaining whether our two stimulating electrodes are stimulating nonoverlapping sets of fibers is

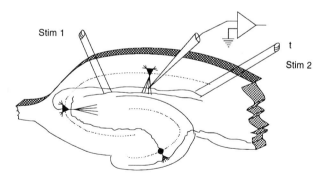

FIG. 3 Stimulating and recording electrode arrangement for examining two independent pathways in the hippocampal slice. Two bipolar stimulating electrodes (stim 1 and stim 2) are placed on either side of an extracellular recording electrode placed in stratum radiatum. The independence of the pathways can be ascertained by examining paired-pulse facilitation (see above).

required. One way to determine pathway independence is to use paired-pulse facilitation (PPF), the phenomenon whereby two closely spaced (25–100 msec) stimuli delivered to the same afferent fibers will result in the enhancement of the postsynaptic response to the second stimulus (7). Thus, we would expect that if S_1 and S_2 are stimulating overlapping sets of fibers, then stimulating one pathway immediately before the other would alter the response of the second pathway. Alternatively, if the pathways are independent then there should be no enhancement of S_1 when it is preceded by S_2, and vice versa.

Intracellular Recording with Microelectrodes

Conventional intracellular recording can be used to monitor the activity of individual neurons. The technique offers the advantage of allowing one to gain intracellular access to a single cell, maintain some control over its somatic membrane potential, inject small substances, and record for prolonged periods of time without substantially dialyzing the postsynaptic neuron. Fine micropipettes are pulled to small tip diameters (~1 μm) and filled with a concentrated salt solution such as KCl(2–3M) or cesium acetate (2 M). The compact distribution of the pyramidal cell bodies in the hippocampal slice makes it easy to position the micropipette. In fact, the quality of the electrode notwithstanding, the initial positioning of the micropipette is probably the most important factor in obtaining a good intracellular recording with ease. The pipette should be aligned parallel to and above the cell body layer. Usually the micromanipulator holding the micropipette is connected to a motorized driver. Initially, the micropipette is advanced with large steps until the top of the slice and the first few cell bodies are encountered. To monitor membrane resistance, small current pulses can be delivered through the pipette. The size of the motorized steps can now be decreased and the capacitance compensation on the electrode can be increased to facilitate penetration. Intracellular recordings are obtained by near simultaneous small steps and "buzzes", transient increases in the capacitance compensation of the electrode. Once a cell is impaled, long duration (1 sec) hyperpolarizing pulses (−0.01 to −0.05 nA) can help "tune-up" the cell by promoting the hyperpolarization of the cell to a healthy resting membrane potential (−60 to −70 mV).

Whole-Cell Voltage Clamp Recording

Whole-cell voltage-clamp recording offers the following advantages over standard intracellular recording: better voltage control of the membrane potential, increased access for introduction of drugs, peptides, etc., and the ability to record and resolve small spontaneous and miniature synaptic events. Recording small synaptic events can be extremely useful in ascertaining the synaptic locus of plasticity induced by synaptic activity or by pharmacological manipulations. Spontaneous synaptic potentials are caused by neurotransmitter release elicited by spontaneous action potentials in presynaptic neurons. Miniature synaptic potentials are caused by spontaneous neurotransmitter release in the absence of presynaptic action potentials (e.g., in the presence of a Na^+ channel blocker). Changes in the size of miniature synaptic events reflect either modifications of postsynaptic receptors or changes in vesicular content. Changes in the frequency of miniature synaptic events may reflect changes in the probability of neurotransmitter release or the addition or subtraction of postsynaptic receptor sites.

We will focus on methods employed in the blind patching technique; individuals interested in visualized recording technique should consult several other sources (8, 9). The blind patching technique involves simply lowering a patch pipette into the pyramidal cell layer of the hippocampus while monitoring the resistance changes across the electrode. A gigaohm seal is obtained on a cell membrane without visualizing the cell and the membrane is ruptured to obtain whole-cell access.

Advancing Pipette and Getting Seals

Before introducing the pipette into the bath or tissue a small amount of positive pressure should be applied (e.g., with a syringe about 0.1–0.5 ml) to avoid any interaction of the pipette with debris or dead cells on the surface of the slice. In thick slices the patch electrode should be advanced 1–2 cell bodies (~50 μm) below the surface of the slice before one seriously begins seeking gigaohm seals. Seals may be obtained in either the current- or voltage clamp recording mode. A small repetitive current or voltage pulse is applied across the pipette to monitor changes in resistance as the electrode is advanced through the slice. The positive pressure can be removed at any point after the dead superficial cells are passed. As the electrode encounters cell bodies, changes in the resistance can be observed. In general the larger the initial change in resistance the better the chances of obtaining a gigaohm seal. If the positive pressure has not been removed it should be removed following this initial jump in resistance. This should further increase the resistance of the seal. Usually, in order to obtain a full gigaohm seal, suction is required. In general the best and most durable seals are those which require the least amount of suction and "playing with" to obtain them. After the gigaohm seal is obtained it is sometimes desirable to allow the seal to stabilize for a few minutes before attempting to rupture or "breakthrough" the patch.

Rupturing is usually accomplished by short puntate episodes of either suction or positive pressure.

In whole-cell recording special consideration should be given to continuous monitoring and compensation of the access resistance. The access resistance represents both the series resistance of the pipette and the residual resistance of the seal and ruptured patch. Large access resistances can impose significant voltage errors into whole-cell recording. Especially dangerous is the possibility that series resistance may change during the course of an experiment leading to apparent changes in EPSC size which are not biological in nature. As such, it is extremely important to continuously monitor the series resistance and compensate as much as possible for the error imposed (9).

Long-Term Potentiation

Long-term potentiation (LTP) is a form of activity-dependent plasticity that is extensively studied in the hippocampal slice preparation (10). Following LTP induction, the strength of synaptic transmission is enhanced for periods of hours. As it is often useful to compare the effects of various synaptic or pharmacological manipulations with LTP a brief description of the methods for inducing LTP is included here (see Fig. 4).

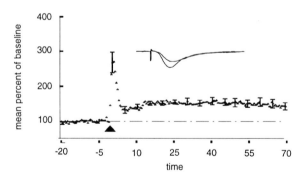

FIG. 4   Example of long-term potentiation produced in hippocampal slices as a result of high-frequency stimulation. Four trains of tetanic stimulation (100 Hz for 1 sec each) were delivered at the time indicated by the triangle. Synaptic strength remained elevated for at least 1 hr following LTP induction. Also shown are two representative fEPSPs before and after LTP induction.

Tetanus-Induced Long-Term Potentiation

Long-term potentiation is initiated when postsynaptic neurons become depolarized, NMDA receptor channels are activated, and Ca^{2+} flows into the postsynaptic neuron (10, 11). One way to induce LTP is to use high-frequency stimulation to release large amounts of neurotransmitter to depolarize the postsynaptic cell. Different laboratories use different types of synaptic stimulation to induce LTP (10). One popular protocol is to deliver 1–4 trains of stimuli at a frequency of 100 Hz for 1 sec each. Another protocol involves so-called theta burst stimulation where short bursts of stimuli are delivered at frequencies approaching the theta frequency (12). It is useful to define LTP as synaptic enhancement that lasts for at least 1 hr to distinguish it from the short-term forms of plasticity that may involve different cellular processes. Long-term depression can be elicited by applying prolonged 1 Hz stimulation (5–15 min) (13, 14).

Pairing-Induced Long-Term Potentiation

Tetanus-induction protocols utilize endogenous neurotransmitter release during high-frequency stimulation to depolarize postsynaptic neurons and induce LTP. With intracellular or whole-cell recording one can inject depolarizing current to substitute for neurotransmitter release (15). This depolarization (to ~0 mV) is coupled with low frequency stimulation (1 Hz) of the Schaffer collaterals to deliver an adequate amount of neurotransmitter to activate NMDA channels. In order to adequately depolarize neurons it is desirable to use a cesium-containing electrolyte (e.g., cesium gluconate or cesium acetate). Cesium diffuses into the cell and blocks K^+ channels (16), allowing you to control the membrane potential of the cell. Following the pairing of depolarization and low frequency stimulation (30–45 sec) the membrane potential is returned to its prepairing value and the normal frequency of synaptic stimulation is resumed.

Conclusions

The hippocampal slice is an extremely useful preparation for studies of synaptic transmission and plasticity. The regular and ordered anatomy of the hippocampus makes extracellular recordings feasible and informative. In addition, intracellular and whole-cell voltage clamp recordings are easy to obtain and allow researchers to control the membrane potential of the cell and to introduce pharmacological tools directly into individual neurons. Lastly, because the hippocampal slice is such a popular preparation, one can take advantage of the many published studies in designing and interpreting experiments.

References

1. R. Lorente De No, *J. Psychol. Neurol.,* 113 (1934).
2. S. Ramon y Cajal, "The Structure of Ammon's Horn." Thomas, Springfield, Illinois, 1968.
3. D. G. Amaral and M. P. Witter, *Neuroscience (Oxford)* **31,** 571 (1989).
4. M. P. Witter, H. J. Groenewegen, F. H. L. D. Silva, et al. *Prog. Neurbiol.* **33,** 161 (1989).
5. E. H. Buhl, K. Halasy, and P. Somogyi, *Nature (London)* **368,** 823 (1994).
6. L. W. Swanson, J. M. Wyss, and W. M. Cowan, *J. Comp. Neurol.* **181,** 681 (1978).
7. T. Manabe, D. J. A. Wyllie, D. J. Perkel, and R. A. Nicoll, *J. Neurophysiol.* **50,** 1451 (1993).
8. F. A. Edwards, A. Konnerth, B. Sakmann, and T. Takahaski, *Pfluegers Arch.* **414,** 600 (1989).
9. R. Sherman-Gold, (ed.). "The Axon Guide." Axon Instruments, Foster City, California, 1993.
10. T. V. P. Bliss, and G. L. Collingridge, *Nature (London)* **361,** 31 (1993).
11. A. U. Larkman and J. J. B. Jack *Curr. Opin. Neurobiol.* **5,** 324 (1995).
12. J. Larson, D. Wong, and G. Lynch, *Brain Res.* **368,** 347 (1986).
13. S. M. Dudek and M. F. Bear, *Proc. Natl. Acad. Sci. U.S.A.* **89,** 4363 (1992).
14. R. M. Mulkey and R. C. Malenka, *Neuron* **9,** 967 (1992).
15. B. Gustafsson, H. Wigstrom, W. C. Abraham, and Y. Y. Huang, *J. Neurosci.* **7,** 774 (1987).
16. F. Bezanilla and C. M. Armstrong, *J. Gen. Physiol.* **70,** 549 (1972).

[27] Redox Congeners of Nitric Oxide, N-Methyl-D-Aspartate Receptors, and Intracellular Calcium Ion

Won-Ki Kim, Jonathan S. Stamler, and Stuart A. Lipton

Introduction

In the central nervous system (CNS), nitric oxide (NO·) and its closely related redox congeners have been reported to be both neuroprotective and neurodestructive. This apparent paradox can be explained, at least in part, by the finding that NO effects on neurons are dependent on the redox state. Moreover, the chemical milieu may influence the redox state of the NO group. These claims are highlighted by the finding that tissue concentrations of cysteine approach 700 μM in settings of cerebral ischemia (1), levels of thiol that would be expected to influence both the redox state of the system and the NO group itself (2).

In neurons, nitric oxide may serve as a neurotransmitter between nerve cells, as a paracrine signal between different types of cells, or as an autocrine signal in a given cell. Nitric oxide is produced following stimulation of N-methyl-D-aspartate (NMDA) receptors and the subsequent influx of Ca^{2+}. This results in stimulation of neuronal nitric oxide synthase. Nitric oxide, induced in this manner, reportedly affects neuronal plasticity, including roles in neurite outgrowth, synaptic transmission, and long-term potentiation (LTP) (3–9). However, excessive activation of NMDA receptors has also been shown to be associated with various neurological disorders, including focal ischemia, epilepsy, trauma, neuropathic pain, and chronic neurodegenerative maladies, such as Parkinson's disease, Huntington's disease, amyotrophic lateral sclerosis, and acquired immunodeficiency syndrome (AIDS) dementia (10).

The activity of the NMDA receptor has been shown to be regulated by several different allosteric modulators which interact with specific sites (11). These include coagonist binding sites for glutamate (or NMDA) and glycine, and a redox modulatory site(s) defined by the existence of critical thiol group(s) (11, 12). Our laboratory and collaborators (13) have identified two cysteine residues in the rat NMDAR1 subunit (Cys-744 and Cys-798) mediating redox modulation of the NMDA receptor–channel complex, although other cysteine residues may also be involved. This redox modulatory site(s) controls the frequency of opening of the receptor-coupled ion channel: when oxidized (probably to a disulfide bond), channel activity is downregulated and when chemically reduced, the channel opens more frequently, allowing additional permeation of Ca^{2+} following agonist binding.

Excess influx of Ca^{2+} is associated with neurotoxicity (10). One possible contributory mechanism is that excessive rises in neuronal Ca^{2+} lead to an increase in constitutive nitric oxide synthase activity, in turn leading to release of nitric oxide which damages surrounding neurons (14, 15). However, several groups have been unable to demonstrate direct toxicity of nitric oxide on cortical neurons and some have even found that NO donor compounds protect against NMDA neurotoxicity (16). Similar discordant effects of nitric oxide have been reported in other tissues. To address this paradox, we studied the NO-related activity by considering its various redox-related states and their distinctive chemistries: NO^+ (nitrosonium ion, with one less electron than $NO\cdot$), $NO\cdot$ (nitric oxide, with one free electron in its outer orbital and, therefore, a radical species), and NO^- (nitroxyl anion, with one additional electron compared to $NO\cdot$) (2).

Nitric Oxide, Superoxide, Peroxynitrite, and Neurotoxicity

The addition of $NO\cdot$ releasing agents, such as S-nitrosocysteine (Cys-NO; 200 μM) or 3-morpholinosydnonimine (SIN-1; 1 mM) to rat cerebrocortical cultures caused dose- and time-dependent neuronal cell killing; neuronal cell damage or death was confirmed by trypan blue exclusion and lactate dehydrogenase (LDH) leakage. The toxicity of Cys-NO was inhibited by preincubation with superoxide dismutase (SOD, 50 U/ml) and catalase (CAT, 50 U/ml), whereas the rate of $NO\cdot$ generation was unaffected or slightly increased (as demonstrated by a nitric oxide sensor electrode, World Precision Instruments, Sarasota, FL). We confirmed this superoxide-dependent toxicity with SIN-1 and the N-hydroxynitrosamine (NONOate) spermine–NO (17, 18). These events are best rationalized as follows: Cys-NO and SIN-1 rapidly liberate nitric oxide ($NO\cdot$), which in turn reacts with endogenous superoxide ($O_2^{\cdot-}$) to induce neuronal damage through formation of peroxynitrite ($OONO^-$).

$$NO\cdot + O_2^{\cdot-} \rightarrow ONOO^- \overset{H^+}{\rightleftharpoons} ONOOH$$
$$ONOOH \rightleftharpoons O{=}N\diagdown_{O}{\cdots}OH \rightarrow NO_3^- + H^+$$

$ONOO^-$ and its degradation products $\rightarrow$ neuronal cell death

According to this explanation, the addition of SOD scavenges $O_2^{\cdot-}$ and is expected to limit the formation of peroxynitrite and its degradation products that cause neuronal cell death (19). This pathway is supported by the direct demonstration of neurotoxic effects of $OONO^-$ on neurons (19). As predicted, SOD–CAT did not prevent the lethal action of $OONO^-$. The neurotoxic effects of peroxynitrite could be due to nitration of tyrosine residues on critical proteins, peroxidation of lipids, and oxidation of sulfhydryls (20, 21). Interestingly, NMDA receptor activation has been reported to lead to production of $NO\cdot$ (14) and O_2^- (22), and thus presumably would generate $OONO^-$, at least under some conditions. Observations that increased nitric oxide synthase (NOS) activity leads to DNA damage and hence stimulation of poly(ADP-ribose) synthase (23), which compromises cell viability, are consistent with the nitrosative and nitrative actions of $OONO^-$. However, neuronal susceptibility to $NO\cdot$ and/or O_2^--induced damage could depend on the nature of the neurons and on the environmental conditions. For example, cerebellar granule cells, known to contain high levels of NOS (24, 25), could be more resistant than other neurons to NO-induced cell death (26). We also favor the possibility that lower levels of $OONO^-$ may serve a signaling function.

Nitric oxide reacts with thiol groups under physiological conditions at only very slow rates. In addition, we found that $NO\cdot$ itself is not primarily responsible for regulation of NMDA receptor activity, as we found no correlation between rates of $NO\cdot$ release from NO donor compounds (measured by the $NO\cdot$ sensor electrode) and the amplitude of NMDA-evoked responses (monitored by whole-cell recording with a patch electrode or by digital calcium imaging with the dye Fura-2). For example, ascorbate (or thiol) markedly enhanced the rate of release of $NO\cdot$ from Cys-NO. Yet, at the same time, these reducing agents markedly diminished the inhibitory effects of Cys-NO on NMDA-evoked currents (19), presumably by eliminating NO^+ equivalents ($NO^+ \xrightarrow{e^-} NO\cdot$) and thus preventing transnitrosation of NMDA receptor thiol at the redox modulatory site(s) (see below). Furthermore, other $NO\cdot$-releasing reagents such as diethylamine–NO and spermine–NO did not alter NMDA-mediated increase in $[Ca^{2+}]_i$. Hence, in our view nitric oxide is not primarily responsible for downregulation of NMDA receptor activity; other redox-related states of the NO group appear to be involved (but see Ref. 27 in which nitric oxide is thought to act at the zinc site of the NMDA receptor because of inhibition of NO effects by EDTA. We note, however, that the transfer of an NO^+ equivalent to a protein thiol group, a reaction termed S-nitrosylation may also depend on catalytic amounts of transition metals. In particular, metals can facilitate nitrosative reactions involving $NO\cdot$ (2). Metal-ion chelators, such as EDTA, prevent these reactions. Therefore, inhibition of the effects of NO donors by metal-ion chelators does not necessarily imply that the NO group itself is acting at a site where metals bind, such as the zinc site).

Nitrosonium (NO+), S-Nitrosylation of NMDA Receptor Redox Sites, and Neuroprotection

We next examined the effects of several NO group donors on NMDA receptor-mediated neurotoxicity. Sodium nitroprusside (SNP), which has NO^+ character, significantly protected rat cortical neurons against NMDA-induced neurotoxicity, whereas SNP alone had no significant affect when added to cultures (19). SNP does not liberate $NO \cdot$ spontaneously in the absence of reductive activation, but will react with thiolate anion (RS^-) on the NMDA redox site(s). A complex is formed in which sulfur is bound to nitrogen; this reaction is best rationalized in terms of S-nitros(yl)ation (Fig. 1).

Downregulation of NMDA receptor activity may occur directly (through formation of the above complex) or by way of subsequent oxidation of the redox site to disulfide (represented by a dashed line in the chemical structure in Fig. 1). Likewise, SNP decreased NMDA receptor activity as demonstrated by whole-cell recording with a patch electrode or by digital calcium imaging with the calcium-sensitive dye Fura-2. In the presence of SOD, Cys-NO also attenuated NMDA-evoked Ca^{2+} influx, a prerequisite for NMDA receptor-mediated neurotoxicity. Not surprisingly therefore, under the same conditions, application of Cys-NO ameliorated NMDA receptor-mediated neurotoxicity. These findings can be explained best by Cys-NO donating NO^+. Thus, facile transfer of an NO^+ equivalent to thiol groups of the NMDA receptor (i.e., heterolytic fission of RS-NO) results in a nitrosothiol derivative of the NMDA receptor, which downregulates receptor activity. Under these conditions, any $NO \cdot$ produced by alternative homolytic cleavage of Cys-NO is prevented from entering the neurotoxic pathway with $O_2 \cdot^-$ by the presence of SOD. This statement should not be construed to imply the

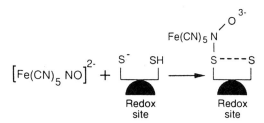

FIG. 1 Mechanism of protection from NMDA receptor-mediated neurotoxicity by sodium nitroprusside (SNP). Redox site indicates one or more thiol groups on the NMDA receptor–channel complex comprising a redox modulatory site that can be S-nitros(yl)ated, possibly facilitating disulfide bond formation. Reprinted from *Neuropharmacology* **33,** S. A. Lipton and J. S. Stamler, Actions of redox-related congeners of nitric oxide at the NMDA receptor, 1229–1233. Copyright 1994, with kind permission from Elsevier Science Ltd, The Boulevard, Langford Lane, Kidlington OX5 1GB, UK.

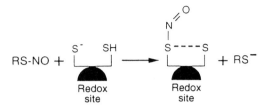

FIG. 2   Mechanism of protection from NMDA receptor-mediated neurotoxicity by *S*-nitrosocysteine as an example of the more general structure, RSNO. In this case the redox modulatory site of the NMDA receptor is transnitrosated from the RSNO. Reprinted from *Neuropharmacology* **33**, S. A. Lipton and J. S. Stamler, Actions of redox-related congeners of nitric oxide at the NMDA receptor, 1229–1233. Copyright 1994, with kind permission from Elsevier Science Ltd, The Boulevard, Langford Lane, Kidlington OX5 1GB, UK.

existence of free nitrosonium ions under our conditions at physiological pH. Rather, a transnitrosation reaction is envisaged that transfers the NO group in the NO^+ form to a thiol group to downregulate NMDA receptor activity, possibly through facilitation of disulfide formation (Fig. 2).

Interconversion of Redox States of the NO Group

One-electron transfer processes are required to interconvert redox-related forms of the NO group (2). Thus, to emphasize further that ambient redox conditions will dictate the potential neuroprotective or neurotoxic effects of a given nitroso- compound, we incubated sodium nitroprusside with excess ascorbate or thiol (cysteine or *N*-acetylcysteine) to show that we could convert nitroprusside to a neurotoxin. These substances act as reducing agents, thereby promoting the evolution of $NO \cdot$ from sodium nitroprusside (i.e., $NO^+ \xrightarrow{e^-} NO \cdot$), which in turn leads to neurotoxicity by reaction with O_2^- (Fig. 3) (19). Furthermore, as predicted, under our conditions SOD protects against this toxic mechanism by scavenging O_2^- and thus attenuating peroxynitrite formation.

S-Nitroso proteins form readily under physiologic conditions and possess biological activities such as endothelial relaxing factor (EDRF)-like effects of vasodilatation and inhibition of platelet aggregation (28). Candidate endogenous species may include RS-NO and iron-nitrosyls (NO^+ equivalents), mimicked in our experiments by the exogenous addition of *S*-nitrosocysteine and SNP, respectively (2). Glutathione is another endogenous thiol that reacts with the NO group to form *S*-nitrosoglutathione, a compound that has been identified *in vivo* (29); this may help explain the effects of glutathione on the redox site of the NMDA receptor (30). Compared to $NO \cdot$, the relative

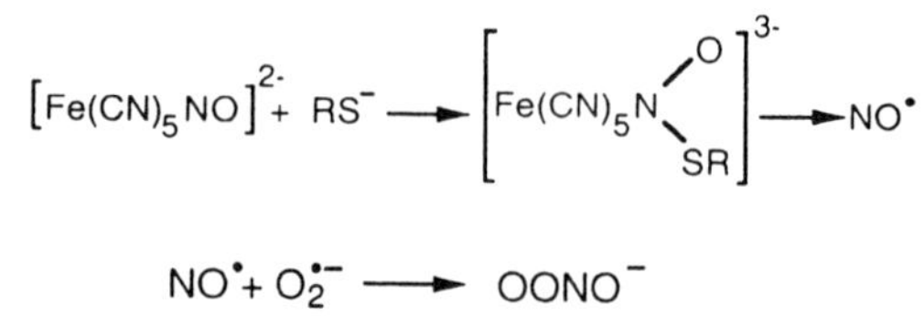

$$[Fe(CN)_5NO]^{2-} + RS^- \longrightarrow \left[Fe(CN)_5N \begin{array}{c} O \\ \diagdown \\ SR \end{array}\right]^{3-} \longrightarrow NO^\bullet$$

$$NO^\bullet + O_2^{\bullet-} \longrightarrow OONO^-$$

FIG. 3 Mechanism of neurotoxicity engendered by sodium nitroprusside (SNP). The NO group of SNP (in the form of NO^+) gains an electron from a donor to liberate nitric oxide ($NO\cdot$). Nitric oxide then reacts with superoxide anion to generate peroxynitrite. Peroxynitrite or one of its breakdown products is neurotoxic (19). Reprinted from *Neuropharmacology* 33, S. A. Lipton and J. S. Stamler, Actions of redox-related congeners of nitric oxide at the NMDA receptor, 1229–1233. Copyright 1994, with kind permission from Elsevier Science Ltd, The Boulevard, Langford Lane, Kidlington OX5 1GB, UK.

stability of RS-NO implies a lack of reactivity of the NO group (NO^+) with superoxide (2). Formation of RS-NO at the redox modulatory site of the NMDA receptor, therefore, may not only downregulate NMDA-associated ion channel activity and consequently harmful Ca^{2+} influx (16), but may also provide a means to avoid neurotoxic reactions of $NO\cdot$ with superoxide.

Nitroglycerin to Downregulate NMDA Receptor Activity and to Ameliorate Neurotoxicity

Based on the above findings, the ideal NO group donor drug would be one that does not rapidly generate $NO\cdot$, but rather would react readily with the critical thiol group(s) of the redox modulatory site(s) of the NMDA receptor to inhibit Ca^{2+} influx. We therefore studied nitroglycerin (NTG) as an exemplary compound. Specifically, this drug does not spontaneously liberate nitric oxide to any significant extent, and it is known to react readily with thiol groups forming derivative thionitrites (RS-NO) or thionitrates (RS-NO$_2$) (Fig. 4) (16, 19).

Using whole-cell recording via patch clamp electrodes and digital calcium imaging with Fura-2, we found that nitroglycerin inhibits NMDA-evoked currents and Ca^{2+} influx (16, 19). Strong evidence that this effect of nitroglycerin is mediated by its reactions with thiol in the above-illustrated manner came from a series of chemical experiments. These studies showed that specific alkylation of thiol groups with N-ethylmaleimide (NEM) completely abrogated the inhibitory effect of nitroglycerin on subsequent NMDA-evoked responses (16).

The finding that nitroglycerin could inhibit NMDA-evoked responses was corroborated by the demonstration that nitroglycerin also significantly ame-

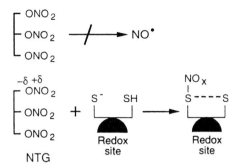

FIG. 4 Mechanism of protection from NMDA receptor-mediated neurotoxicity by nitroglycerin (NTG). Nitroglycerin does not by itself generate nitric oxide ($NO\cdot$). Instead, NMDA receptor activity is downregulated by NTG via transfer of NO_x (where x is 1 or 2) to a redox modulatory site of the receptor, possibly facilitating disulfide bond formation. Reprinted from *Neuropharmacology* **33**, S. A. Lipton and J. S. Stamler, Actions of redox-related congeners of nitric oxide at the NMDA receptor, 1229–1233. Copyright 1994, with kind permission from Elsevier Science Ltd, The Boulevard, Langford Lane, Kidlington OX5 1GB, UK.

liorates NMDA-induced neuronal killing in cerebrocortical cultures (16, 19). In addition, preliminary data suggest that high doses of nitroglycerin are neuroprotective in rat models of focal ischemia under conditions of nearly constant systemic blood pressure and cerebral blood flow (31). These parameters are held stable by inducing tolerance to the systemic effects of nitroglycerin through chronic transdermal application, or by intravenous infusion of a pressor agent concurrently with nitroglycerin.

Nitroxyl Anion to Downregulate NMDA Receptor Activity, Ameliorating Neurotoxicity

So far, we have shown that nitrosonium cation equivalents (NO^+) downregulate NMDA receptor activity by reaction with RS^- at a redox modulatory site of the NMDA receptor (19). More recently, we have extended these observations by studying the effects of nitroxyl anion (NO^-), which contains one additional electron compared to $NO\cdot$. Under physiological conditions, NO^- can be synthesized by the reaction of $NO\cdot$ with SOD[Cu(I)] (32). Nitroxyl anion generated enzymatically in this fashion or by the addition of exogenous NO^- donors was applied to rat cortical neurons in culture while monitoring their NMDA-evoked responses during patch-clamp recording or digital Ca^{2+} imaging with Fura-2.

The NMDA-evoked responses were inhibited by three distinct NO^- donors, and this effect was prevented by preincubation with the reducing re-

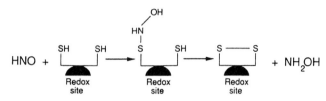

FIG. 5   Mechanism of the action of NO⁻ to oxidize sulfhydryls of a redox modulatory site of the NMDA receptor via an RSNHOH intermediate and thus downregulate receptor activity. This prevents the excessive influx of Ca^{2+} and subsequent neuronal injury. Reprinted from *Neuropharmacology* **33,** S. A. Lipton and J. S. Stamler, Actions of redox-related congeners of nitric oxide at the NMDA receptor, 1229–1233. Copyright 1994, with kind permission from Elsevier Science Ltd, The Boulevard, Langford Lane, Kidlington OX5 1GB, UK.

agent dithiothreitol (DTT). Moreover, similar concentrations of the NO⁻ donors consistently ameliorated NMDA receptor-mediated neurotoxicity to a significant degree. These effects can be explained by the known reaction of NO⁻ (in the singlet state) with thiol, which leads to sulfhydryl oxidation to disulfide, through intermediate formation of RSNHOH, in this case at a redox modulatory site of the NMDA receptor (Fig. 5).

Nitroxyl anion produced by SOD[Cu(I)] appears to be in the triplet state (a lower energy state of the paired electrons in the outer molecular orbital than the singlet state). This species reacts preferentially with oxygen (rather than thiol) to generate peroxynitrite. In preliminary studies, using both patch-clamp recording and digital calcium imaging, we have shown that peroxynitrite decreased NMDA receptor responses by oxidizing thiol group(s) of the redox modulatory site(s) of the NMDA receptor. Indeed, incubation of NO⁻ donors with SOD[Cu(I)], downregulated NMDA–receptor activity, an effect best explained by formation of OONO⁻.

Conclusion

In summary, the actions of the NO group are correlated with its redox state. Nitric oxide appears to be associated, at least in part, with neuronal injury caused by excessive activation of the NMDA receptor. In this scenario peroxynitrite forms from the reaction of nitric oxide (NO ·) with superoxide anion ($O_2^{\cdot -}$). Drugs containing the NO group in alternative redox states may prevent NMDA receptor-mediated neurotoxicity. This mechanism appears to involve S-nitrosylation or oxidation of critical thiols to disulfide bonds in the redox modulatory site(s) of the NMDA receptor to downregulate channel activity. Nitric oxide itself reacts only very slowly with thiol (-SH) groups

(33–35), consistent with the lack of effect of NO · on NMDA receptor-mediated responses under physiological conditions. It is inferred that the alternative redox-activated states, NO^+ and NO^-, are likely to be the species directly reactive with thiol groups. Under certain conditions, neuroprotection may be achieved by limiting peroxynitrite formation from NO ·, for example with nitric oxide synthase inhibitors and superoxide scavengers, or by fostering NO group transfer (of NO^+ or NO^-) to the critical thiol groups of the redox modulatory site(s) of the NMDA receptor. It is becoming increasingly evident that in addition to NMDA receptors, biological activities of many other proteins containing critical cysteine residues can be regulated by S-nitrosylation and oxidation, in a sense similar to the type of control exerted by phosphorylation of critical tyrosine or serine groups (28). This chemical reaction may be a new and ubiquitous pathway for the molecular control of protein function by potential reactive sulfhydryl centers.

References

1. A. Slivka and G. Cohen, *Brain Res.* **608,** 33 (1993).
2. J. S. Stamler, D. J. Singel, and J. Loscalzo, *Science* **258,** 1898 (1992).
3. G. A. Bohme, C. Bon, J. M. Stutzmann, A. Doble, and J. C. Blanchard, *Eur. J. Pharmacol.* **199,** 379 (1991).
4. T. J. O'Dell, R. D. Hawkins, E. R. Kandel, and O. Arancio, *Proc. Natl. Acad. Sci. U.S.A.* **88,** 11285 (1991).
5. E. M. Schuman and D. V. Madison, *Science* **254,** 1503 (1991).
6. D. T. Hess, S. I. Patterson, D. S. Smith, and J. H. Skene, *Nature (London)* **366,** 562 (1993).
7. M. Zhuo, S. A. Small, E. R. Kandel, and R. D. Hawkins, *Science* **260,** 1946 (1993).
8. P. R. Montague, C. D. Gancayco, M. J. Winn, R. B. Marchase, and M. J. Friedlander, *Science* **263,** 973 (1994).
9. E. M. Schuman and D. V. Madison, *Science* **263,** 532 (1994).
10. S. A. Lipton and P. A. Rosenberg, *N. Engl. J. Med.* **330,** 613 (1994).
11. S. A. Lipton, *Trends Neurosci.* **16,** 527 (1993).
12. E. Aizenman, S. A. Lipton, and R. H. Loring, *Neuron* **2,** 1257 (1989).
13. J. M. Sullivan, S. F. Traynelis, H. S. V. Chen, W. Escobar, S. F. Heinemann, and S. A. Lipton, *Neuron* **13,** 929 (1994).
14. V. L. Dawson, T. M. Dawson, E. D. London, D. S. Bredt, and S. H. Snyder, *Proc. Natl. Acad. Sci. U.S.A.* **88,** 6368 (1991).
15. V. L. Dawson, T. M. Dawson, D. A. Bartley, G. R. Uhl, and S. H. Snyder, *J. Neurosci.* **13,** 2651 (1993).
16. S. Z. Lei, Z. H. Pan, S. K. Aggarwal, H. S. Chen, J. Hartman, N. J. Sucher, and S. A. Lipton, *Neuron* **8,** 1087 (1992).
17. M. Feelisch, J. Ostrowski, and E. Noack, *J. Cardiovasc. Pharmacol.* **14**(Suppl. 11), S13 (1989).

18. C. M. Maragos, D. Morley, D. A. Wink, T. M. Dunams, J. E. Saavedra, A. Hoffman, A. A. Bove, L. Isaac, J. A. Hrabie, and L. K. Keefer, *J. Med. Chem.* **34,** 3242 (1991).
19. S. A. Lipton, Y. B. Choi, Z. H. Pan, S. Z. Lei, H. S. Chen, N. J. Sucher, J. Loscalzo, D. J. Singel, and J. S. Stamler, *Nature (London)* **364,** 626 (1993).
20. R. Radi, J. S. Beckman, K. M. Bush, and B. A. Freeman, *Arch. Biochem. Biophys.* **288,** 481 (1991).
21. R. Radi, J. S. Beckman, K. M. Bush, and B. A. Freeman, *J. Biol. Chem.* **266,** 4244 (1991).
22. S. Pietri, M. Culcasi, and J. Bockaert, *Nature (London)* **364,** 535 (1993).
23. J. Zhang, V. L. Dawson, T. M. Dawson, and S. H. Snyder, *Science* **263,** 687 (1994).
24. J. Garthwaite, *Trends Neurosci.* **14,** 60 (1991).
25. D. S. Bredt, P. M. Hwang, and S. H. Snyder, *Nature (London)* **347,** 768 (1990).
26. D. S. Bredt and S. H. Snyder, *Neuron* **8,** 3 (1992).
27. L. Fagni, M. Olivier, M. Lafon-Cazal, and J. Bockaert, *Mol. Pharmacol.* **47,** 1239 (1995).
28. J. S. Stamler, D. I. Simon, J. A. Osborne, M. E. Mullins, O. Jaraki, T. Michel, D. J. Singe, and J. Loscalzo, *Proc. Natl. Acad. Sci. U.S.A.* **89,** 444 (1992).
29. B. Gaston, J. Reilly, J. M. Drazen, J. Fackler, P. Ramdev, D. Arnelle, M. E. Mullins, D. J. Sugarbaker, C. Chee, D. J. Singel, J. Loscalzo, and J. S. Stamler, *Proc. Natl. Acad. Sci. U.S.A.* **90,** 10957 (1993).
30. N. J. Sucher and S. A. Lipton, *J. Neurosci. Res.* **30,** 582 (1991).
31. S. Sathi, P. Edgecomb, S. Warach, K. Manchester, T. Donaghey, P. E. Stiege, F. E. Jensen, and S. A. Lipton, *Soc. Neurosci. Abstr.* **19,** 849 (1993).
32. M. E. Murphy and H. Sies, *Proc. Natl. Acad. Sci. U.S.A.* **88,** 10860 (1991).
33. K. R. Hoyt, L. H. Tang, E. Aizenman, and I. J. Reynolds, *Brain Res.* **592,** 310 (1992).
34. W. A. Pryor and J. W. Lightsey, *Science* **214,** 435 (1981).
35. W. A. Pryor, D. F. Church, C. K. Govinden, and J. W. Lightsey, *J. Org. Chem.* **47,** 156 (1982).

[28] Cytosolic and Nuclear Calcium Imaging by Confocal Microscopy

Michel Burnier, Gabriel Centeno, and Hans R. Brunner

Introduction

Cytosolic free calcium serves as an ubiquitous intracellular second messenger system to couple a stimulus to a response in various excitable and nonexcitable tissues. Thus, changes in intracellular calcium concentrations have been implicated as a critical regulatory mechanism in cellular activities such as cell contraction, differentiation, and proliferation (1, 2). Cytosolic free calcium is the intracellular messenger system for many hormones and growth factors. Moreover, increases in intracellular calcium are associated with the activation of enzymes such as the neuronal and endothelial constitutive Ca^{2+}-calmodulin nitric oxide (NO) synthases, with the stimulation of DNA synthesis and with gene activation (2–5). In neurons, variations in intracellular calcium concentrations are also coupled to several activities including the release of neurotransmitters, the regulation of membrane excitability, gene expression, and the modulation of synaptic plasticity (6, 7).

Several methods have been used to measure cytosolic free calcium concentration. The first approaches often implied the disruption of the cells either to microinject a calcium indicator or to insert a calcium-sensitive microelectrode. These techniques were restricted to large cells. Owing to the development and availability of fluorescent calcium-sensitive dyes (Quin-2, Fura-2, or Fluo-3) and their cell-penetrating acetoxymethyl esters, the determination of intracellular free calcium has been considerably simplified and has become possible in many kinds of cells (8–10). In addition to measuring intracellular calcium in cell suspensions in cuvettes, the combined use of these new fluorescent indicators with video microscopic techniques has improved the analysis by allowing a more precise assessment of the distribution of the instantaneous changes in intracellular calcium concentration in single cells. With the single cell observations, it was noted that the changes in intracellular free calcium are often transient and may be spatially nonhomogeneous suggesting thereby that calcium cycling in a target cell may represent an important, spatially and temporally distinct messenger function (11).

The development of fluorescence imaging by video microscopy has afforded new opportunities to localize the changes in intracellular free calcium. Yet, video-imaging is limited by the fact that it produces an imperfect image of the plane of the specimen and that the recorded fluorescence is coming from the plane of focus as well as from out-of focus regions of the specimen. Thus, unless out-of-focus informations are removed computationally using complex deconvolution algorithms, it is difficult to determine precisely the intracellular origin of the fluorescence (12). This technical limitation has been resolved by confocal microscopy which allows one to obtain fluorescence images in which out-of-focus blur is very weak (13). The purpose of this chapter is to review the advantages and limitations of confocal microscopy to measure intracellular free calcium.

Principles of Confocal Microscopy

The general principles of confocal scanning optical microscopy are presented in Fig. 1 and 2 (13, 14). With the conventional fluorescence microscopy, a two-dimensional optical image of the specimen is produced by collecting the fluorescence emitted throughout its whole depth and not just at the plane in which the objective lens is focused. With confocal microscopy, a conical illuminating beam of light focused to a point by a conventional objective lens is scanned through the sample as shown in Figure 1. Figure 2 illustrates the principle in more details: the excitatory light, generally emitted by an argon laser, passes from the illuminating pinhole through an excitation filter. It is then reflected by a dichroic mirror and is focused by the microscope objective to a diffraction limited spot at the focal plane within the specimen. Fluorescence emissions produced within the focused plane as well as within the illuminated cones above and below it, are collected by the objective and pass through the dichroic mirror and the emission filter (not indicated). However, only the emissions from the in focus volume (voxel) are able to

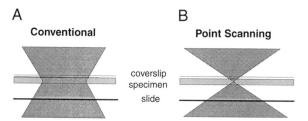

FIG. 1   Schematic representation of the illumination experienced by a specimen during conventional fluorescence microscopy (A) and single-point illumination at the focal plane by a cone of light focused to a diffraction limited point in a unitary beam laser scanning optical microscope (B). Adapted from Ref. 13.

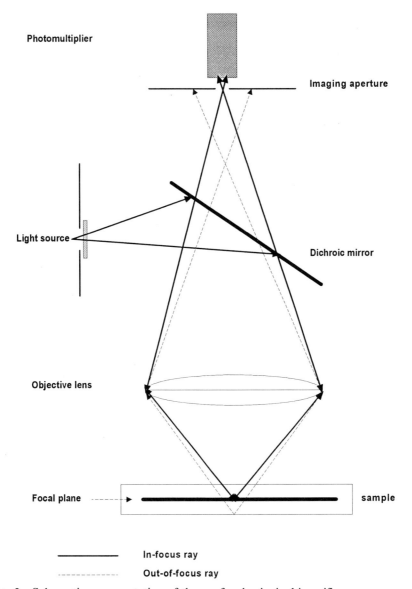

Photomultiplier

Imaging aperture

Light source

Dichroic mirror

Objective lens

Focal plane ⤑ sample

——————— In-focus ray

- - - - - - - - - - - Out-of-focus ray

FIG. 2   Schematic representation of the confocal principal in epifluorescence scanning optical microscopy. Adapted from Ref. 13. See text for detailed description.

cross the imaging aperture to be detected by the photomultiplier. The out-of-focus fluorescence emissions do not pass through the pinhole of the imaging aperture and therefore do not contribute to the final confocal image.

Advantages and Limits of Confocal Microscopy

It is obvious from these principles that confocal microscopy offers some important advantages over conventional fluorescence microscopy. The first is the elimination of the out-of-focus blur. However, the thickness of the optical section which is a function of both the lens and the pinhole aperture is critical for the performance of confocal microscopy. Indeed, if the size of the pinhole is too large, the out-of-focus reflections are able to reach the photodetector and the depth discrimination is completely abolished. The second advantage is the capability for noninvasive optical sectioning of intact and even living specimens both in the x, y and in the x, z planes thus providing optical sections parallel to the optical axis of the microscope. Because the thickness of the section is often in the micron range, the resolution in the Z axis is generally less than in the X and Y axes. These sections can than be used to reconstruct three-dimensional cellular tomographic images. Another advantage of confocal microscopy over conventional fluorescence micros-copy is the higher axial and lateral resolution which is improved with the use of an objective with a high numeric aperture ($NA > 1.2$).

These characteristics of confocal microscopy are also important for cal-cium imaging. The narrower depth of field and the improved lateral resolution allow a better visualization of the intracellular details and thus a better representation of the distribution of intracellular calcium. Because a small and relatively constant volume of the cell is sampled throughout the focal plane, measurement errors due to path length variations are minimized and photobleaching of the entire sample is also relatively low. Even though the light levels of fluorescence produced by the monitoring of free calcium concentration in thin sections are low, the use of sensitive photomultiplier tubes allow to load the specimens with relatively low concentrations of the dyes. With a low dye concentration in the cytosol, the cytosolic free calcium is not buffered. Thus the dye interferes minimally with the parameter under investigation. One possible limitation of confocal microscopy is the acquisi-tion speed which is a constraint of the number of pixels per second that can be scanned. For this reason, the technique is most appropriate for slowly developing (several seconds) calcium stages. However, the temporal resolu-tion can be improved considerably by reducing the number of scanned lines. A single line can be chosen in the cell and scanned repeatedly with a variable frequency up to 500 Hz leading to a millisecond time resolution (15, 16). The acquired lines are then arranged downward in a sequential order to build up a line-scan image.

Commercially available laser-scanning confocal microscopes have generally been developed with an argon laser as the source of light. The laser provides a green (514 nm) and a blue (488 nm) excitation line which allows one to use only some of the calcium indicators such as Fluo-3. In contrast to Fura-2 which is characterized by a shift in its excitation spectrum on binding with calcium, the binding of Fluo-3 to calcium results in an increase of the green fluorescence emission up to 40-fold (10). As discussed below, ratiometric methods for calibration cannot be used.

To be able to use the various calcium indicators that yield a reliable measurement of changes in intracellular calcium by a ratiometric method, UV confocal laser scanning microscopes have been developed by modification of the commercially available microscopes (17, 18). These systems appear to have lateral and axial resolutions suitable for applications to physiological and morphological studies using UV-activated fluorophores. Another approach to ratio-metric calcium measurements using conventional confocal laser-scanning microscopy has been proposed by Lipp and Niggli who loaded their cardiac myocytes with two calcium indicators (Fluo-3 and Fura Red) with excitation spectra in the visible range of wavelengths (16). When intracellular free calcium rises, Fluo-3 shows an increase in fluorescence whereas that of Fura Red decreases. The Fluo-3/Fura Red ratio also allows one to obtain reliable values of intracellular free calcium and obviates the potential problem of changes in intracellular dye concentration. However, for this method to be valid, the two dyes must have the same intracellular distribution.

Fluorescent Calcium Indicators

One of the major technical breakthroughs for measuring or imaging intracellular calcium has been the development of fluorescent dye molecules such as Fura-2, Indo-1, or Fluo-3 containing a calcium binding site modeled on the well-known chelator EGTA (8–10). These molecules have a much higher affinity for calcium than for magnesium, the most important competing divalent ion. This selectivity is crucial because free cytosolic magnesium concentration is about 10,000-fold greater than that of calcium. The early calcium indicators (Quin-2, Fura-2, and Indo-1) require excitation at UV wavelengths near the cutoff for transmission through glass. They were therefore unsuitable for conventional confocal microscopes which contain a laser excitation source in the visible range that is less expensive than an UV laser.

Fluo-3, the calcium indicator most frequently used in confocal microscopy, responds to calcium with an increase in fluorescence intensity without a shift in excitation or emission. This indicator has the advantage of being excitable in the visible wavelengths (500–505 nm) but it lacks any pair of wavelengths for ratioing. Thus, an increase in fluorescence could be the result of a change in intracellular dye concentration as well as the expression of a real increase in cytosolic free calcium. The potential problem of variations in intracellular dye concentration is particularly relevant when studying cells with shapes that vary during the investigation as it is the case in contracting heart or smooth muscle cells. A change in intracellular dye concentration may also occur if the dye leaks from the cell or bleaches rapidly. Fluo-3 has a weak calcium binding capacity (K_d of about 400 nM) which allows one to measure higher calcium values. However, like other dyes, it can easily be saturated, and one is therefore not able to distinguish between calcium levels above several micromolar. This means also that if cells are loaded with a large amount of the dye, it is easy to buffer all variations in intracellular calcium concentration.

The main advantage of these new dyes is that intact cells or tissues can be loaded with the indicator simply by incubating the specimen with the esterified form of the dye. The lipophilic esters penetrate cells rapidly and are cleaved by cellular esterases. Once cleaved, the hydrophilic free form of the indicator is trapped within the cell. In some tissues, problems of dye loading have been reported due to incomplete ester cleavage, extrusion of the dye by anion transport mechanisms, endocytosis, or compartmentalization of the fluorescence into organelles. Endocytosis and compartmentalization can be reduced by careful solubilization of the dye esters with the detergent pluronic acid or by lowering the incubation temperature (19). Anion transport can be prevented with the addition of probenecid or sulfinpyrazone (20). If dye loading remains problematic, the free form of the indicator can be introduced directly into the cytosol by permeabilization of the cells, by microinjection, by iontophoresis from a sharp intracellular electrode, or by using the patch-clamp technique (15, 16, 21). With these approaches, the dye loading is effective but the possibility exists that the cellular homeostasis is perturbed.

Sequestration of the calcium indicators in the various intracellular organelles is an important aspect to consider when evaluating the changes in intracellular fluorescence induced by various cell stimulations. There is indeed some evidence that all calcium dyes do not necessarily localize in the cytosol only and differences in electric charges may perhaps account for the distribution of certain dyes in the nucleus. Thus, rhodamine appears to accumulate preferentially in mitochondria and has been used for these properties as a mitochondria-specific stain or to monitor mitochondrial membrane potential in living cells (22). In certain cells, the Fura-2 fluorescence also accumulates in association with mitochondria (23). As far as Fluo-3 is concerned, there is often a marked fluorescence observed in the nucleus which would suggest either an increased concentration of Fluo-3 in the nucleus or an increased fluorescence intensity when the dye is located in the nucleus due in part to autofluorescence of the intranuclear components (24). A difference in Fluo-3 fluorescence in the cytosol and in the nucleus may have some important methodological consequences. The first is that it becomes mandatory to calibrate separately the nucleus and the cytoplasm. In addition, one must be very prudent with the interpretation of the pseudocolor scale provided by the computer since one color may not necessarily indicate a comparable amount of calcium in the nucleus and in the cytosol.

Nuclear Calcium Signaling

While performing single cell calcium imaging using either video microscopy or confocal laser scanning microscopy, several investigators have found an increase in nuclear fluorescence at rest and in stimulated conditions as illustrated in Fig. 3 (9, 15, 16, 24–33). The finding of a nuclear/cytosolic calcium gradient in neuronal as well as in nonneuronal cells is the source of many important questions regarding not only the mechanism of the calcium gradient but also its physiological role. The permeability of the nuclear envelope is indeed of major interest because calcium is likely to regulate several physiological functions in the nucleus such as gene expression through a variety of calcium binding proteins including protein kinase C and calmodulin, which are constitutively present in the nucleus, or other calcium response elements (34). Calcium may also contribute to some pathological events by activation of proteases and endonucleases that play, for example, a role in cell death.

Today, there is increasing evidence indicating that the changes in nuclear fluorescence observed with the various calcium indicators are not merely technical artifacts but represent true variations in intranuclear calcium concentrations (32). However, there appears to be a large heterogeneity of the calcium distribution at rest or during stimulation depending on the cell type as well as on the stimulus or the experimental conditions. After stimulation, the nuclei of some cells seem to be protected from the changes in cytosolic calcium (25, 33) whereas in other cell types the changes in intranuclear calcium either follow or exceed those occurring in the cytoplasm (for review see Ref. 32). These very different behaviors suggest that the nucleus does not necessarily follow the changes in the cytoplasm passively. Although the mechanisms of the changes in intranuclear calcium are still under investigation, there is convincing evidence today that the changes in intranuclear calcium are linked both to a diffusion of the calcium ions released from the cytosolic stores through the permeable nuclear pores and to an activation of calcium transport systems leading to a calcium release from the nuclear envelope (35–39). In this respect, studies performed in isolated nuclei have demonstrated both an ATP-dependent uptake of calcium and a release of the accumulated calcium by inositol 1,4,5-trisphosphate in the nuclear envelope (36–39). The nuclear membrane thus possesses several elements allowing regulation of the nuclear pore permeability and hence the intranuclear processes.

Conclusions

The concomitant development of fluorescent calcium indicators and confocal laser scanning microscopy has improved the study of the intracellular events considerably occuring at rest or after cell stimulation. The single-cell calcium imaging approach has not only allowed monitoring and visualization of the changes in cytosolic free calcium concentration, it has also revealed the importance of the variations in calcium occurring in other subcellular sites. Thus, the rapidly growing literature evaluating the regulation of intranuclear calcium in various physiological and pathological conditions using confocal microscopy is an evident reflexion of the incredible potential of this new methodology. The possibilities of confocal microscopy are however not limited to the determination of intracellular calcium. Numerous other fluorescent indicators exist to monitor other intracellular ions such as H^+, Mg^{2+}, Na^+, or different second messenger systems like cyclic AMP or protein kinase C (40, 41).

Acknowledgment

This work was supported by a grant of the Fonds National suisse de la Recherche Scientifique (Grant No. 32-32535.91)

References

1. W. H. Moolenaar, L. H. K. Defize, and S. W. de Laat, *J. Exp. Biol.* **124,** 359 (1986).
2. H. Rasmussen and P. Q. Barrett, *Physiol. Rev.* **64,** 938 (1984).
3. A. L. Boynton, J. F. Whitfield, and J. P. MacManus, *Biochem. Biophys. Res. Commun.* **95,** 745 (1980).
4. B. A. White, *J. Biol. Chem.* **260,** 1213 (1985).
5. C. Nathan and Q. W. Xie, *Cell (Cambridge, Mass.)* **78,** 915 (1994).
6. H. Bading, D. D. Ginty, and M. E. Greenberg, *Science* **260,** 181 (1993).

7. J. Wang, J. J. Renger, L. C. Griffith, R. J. Greenspan, and C. F. Wu, *Neuron* **13,** 1373 (1994).

8. G. Grynkiewcz, M. Poenie, and R. Y. Tsien, *J. Biol. Chem.* **260,** 3440 (1985).

9. J. P. Y. Kao, A. T. Harootunian, and R. Y. Tsien, *J. Biol. Chem.* **264,** 8179 (1989).

10. A. Minta, J. P. Y. Kao, and R. Y. Tsien, *J. Biol. Chem.* **264,** 8171 (1989).

11. M. J. Berridge and R. F. Irvine, *Nature (London)* **341,** 197 (1989).

12. F. S. Fay, W. Carrington, and K. E. Fogarty, *J. Microsc.* **153,** 133 (1989).

13. D. Shotton and N. White, *Trends Biol. Sci.* **14,** 435 (1989).

14. G. J. Brakenhoff, P. Blom, and P. Barends, *J. Microsc.* **117,** 219 (1979).

15. A. Hernandez-Cruz, F. Sala, and P. R. Adams, *Science* **24,** 858 (1990).

16. P. Lipp and E. Niggli, *Cell Calcium* **14,** 359 (1993).

17. C. Bliton, J. Lechleiter, and D. E. Clapham, *J. Microsc.* **169,** 15 (1993).

18. K. Kuba, S. Y. Hua, and T. Hayashi, *Cell Calcium* **16,** 205 (1994).

19. A. Malgaroli, D. Milani, J. Meldolesi, and T. Pozzan, *J. Cell Biol.* **105,** 2145 (1987).

20. F. DiVirgilio, T. H. Steinberg, J. A. Swanson, and S. C. Silverstein, *J. Immunol.* **140,** 915 (1988).

21. B. J. Bacskai, P. Wallen, V. Lev-Ram, S. Grillner, and R. Y. Tsien, *Neuron* **14,** 19 (1995).

22. L. V. Johnson, M. L. Walsh, B. J. Bockus, and L. B. Chen, *J. Cell Biol.* **88,** 526 (1981).

23. S. F. Steinberg, J. P. Bilezikian, and Q. Al-Awqati, *Am. J. Physiol.* **253,** C744 (1987).

24. M. Burnier, G. Centeno, E. Burki, and H. R. Brunner, *Am. J. Physiol.* **266,** C1118 (1994).

25. D. A. Williams, K. E. Fogarty, R. Y. Tsien, and F. S. Fay, *Nature (London)* **318,** 558 (1985).

26. R. V. Yelamarti, B. A. Miller, R. C. Scaduto, F. T. S. Yu, D. L. Tillotson, and J. Y. Cheung, *J. Clin. Invest.* **85,** 1799 (1990).

27. R. W. Tucker and F. S. Fay, *Eur. J. Cell Biol.* **51,** 120 (1990).

28. Y. Horikoshi, T. Furuno, R. Teshima, J. Sawada, and M. Nakanishi, *Biochem. J.* **304,** 57 (1994).

29. S. L. Mironov and A. Hermann, *J. Biolumin. Chemilumin.* **9,** 233 (1994).

30. I. Gillot and M. Whitaker, *Cell Calcium* **16,** 269 (1994).

31. J. Carroll, K. Swann, D. Whittingham, and M. Whitaker, *Development (Cambridge, UK)* **120,** 3507 (1994).

32. B. Himpens, H. de Smedt, and R. Casteels, *Cell Calcium* **16,** 239 (1994).

33. F. A. Al-Mohanna, K. W. T. Caddy, and S. R. Bolsover, *Nature (London)* **367,** 745 (1994).

34. M. Sheng, M. A. Thompson, and M. E. Greenberg, *Science* **252,** 1427 (1991).

35. D. M. O'Malley, *J. Neurosci.* **14**(10), 5741 (1994).

36. P. Nicotera, D. J. McConkey, D. P. Jones, and S. Orrenius, *Proc. Natl. Acad. Sci. U.S.A.* **86,** 453 (1989).

37. O. V. Gerasimenko, J. V. Gerasimenko, A. V. Tepikin, and O. H. Petersen, *Cell (Cambridge, Mass.)* **80,** 439 (1995).

38. A. N. Malviya, P. Rogue, and G. Vincendon, *Proc. Natl. Acad. Sci. U.S.A.* **87,** 9270 (1990).

39. N. Matter, M. F. Ritz, S. Freyermuth, P. Rogue, and A. N. Malviya, *J. Biol. Chem.* **268,** 732 (1993).
40. S. R. Adams, A. T. Harootunian, Y. I. Buechler, S. S. Taylor, and R. Y. Tsien, *Nature (London)* **349,** 694 (1991).
41. C. S. Chen and M. Poenie, *J. Biol. Chem.* **268,** 15812 (1993).

[29] Effects of Nitrogen Monoxide on Cellular Iron Metabolism

Des R. Richardson and Prem Ponka

Introduction

Nitrogen Monoxide Mediates Many Biological Effects by Binding to Iron

Virtually every field in physiology has been influenced by the discovery that nitrogen monoxide (NO) has an important role as a biological messenger molecule. Nitrogen monoxide is a small, relatively unstable, potentially toxic, diatomic gas that is produced in a wide variety of mammalian cells (1–7). In biological systems it is likely that NO may exist in a number of redox-related states (see Section 1.3 below), and in the context of this article, the term "nitrogen monoxide (NO)" is used generically to encompass all these species.

Nitrogen monoxide has been shown to have two principal functions in cells, namely servoregulation and cytotoxicity. In considering servoregulation, NO is produced in small amounts under physiological conditions and mediates vasorelaxation, regulates blood pressure, controls the adhesion and aggregation of platelets and neutrophils, and is also involved in neurotransmission. Most of these actions are mediated through the binding of NO to iron (Fe) in the heme prosthetic group of soluble guanylate cyclase (3). The role of Fe in mediating the functions of NO is also apparent when examining the cytotoxic effects of the molecule. The cytotoxic functions of NO are observed when it is produced in much larger amounts by activated macrophages, hepatocytes, and other cells following their exposure to cytokines or microbial products such as lipopolysaccharide. Nitrogen monoxide produced via such high output systems inhibits the proliferation of intracellular pathogens and tumor cells. These cytotoxic effects can be explained by the reactivity of NO and Fe in the active sites of critically important macromolecules (8–16). The significance of the interaction of NO and Fe in mediating cytotoxicity will be discussed further in Section 1.4.

Before discussing in detail the interaction of NO with Fe and its important implications, we will briefly describe the way cells obtain and metabolize Fe.

Methods in Neurosciences, Volume 31

Cellular Iron Metabolism: Iron Is an Obligate Requirement for Life

Iron is indispensible for both cellular growth and proliferation, as it plays critical roles in many essential metabolic processes (17). Iron is essential for the activity of many ubiquitous enzymes such as the cytochromes, hemoglobin, myoglobin, ribonucleotide reductase, and aconitase (18, 19). In mammals, Fe is transported by the plasma glycoprotein, transferrin (Tf), which has a molecular mass of 80 kDa and contains two specific high affinity Fe(III) binding sites (for review see Ref. 20). The initial event in cellular Fe acquisition from Tf is the binding of the protein to its specific transferrin receptor (TfR) on the cell membrane. The TfR is then internalized within an endocytotic vesicle and the Fe is released from Tf by intravesicular acidification. Iron is then transported to intracellular sites by largely unknown mechanisms, and the Fe-free apo-Tf attached to the TfR returns to the cell surface where Tf is released (20, 21). In some normal and neoplastic cell types other processes apart from the classic receptor-mediated endocytosis mechanism can also remove Fe from Tf (22–26). In melanoma cells this second process increases after saturation of the TfR and is consistent with adsorptive pinocytosis (27).

After Fe has been released from Tf, it enters a poorly characterized compartment known as the intracellular Fe pool (28). The Fe in this pool has been suggested to be bound to low molecular weight ligands such citrate, amino acids, ascorbate, ATP, and sugars (28). However, despite numerous studies, there is little evidence to implicate a role for low molecular weight ligands in intracellular Fe transport (29–31), and further characterization of this compartment is essential. The Fe that enters the hypothetical intracellular pool can be used for the synthesis of heme in the mitochondrion or can be incorporated into nonheme Fe-containing proteins such as ribonucleotide reductase and iron-sulfur (Fe-S) proteins. Iron in excess of that required for metabolic functioning is then stored in ferritin. Ferritin is a ubiquitous protein whose function is the sequestration and storage of Fe. Mammalian ferritin consists of a protein shell having a molecular mass of 450 kDa and is composed of 24 subunits. Ferritin shells can contain up to 4500 atoms of Fe per molecule (32).

Cellular Fe uptake via the Tf–TfR pathway and intracellular Fe storage in ferritin are coordinately regulated through a feedback control mechanism. Under conditions of low Fe supply, TfR expression (a prerequisite for Fe uptake by cells) is increased and ferritin synthesis is decreased, and these responses represent a control mechanism that maintains Fe homeostasis. An exactly opposite scenario occurs under conditions of high Fe levels. The regulation depends on specific structural elements located in the 5′ untranslated region (UTR) of ferritin mRNA and 3′-UTR of TfR mRNA.

These iron-responsive elements are highly conserved, and constitute binding sites for a cytoplasmic factor known as the iron regulatory protein (IRP, formerly called IRF, IRE-BP, FRP, or p90; reviewed in Refs. 33 and 34). Interestingly, IRP was more recently shown to be identical to the Fe-S protein, cytoplasmic aconitase (35). Cells depleted of Fe contain higher levels of activated IRP, which then forms complexes with the IRE. When the IRP binds to IRE of ferritin mRNA, it represses its translation, whereas IRP binding to IREs of TfR mRNA protects the message against degradation (33, 34). In Fe-replete cells, the RNA-binding activity of IRP is decreased and ferritin mRNA is translated while TfR mRNA is degraded. In general, IRP is a regulator of cellular Fe metabolism that monitors intracellular Fe concentrations and maintains them at appropriate levels. To understand the effects of NO on Fe metabolism, it is relevant to discuss the chemistry of NO in terms of defining possible intracellular targets.

Redox-Related Species of Nitrogen Monoxide and Biological Targets

Stamler *et al.* (4) have pointed out that NO can exist in distinct redox-related states which have different reaction specificities. These redox-related forms of NO include the nitrosonium cation (NO^+), nitric oxide ($NO\cdot$), and the nitroxyl anion (NO^-). The different targets with which these redox-related species interact may explain the rich diversity of functions ascribed to NO (4).

From a biological standpoint, the important reactions of nitric oxide ($NO\cdot$) are those with oxygen, superoxide (O_2^-), and transition metal ions (4). The biological half-life of $NO\cdot$ is in the order of seconds, and under physiological conditions, $NO\cdot$ reacts rapidly with O_2^- to form peroxynitrite ($ONOO^-$). Once $ONOO^-$ is protonated it rapidly decomposes, generating a strong oxidant, probably the hydroxyl radical [$HO\cdot$; Eq. (1)]. The $HO\cdot$ can induce lipid peroxidation and sulfhydryl oxidation resulting in cytotoxicity (36–38). The reactions shown below in Eq. (1) are well established and there is evidence to suggest that they may occur *in vivo* (36, 39).

$$NO\cdot + O_2^- \rightarrow ONOO^- + H^+ \rightleftharpoons ONOOH \rightleftharpoons HO\cdot + NO_2\cdot \rightarrow NO_3^- + H^+ \quad (1)$$

Stamler and colleagues have reported that $ONOO^-$ can mediate the S-nitrosylation of a critical sulphydryl group in glyceraldehyde-3-phosphate dehydrogenase (40). These data could indicate that under some circumstances, $ONOO^-$ may have some physiologically relevant regulatory functions.

Nitric oxide also readily forms ligand–metal complexes with transition metal ions, especially Fe in heme- or nonheme-containing proteins (4). Much interest has focused on the reactivity of NO· with Fe-S centers in proteins such as aconitase (8, 9), the Fe-S proteins of the mitochondrial respiratory system (41) and ferrochelatase (42, 42a). The inhibition of ferrochelatase activity by NO could suggest that this molecule may have a role in regulating heme synthesis, since ferrochelatase is involved in the insertion of Fe(II) into protoporphyrin to form heme (19). The interaction of NO with the heme center in guanylate cyclase activates the enzyme (3, 7, 11) and results in smooth muscle relaxation. In contrast, the binding of NO to the heme pros-thetic group of cytochrome P450 inhibits its enzymic activity (43). Interest-ingly, NO inhibits NOS activity by binding to the heme center of this protein, such that the product of this enzyme is involved in regulating its activity (44). Nitrogen monoxide also reacts rapidly with the Fe(II) of the heme center in oxyhemoglobin to form methemoglobin [Fe(III)–hemoglobin] and nitrate. This is a very important physiological reaction which serves to inacti-vate NO once it reaches the bloodstream.

Traylor and Sharma (11) have conducted a theoretical analysis of what makes NO· so suitable in fulfilling its biochemical functions. They have concluded that NO· is a unique ligand as to its reaction properties with heme and nonheme Fe proteins. One of these important properties of NO· is its very high affinity for Fe(II) heme. Moreover, NO· is uniquely capable, as compared to other ligands, of binding Fe in heme in the absence of proximal ligands (e.g., imidazole). In fact, the binding constant for the proximal imidaz-ole to Fe is decreased after NO· binds to Fe in heme, leading to the displace-ment of this residue and some change in protein conformation (11). It is likely that the breaking of the bond between this imidazole residue and the Fe in heme represents a mechanism by which NO· activates guanylate cyclase (3). However, more recent data suggest that the activation of guanylate cyclase by NO· involves displacement of both proximal and distal histidine ligands (45).

The nitrosonium cation (NO^+) may react with amide, carboxyl, and hy-droxyl groups as well as aromatic compounds (4). However, the most import-ant reaction of NO^+ under physiological conditions is its reaction with thiols (4). Nitrosothiols with biological activity may be those derived from cysteine, glutathione, and proteins such as albumin. *In vivo,* it may be that these compounds act to transfer NO^+ to thiol-containing proteins (46). S-nitrosyla-tion may control the biological activity of proteins, having functions akin to those mediated by phosphorylation (38, 46, 47). Also, *S*-nitrosothiols are relatively stable toward O_2, and nitroso proteins detected in plasma such as S-nitrosylated albumin could possibly serve as a source and a sink of NO (48, 49). Similarly, Fe-nitrosyl complexes with NO^+ character have been proposed as another possibility for intracellular NO storage (50). It is also possible that nitric oxide synthase could produce both NO· and NO⁺ depending on the redox milieu inside the cell (4).

Very little is known about the biological significance of NO⁻, although Butler *et al.* (39) have suggested that a physiological role of this species is feasible. Bastian *et al.* (51) have obtained evidence that NO⁻ can release Fe from horse spleen ferritin *in vitro*. These latter authors have suggested that NO⁻ mediated release of Fe from ferritin may be responsible for the Fe release observed from tumor target cells after exposure to NO-generating macrophages (12). It is also of interest that NO⁻ can bind to reduced cytochrome *d* (52), and also to Fe(III) heme bound to protein (53).

Importance of Nitric Oxide–Iron Interactions in Cytotoxic Effector Mechanisms

As discussed in the previous section, many of the biological effects of NO are mediated through its avid binding to Fe, and this results in a wide variety of metabolic changes, including modulation of protein activity (3, 54–56), mobilization of Fe from ferritin (51, 57), and intracellular Fe release (12, 13). Considering the role of NO in Fe release, Hibbs and others have shown that activated macrophages produce reactive nitrogen intermediates, including NO (58–61). Cocultivation of activated macrophages with tumor target cells results in the inhibition of target cell DNA synthesis and mitochondrial respiration, and in addition a concomitant loss of a large fraction of intracellular Fe (12, 13). Hibbs and associates (12) suggested that the inhibition of DNA synthesis and mitochondrial respiration may be related to the loss of Fe, because the functional integrity of these metabolic pathways are dependent on the activities of proteins containing Fe centers. These proteins include the rate-limiting enzyme of DNA synthesis, ribonucleotide reductase (14–16), NADH ubiquinone oxidoreductase, and succinate-ubiquinone oxidoreductase of the electron transport chain (62).

Studies by Drapier and Hibbs (8) demonstrated that mitochondrial aconitase (an Fe-S protein) is inhibited in NO-generating cells, and that this inhibition occurred simultaneously with the inhibition of DNA synthesis. It was suggested that the efflux of Fe observed in previous studies was due to the release of Fe from aconitase (8). Further support for the role of an interaction between NO and Fe in macrophage-induced cytostasis have come from investigations demonstrating stable Fe-nitrosyl complexes in activated macrophages (63, 64) and their target cells (65, 66). The Fe-nitrosyl complexes can be detected by electron paramagnetic resonance spectroscopy because they produce a signal at $g = 2.039$, typical of Fe-S complexes (67–70). Vanin (69) have determined that these complexes have the general chemical formula

Fe(RS)$_2$(NO)$_2$ (i.e., Fe–dithiol–dinitrosyl complex). In addition, Bastian *et al.* (66) have reported that murine tumors exhibit EPR-signals related to the formation of not only Fe(RS)$_2$(NO)$_2$ complexes but also two distinct heme–nitrosyl complexes. These two heme–nitrosyl species differ by having their Fe either five or six coordinate (66). Hence, the identification of Fe-nitrosyl complexes in tumor cells links activated macrophage-induced inhibition of Fe-containing enzymes with NO biosynthesis. Lee *et al.* (71) have identified EPR-active Fe-nitrosyl complexes in ferritin, a result that can perhaps explain an earlier observation that NO can mediate Fe release from ferritin *in vitro* (57). Moreover, it has been postulated that ferritin may act as an intracellular store for NO (71), an intriguing suggestion which would have broad physiological significance. Vanin (69) has suggested that cells exposed to NO release Fe complexed to this molecule and also to thiol ligands such as cysteine. In contrast to the studies of Hibbs and others (8, 63, 64), Vanin and associates (70) have suggested that the Fe released may be derived from loosely bound pools of nonheme Fe ("free" Fe) instead of mitochondrial Fe centers. Considering the arguments and results described above, it could be suggested that NO may be diffusing into cells, ligating to the Fe centers of proteins, removing Fe, and then diffusing from the cell in the form of Fe-nitrosyl complexes.

The extensive interaction of NO with intracellular Fe-containing molecules has prompted us to investigate its effects on the Fe metabolism of neoplastic cells. First, we thought that it was important to examine whether NO by itself could remove Fe from cells prelabeled with ^{59}Fe-Tf. This is because previous studies that demonstrated marked release of ^{59}Fe from target cells in the presence of NO-generating macrophages were done after labeling cells with the nonphysiologically relevant Fe complex, ^{59}Fe-citrate (12). Hence, it was unknown whether the Fe released was derived from a physiologically relevant intracellular Fe pool. Under physiological circumstances, most Fe is bound to Tf which donates its Fe to cells via an interaction with the specific TfR (20). Second, since Fe is essential for cellular growth and proliferation, examination of the effect of NO on Fe uptake from Tf was of interest. Third, the effects of redox-related species of NO on IRP activation, TfR expression, and Fe uptake were also examined.

Methods

Nitrogen Monoxide-Generating Compounds

The NO· generating compound used was *S*-nitro-*N*-acetylpenicillamine (SNAP) (72) and the NO·/ONOO$^-$ donor was 3-morpholinosydnonimine (SIN-1) (38). The SNAP was synthesized by the method of Field *et al.* (73)

and SIN-1 and its inactive analog SIN-1C were kind gifts from Dr. Rainer
Henning (Cassella, A. G., Frankfurt, Germany). The NO^+ generator, sodium
nitroprusside (SNP) (4, 38), was obtained from Sigma (St. Louis, MO).

Cells

The human melanoma cell line, SK-Mel-28, and the human erythroleukemia
cell line, K562, were obtained from the American Type Culture Collection
(Rockville, MD). The MDW4 cell line was derived from a DBA-2 mouse
leukemia and was kindly provided by Dr. W. Lapp (McGill University,
Department of Physiology, Montréal, Canada).

Preparation of $^{59}Fe,^{125}I$-Transferrin

Human apo-Tf was obtained from Boehringer Mannheim (Indianapolis, IN)
and was labeled with ^{59}Fe using a ferric nitrilotriacetate (NTA) complex in
a molar ratio of 1 Fe : 10 NTA via the method of Hemmaplardh and Morgan
(74). Radioiodine labeling of Tf was performed by the iodine monochloride
method of McFarlane (75) to give a specific activity of approximately 0.8
μCi ^{125}I per nmol of Tf. Unbound ^{59}Fe and ^{125}I were removed by gel filtration
through Sephadex G-25 (Pharmacia, Uppsala, Sweden) followed by vacuum
dialysis against at least four changes of 0.15 M NaCl adjusted to pH 7.4 with
sodium bicarbonate (1.4% w/v).

Iron Metabolism, Experimental Procedure

Iron and Transferrin Uptake by Cells

The effect of NO-generating compounds on ^{59}Fe- and ^{125}I-Tf uptake was
investigated by incubating cells with $^{59}Fe,^{125}I$-Tf (1.25 μM), 20 mM HEPES
(pH 7.4), and the NO-generating compounds. After this incubation the cells
were washed four times with ice-cold Hanks' balanced salt solution (BSS)
and then incubated with pronase (1 mg/ml) for 30 min at 4°C to separate
membrane-bound (pronase-sensitive) from internalized (pronase-resistant)
radioactivity, as described (23, 76).

Iron and Transferrin Release from Cells

The effect of NO-generating agents on ^{59}Fe release was examined by labeling
cells with $^{59}Fe,^{125}I$-Tf (1.25 μM) or ^{59}Fe-citrate (molar ratio of Fe : citrate =
1 : 100; [Fe] = 2.5 μM) for 2–24 hr at 37°C. After the cells were labeled they

were then washed three times with prewarmed medium and then incubated three times for 20 min at 37°C. This extensive washing procedure was implemented because, once released, the postulated NO-^{59}Fe complex can then possibly be bound by ^{125}I-apo-Tf released from the cells. The ^{59}Fe-Tf so formed can donate the released ^{59}Fe back to the cell resulting in no net ^{59}Fe release. Hence, it is important to remove Tf from the cells before the addition of NO-generating agents. Washed cells were then reincubated for 2–24 hr at 37°C in MEM or MEM containing NO-generating agents. Next, the cells and supernatant were separated via centrifugation and the radioactivity measured in each (77).

Measurement of Iron Regulatory Protein RNA-Binding Activity by the Gel-Retardation Assay

The gel-retardation assay was used to measure the interaction between IRP and IRE by established techniques (78, 79). Briefly, after incubation with RPMI alone (control) or various NO$^+$/NO· producing agents, 5×10^6 K562 cells were washed with ice-cold PBS and lysed at 4°C in 100 μl of extraction buffer [10 mM HEPES pH 7.5, 3 mM MgCl$_2$, 40 mM KCl, 5% glycerol, 1 mM dithiothreitol, and 0.2% Nonidet P-40 (NP-40)]. After lysis, the samples were centrifuged at 10,000 g for 3 min to remove nuclei. Samples of the cytoplasmic extracts are diluted to a protein concentration of 100 μg/ml in lysis buffer without Nonidet P-40, and 2 μg aliquotes were analyzed for IRP by incubation with 0.1 ng of ^{32}P-labeled pSPT-fer RNA transcript (provided kindly by Dr. Lukas Kühn, Swiss Institute for Experimental Cancer Research) (79). RNA was transcribed *in vitro* from linearized plasmid templates by T7 RNA polymerase in the presence of [α-^{32}P]CTP. To form RNA–protein complexes, cytoplasmic extracts were incubated for 10 min at room temperature with 0.1 ng of labeled RNA. Unprotected probe was degraded by incubation with 1 unit of RNase T1 for 10 min, and heparin (5 mg/ml) was then added for another 10 min to exclude nonspecific binding. RNA–protein complexes were analyzed in 6% nondenaturing polyacrylamide gels as described by Konarska and Sharp (80). In parallel experiments, samples were treated with 2% 2-mercaptoethanol prior to the addition of the RNA probe. Autoradiographs were quantitated by scanning densitometry.

Northern Blot Analysis

Northern blot analysis was performed by established techniques (81).

Results

Effect of Nitrogen Monoxide-Generating Agents on Iron Uptake from Transferrin

Because the uptake of Fe from Tf is essential for cellular proliferation (17), we investigated the effect of NO-generating agents on Fe uptake from this protein (77). Surprisingly, the NO^+-generating agent, SNP, appreciably reduced the uptake of ^{59}Fe without affecting Tf uptake (Fig. 1). Similar effects on Fe uptake were also found in the presence of the NO· generator, SNAP, or the NO·/ONOO$^-$ generator SIN-1 in the presence of superoxide dismutase (77). However, dialysis experiments demonstrated that NO did not directly remove Fe from Tf (77). Previous studies have shown that NO can ligate to the Fe-binding site of lactoferrin (82), and since the Fe-binding site of serum Tf is very similar (83), it is possible that an NO–Fe–Tf complex may form. Because endosomal acidification is involved in releasing Fe from Tf, the formation of an NO–Fe–Tf complex may affect the lability of the Fe-binding

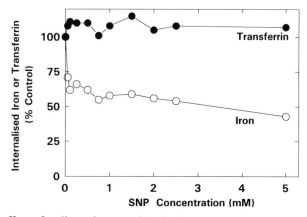

FIG. 1   The effect of sodium nitroprusside (SNP) concentration on internalized ^{59}Fe- and ^{125}I-transferrin uptake from $^{59}Fe,^{125}I$-transferrin by SK-MEL-28 melanoma cells. Cells were incubated with medium containing $^{59}Fe,^{125}I$-transferrin (1.25 μM) and various concentrations of SNP (0.05–5 mM) for 4 hr at 37°C. After this incubation period the cells were washed and incubated with pronase (1 mg/ml) for 30 min at 4°C to separate internalized from membrane-bound radioactivity. The results are the means of duplicate and triplicate determinations in a typical experiment from a total of three experiments performed. For each experimental point the standard deviation did not exceed 5%. [From: D. R. Richardson, V. Neumannova, and P. Ponka, *Biochim. Biophys. Acta* **1266,** 250 (1995) with permission].

site to acidification. Consequently, this could result in the decreased Fe uptake observed and further studies have been initiated to examine this hypothesis.

Effect of Nitrogen Monoxide-Generating Agents on Iron Release from Prelabeled Cells

The present studies using three cell lines (SK-Mel-28, K562, and MDW4 cells) found that the NO-generating agents (SNP, SNAP, and SIN-1) could not remove appreciable amounts of ^{59}Fe from cells which had been prelabeled with ^{59}Fe-Tf or ^{59}Fe-citrate (77). In contrast, these NO-producing agents could markedly inhibit [^{3}H]thymidine incorporation (Fig. 2), suggesting that Fe release was not a prerequisite for inhibiting proliferation (77). It is possible that other factors released by the macrophage or the direct physical contact of the macrophage with target cell may be required to facilitate the Fe release observed by Hibbs et al. (12). Alternatively, it could be that only certain cell types are sensitive to NO-mediated Fe release.

Effect of Nitrogen Monoxide on Iron-Regulatory Protein

As described previously, mitochondrial aconitase is one Fe-S protein the activity of which is modulated by NO, and IRP which is present in the cytosol has a high homology to mitochondrial aconitase. Therefore, the biological activity of IRP may be modulated by NO, and in fact, Drapier et al. (84) and Weiss et al. (85) have shown that NO can activate IRP RNA-binding activity. Considering the possible target sites of NO on the IRP molecule, it has been shown that in Fe replete cells IRP possesses a cubane 4Fe-4S cluster which prevents IRE binding, and in this state the protein displays aconitase activity (86–88). In contrast, in cells depleted of Fe, the Fe-S cluster is not present, and under these conditions IRP can bind to the IRE of mRNA. In addition, free sulfhydryl groups, such as cysteine-437, have also been shown to regulate the binding of IRP to the IRE (88, 89). However, the relative roles of the Fe-S cluster and sulfhydryl group(s) in IRP regulation are not well understood.

The finding that NO can increase the binding of IRP to the IRE has prompted Drapier et al. (84) and Weiss et al. (85) to suggest that NO may react directly with the Fe-S cluster of IRP, leading to a loss of the cluster and an increase in RNA-binding activity. However, two studies have sug-

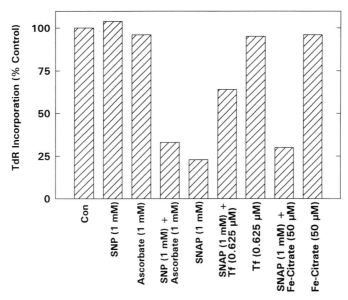

FIG. 2 The effect of the nitrosonium ion (NO⁺) generator, sodium nitroprusside (SNP), and the nitric oxide (NO·) generator, S-nitroso-N-acetylpenicillamine (SNAP), on the incorporation of [³H]thymidine by K562 cells. Cells were incubated for 18 hr at 37°C in the presence of either SNP (1 mM), ascorbate (1 mM), SNP (1 mM) plus ascorbate (1 mM), SNAP (1 mM), SNAP (1 mM) plus unlabeled diferric Tf (0.625 μM), unlabeled diferric Tf (0.625 μM), SNAP (1 mM) plus 50 μM ferric citrate, or 50 μM ferric citrate only. The cells were then washed and exposed to 10 μCi/ml of [³H]thymidine for 1 hr at 37°C. After this labeling period the cells were washed, incubated with 20% trichloroacetic acid (TCA) for 1 hr at 4°C, and then washed twice in ice-cold 10% TCA. The results are means of triplicate determinations in a typical experiment of two experiments performed. For each experimental condition the standard deviation did not exceed 5%. [From: D. R. Richardson, V. Neumannova, and P. Ponka, *Biochim. Biophys. Acta* **1266,** 250 (1995) with permission].

gested that it is not NO but ONOO⁻ which reacts with the 4Fe-4S cluster of mitochondrial aconitase and IRP (90, 91). Although Drapier *et al.* (84) and Weiss *et al.* (85) demonstrated an increase in IRP RNA-binding activity after exposure to NO, they did not consider the effects of congeners of NO on IRP-binding activity, nor the effects of these species on TfR expression and Fe uptake from Tf.

Studies by the authors using K562 cells have shown that the NO⁺ producing agent, SNP (4), decreased IRP activation (Fig. 3A), TfR mRNA (Fig. 3B), TfR number and Fe uptake (25). In contrast, the NO· generator, SNAP (72),

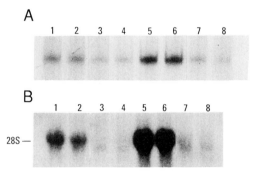

FIG. 3   Changes in (A) the RNA-binding activity of IRP and (B) the transferrin receptor mRNA levels in K562 cells incubated for 18 hr with RPMI (untreated control, lane 1) or the nitrosonium ion (NO$^+$) generator, sodium nitroprusside (SNP: 0.1 mM, 1 mM, and 5 mM; lanes 2, 3, 4, respectively), desferrioxamine (DFO, 50 μM, lane 5) or DFO (50 μM) plus SNP (0.1 mM, 1 mM and 5 mM; lanes 6, 7, 8, respectively) [From: D. R. Richardson, V. Neumannova, E. Nagy, and P. Ponka, *Blood* **86,** 3211 (1995) with permission].

increased IRP activation (Fig. 4A), and TfR mRNA (Fig. 4B) but only slightly increased TfR number and Fe uptake (25). Considering these opposing results, regulation of IRP by NO could occur at two different sites on the protein. One target site may be the sulfhydryl group of cysteine-437, whereas

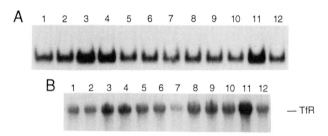

FIG. 4   Effect of the NO· generator, *S*-nitroso-*N*-acetylpenicillamine (SNAP) and the NO·/peroxynitrite (ONOO$^-$) generator, 3-morpholinosydnonimine (SIN-1) on (A) active IRP levels and (B) transferrin receptor mRNA levels in K562 cells. Cells were incubated for 18 hr with RPMI-1640 only (control, lane 1), SNAP (0.1 mM, 0.5 mM, 1 mM; lanes 2, 3, and 4, respectively), SIN-1 (0.1 mM, 0.5 mM, 1.0 mM; lanes 5, 6, 7, respectively), SIN-1 (0.1 mM, 0.5 mM, 1.0 mM) plus superoxide dismutase (SOD; 500 U/ml) and catalase (CAT; 500 U/ml) (lanes 8, 9, and 10, respectively), desferrioxamine (0.1 mM; lane 11), SOD plus CAT (lane 12). [From: D. R. Richardson, V. Neumannova, E. Nagy, and P. Ponka, *Blood* **86,** 3211 (1995). Used with permission].

a second site may be the dynamic Fe-S cluster. Regarding these two sites, it has been suggested that NO· and NO$^+$ can react at different sites on protein molecules (4), NO· via direct coordination to Fe centers and NO$^+$ via S-nitrosylation of thiol groups. It is conceivable that these redox forms of NO may have distinct roles in regulating RNA-binding activity of IRP (25).

It has been demonstrated that S-nitrosylation of cysteine sulphydryl groups by NO$^+$ may aid disulfide bond formation (38, 46, 92). The NO$^+$ generated by SNP may react with the sulfhydryl group of cysteine-437, resulting in disulfide bridging between this residue and cysteine 503 or 506, which has been shown to inhibit IRP binding to the IRE (88). Therefore, the decrease in IRP RNA-binding activity observed in our experiments in the presence of SNP could be explained by this reaction. However, further studies are required to directly determine the possible target sites of NO$^+$ on this protein. The increase in IRP RNA-binding activity in K562 cells in the presence of SNAP may be due to the reaction of NO· directly with the Fe-S cluster, resulting in a loss of the cluster and an increase in RNA binding, as proposed by Drapier *et al.* (84) and Weiss *et al.* (85). Alternatively, the NO· produced from SNAP may be converted to ONOO$^-$ intracellularly which may then interact with the Fe-S cluster. It is of interest that despite IRP activation induced by SNAP, there was only a slight increase in TfR number and Fe uptake. Further studies are necessary to determine whether intracellular generation of NO by NOS can result in changes in TfR number and Fe uptake.

Conclusions

Discovery of NO and its interaction with intracellular Fe-containing molecules has revealed previously unexpected aspects of cellular Fe metabolism. It can be predicted that NO, by virtue of its high affinity for Fe in protein targets including the IRP, is likely to dramatically affect cellular Fe metabolism as a whole. Hence, cellular Fe homeostasis could be regulated not only by Fe levels per se but also by other factors which manifest themselves, such as those that occur during inflammation, for example. On the other hand, it is tempting to speculate that the intracellular Fe level may be one factor affecting NO production. This idea is endorsed by work showing that the Fe(III) chelator, desferrioxamine, can markedly increase the transcription of NO synthase, whereas it was decreased by the addition of Fe salts (93). Considering the high affinity of NO for Fe, and its interaction with key cellular proteins containing Fe, further vigorous research is needed to define an interesting and important link between NO and Fe metabolism.

Acknowledgments

This work was supported by grants from the Medical Research Council of Canada (P.P. and D.R.R.) and a Terry Fox New Investigator Award from the National Cancer Institute of Canada (to D.R.R.). D.R.R. is a Medical Research Council of Canada Scholar.

References

1. M. A. Marletta, *Trends Biochem. Sci.* **14,** 488 (1989).
2. S. Moncada, R. M. J. Palmer, and E. A. Higgs, *Pharmacol. Rev.* **43,** 109 (1991).
3. L. J. Ignarro, *Blood Vessels* **28,** 67 (1991).
4. J. S. Stamler, D. J. Singel, and J. Loscalzo, *Science* **258,** 1898 (1992).
5. S. Moncada and E. A Higgs, *N. Engl. J. Med.* **329,** 2002 (1993).
6. R. G. Knowles and S. Moncada, *Biochem. J.* **298,** 249 (1994).
7. D. S. Bredt and S. H. Snyder, *Annu. Rev. Biochem.* **63,** 175 (1994).
8. J.-C. Drapier and J. B. Hibbs, Jr., *J. Clin. Invest.* **78,** 790 (1986).
9. J.-C. Drapier and J. B. Hibbs, Jr., *J. Immunol.* **140,** 2829 (1988).
10. Y. Henry, C. Ducrocq, J.-C. Drapier, D. Servent, C. Pellat, and A. Guissani, *Eur. Biophys. J.* **20,** 1 (1991).
11. T. G. Traylor and V. S. Sharma, *Biochemistry* **31,** 2847 (1992).
12. J. B. Hibbs, Jr., R. R. Taintor, and Z. Vavrin, *Biochem. Biophys. Res. Commun.* **123,** 716 (1984).
13. J. B. Hibbs, Jr., R. R. Taintor, Z. Vavrin, and E. M. Rachlin, *Biochem. Biophys. Res. Commun.* **157,** 87 (1988).
14. M. Lepoivre, B. Chenais, A. Yapo, G. Lemaire, L. Thelander, and J.-P. Tenu, *J. Biol. Chem.* **265,** 14143 (1990).
15. M. Lepoivre, F. Fieschi, J. Coves, L. Thelander, and M. Fontecave, *Biochem. Biophys. Res. Commun.* **179,** 442 (1991).
16. N. S. Kwon, D. J. Stuehr, and C. F. Nathan, *J. Exp. Med.* **174,** 761 (1991).
17. L. C. Kühn, H. M. Schulman, and P. Ponka, *in* "Iron Transport and Storage" (P. Ponka, H. M. Schulman, and R. C. Woodworth, eds.), p. 149. CRC Press, Boca Raton, Florida, 1990.
18. R. Cammack, J. M. Wrigglesworth, and H. Baum, *in* "Iron Transport and Storage" (P. Ponka, H. M. Schulman, and R. C. Woodworth, eds.), p. 17. CRC Press, Boca Raton, Florida, 1990.
19. P. Ponka, H. M. Schulman, and T. Cox, *in* "Biosynthesis of Heme and Chlorophylls" (H. A. Dailey, ed.), p. 393. McGraw-Hill, New York, 1990.
20. E. H. Morgan, *Mol. Aspects Med.* **4,** 1 (1981).
21. R. D. Klausner, G. Ashwell, J. Van Renswoude, J. B. Harford, and K. R. Bridges, *Proc. Natl. Acad. Sci. U.S.A.* **80,** 2263 (1983).
22. M. A. Page, E. Baker, and E. H. Morgan, *Am. J. Physiol.* **246,** G26 (1984).
23. D. R. Richardson and E. Baker, *Biochim. Biophys. Acta* **1053,** 1 (1990).
24. D. R. Richardson and E. Baker, *J. Biol. Chem.* **267,** 13972 (1992).

25. D. R. Richardson, V. Neumannova, E. Nagy, and P. Ponka, *Blood* **86,** 3211 (1995).
26. D. R. Richardson and P. Ponka, *Biochim. Biophys. Acta* **1269,** 105 (1995).
27. D. R. Richardson and E. Baker, *J. Cell. Physiol.* **161,** 160 (1994).
28. A. Jacobs, *Blood* **50,** 433 (1977).
29. I. Romslo, *in* "Iron in Biochemistry and Medicine" (A. Jacobs and M. Worwood, eds.), Vol. 2, p. 325. Academic Press, London, 1980.
30. D. Vyoral, A. Hradilek, and J. Neuwirt, *Biochim. Biophys. Acta* **1137,** 148 (1992).
31. D. R. Richardson, P. Ponka, and D. Vyoral, *Blood* **87,** 3477 (1996).
32. P. M. Harrison, S. C. Andrews, P. J. Artymiuk, G. C. Ford, D. M. Lawson, J. M. A. Smith, A. Treffry, and J. L. White, *in* "Iron Transport and Storage" (P. Ponka, H. M. Schulman, and R. C. Woodworth, eds.), p. 81. CRC Press, Boca Raton, Florida, 1990.
33. L. C. Kühn, *Br. J. Haematol.* **47,** 183 (1992).
34. R. D. Klausner, T. A. Rounault, and J. B. Harford, *Cell (Cambridge, Mass.)* **72,** 19 (1993).
35. A. Constable, S. Quick, N. K. Gray, and M. W. Hentze, *Proc. Natl. Acad. Sci. U.S.A.* **89,** 4554 (1992).
36. J. S. Beckman, T. W. Beckman, J. Chen, P. A. Marshall, and B. A. Freeman, *Proc. Natl. Acad. Sci. U.S.A.* **87,** 1620 (1990).
37. R. Radi, J. S. Beckman, K. M. Bush, and B. A. Freeman, *Arch. Biochem. Biophys.* **288,** 481 (1991).
38. S. A. Lipton, Y. B. Choi, S. H. Pan, S. Z. Lei, H. S. V. Chen, N. J. Sucher, J. Loscalzo, D. J. Singel, and J. S. Stamler, *Nature (London)* **364,** 626 (1993).
39. A. R. Butler, F. W. Flitney, and D. L. H. Williams, *Trends Pharmacol. Sci.* **16,** 18 (1995).
40. S. Mohr, J. S. Stamler, and B. Brüne, *FEBS Lett.* **348,** 223 (1994).
41. J. Stadler, T. R. Billiar, R. D. Curran, D. J. Stuehr, J. B. Ochoa, and R. L. Simmons, *Am. J. Physiol.* **260,** C910 (1991).
42. Y.-M. Kim, H. A. Bergonia, C. Muller, B. R. Pitt, W. D. Watkins, and J. R. Lancaster, Jr., *J. Biol. Chem.* **270,** 5710 (1995).
42a. V. M. Sellers, M. K. Johnson, and H. A. Dailey, *Biochemistry* **35,** 2699 (1966).
43. O. G. Khartenko, S. S. Gross, A. R. Rifkind, and J. R. Vane, *Proc. Natl. Acad. Sci. U.S.A.* **90,** 11147 (1993).
44. J. M. Griscavage, J. M. Fukuto, Y. Komori, and L. J. Ignarro, *J. Biol. Chem.* **269,** 21644 (1994).
45. A. E. Yu, S. Hu, T. G. Spiro, and J. N. Burstyn, *J. Am. Chem. Soc.* **116,** 4117 (1994).
46. J. S. Stamler, *Cell (Cambridge, Mass.)* **78,** 931 (1994).
47. S. H. Snyder, *Nature (London)* **364,** 577 (1993).
48. J. S. Stamler, D. I. Simon, J. A. Osborne, M. E. Mullins, O. Jaraki, T. Michel, D. J. Singel, and J. Loscalzo, *Proc. Natl. Acad. Sci. U.S.A.* **89,** 444 (1992).
49. J. S. Stamler, O. Jaraki, J. Osborne, D. I. Simon, J. Keaney, J. Vita, D. Singel, C. R. Valeri, and J. Loscalzo, *Proc. Natl. Acad. Sci. U.S.A.* **89,** 7674 (1992).
50. A. Mulsch, P. Mordvintcev, A. F. Vanin, and R. Busse, *FEBS Lett.* **294,** 252 (1991).

51. N. R. Bastian, M. J. P. Foster, C. L. Bello III, D. V. Kinikini, T. J. Smith, and J. B. Hibbs, Jr., *Endothelium* **3**, S12 (1995).

52. F. T. Bonner, M. N. Hughes, R. K. Poole, and R. I. Scott, *Biochim. Biophys. Acta* **25**, 133 (1991).

53. D. A. Bazylinski and T. C. Hollocher, *J. Am. Chem. Soc.* **107**, 7982 (1985).

54. J. Zhang, V. L. Dawson, T. M. Dawson, and S. H. Snyder, *Science* **263**, 687 (1994).

55. D. Salvemini, T. P. Misko, J. L. Masferrer, K. Seibert, M. G. Currie, and P. Needleman, *Proc. Natl. Acad. Sci. U.S.A.* **90**, 7240 (1993).

56. L. J. Ignarro, *Pharmacol. Toxicol.* **67**, 1 (1990).

57. D. W. Reif and R. D. Simmons, *Arch. Biochem. Biophys.* **283**, 537 (1990).

58. J. B. Hibbs, Jr., R. R. Taintor, and Z. Vavrin, *Science* **235**, 473 (1987).

59. D. J. Stuehr and C. F. Nathan, *J. Exp. Med.* **169**, 1543 (1989).

60. R. Keller, M. Geiges, and R. Keist, *Cancer Res.* **50**, 1421 (1990).

61. C. J. Lowenstein and S. H. Snyder, *Cell (Cambridge, Mass.)* **70**, 705 (1992).

62. J. B. Hibbs, Jr., R. R. Taintor, Z. Vavrin, D. L. Granger, J. C. Drapier, I. J. Amber, and J. Lancaster, Jr., *in* "Nitric Oxide from L-Arginine: A Bioregulatory System" (S. Moncada and E. A. Higgs, eds.), p. 189. Elsevier, Amsterdam, 1990.

63. J. R. Lancaster and J. B. Hibbs, Jr., *Proc. Natl. Acad. Sci. U.S.A.* **87**, 1223, (1990).

64. C. Pellat, Y. Henry, and J. C. Drapier, *Biochem. Biophys. Res. Commun.* **166**, 119 (1990).

65. J. C. Drapier, C. Pellat, and Y. Henry, *J. Biol. Chem.* **266**, 10162 (1991).

66. N. R. Bastian, C. Y. Yim, J. B. Hibbs, Jr., and W. E. Samlowski, *J. Biol. Chem.* **269**, 5127 (1994).

67. B. Commoner, J. C. Woloum, B. H. Senturia, Jr., and J. L. Ternberg, *Cancer Res.* **30**, 2091 (1970).

68. R. W. Chiang, J. C. Woolum, and B. Commoner, *Biochim. Biophys. Acta* **257**, 452 (1972).

69. A. F. Vanin, *FEBS Lett.* **289**, 1 (1991).

70. A. F. Vanin, G. B. Men'shikov, I. A. Moroz, P. I. Mordvintcev, V. A. Serezhenkov, and D. S. Burbaev, *Biochim. Biophys. Acta* **1135**, 275 (1992).

71. M. Lee, P. Arosio, A. Cozzi, and N. D. Chasteen, *Biochemistry* **33**, 3679 (1994).

72. M. Feelisch, *J. Cardiovasc. Pharmacol.* **17**, 525 (1991).

73. L. R. V. Field, R. V. Dilts, R. Ravichandran, P. G. Lenhert, and G. E. Carnahan, *J. Chem. Soc. Trans.* **1**, 249 (1978).

74. D. Hemmaplardh and E. Morgan, *Int. J. Appl. Radiat. Isot.* **27**, 89 (1976).

75. A. S. McFarlane, *Nature (London)* **182**, 58 (1958).

76. M. Karin and B. Mintz, *J. Biol. Chem.* **256**, 3245 (1981).

77. D. R. Richardson, V. Neumannova, and P. Ponka, *Biochim. Biophys. Acta* **1266**, 250 (1995).

78. E. A. Leibold and H. N. Munro, *Proc. Natl. Acad. Sci. U.S.A.* **85**, 2171 (1988).

79. E. W. Müllner, B. Neupert, and L. C. Kühn, *Cell (Cambridge, Mass.)* **58**, 373 (1989).

80. M. M. Konarska and P. A. Sharp, *Cell (Cambridge, Mass.)* **46**, 845 (1986).

81. J. Sambrook, E. F. Fritsch, and T. Maniatis, *in* "Molecular Cloning: A Labora-

tory Manual.'' Cold Spring Harbor Laboratory, Cold Spring Harbor, New York, 1989.

82. A. J. Carmichael, L. Steelgoodwin, B. Gray, and C. M. Arroyo, *Free Radical Res. Commun.* **19,** S201 (1993).

83. B. F. Anderson, H. M. Baker, E. J. Dodson, G. E. Norris, S. V. Rumball, and E. N. Baker, *Proc. Natl. Acad. Sci. U.S.A.* **84,** 1769 (1987).

84. J.-C. Drapier, H. Hirling, J. Wietzerbin, P. Kaldy, and L. C. Kühn, *EMBO J.* **12,** 3643 (1993).

85. G. Weiss, B. Goossen, W. Doppler, D. Fuchs, K. Pantopoulos, G. Werner-Felmayer, H. Wachter, and M. W. Hentze, *EMBO J.* **12,** 3651 (1993).

86. D. J. Haile, T. A. Rouault, J. B. Harford, M. C. Kennedy, G. A. Blondin, H. Bienert, and R. D. Klausner, *Proc. Natl. Acad. Sci. U.S.A.* **89,** 11735 (1992).

87. A. Emery-Goodman, H. Hirling, L. Scarpellino, B. Henderson, and L. C. Kühn, *Nucleic Acids Res.* **21,** 1457 (1993).

88. H. Hirling, B. R. Henderson, and L. C. Kühn, *EMBO J.* **13,** 453 (1994).

89. C. C. Philpott, D. Haile, T. A. Rouault, and R. D. Klausner, *J. Biol. Chem.* **268,** 17655 (1993).

90. A. Hausladen and I. Fridovich, *J. Biol. Chem.* **269,** 29405 (1994).

91. L. Castro, M. Rodriguez, and R. Radi, *J. Biol. Chem.* **269,** 29409 (1994).

92. S. Z. Lei, Z.-H. Pan, S. K. Aggarwal, H.-S. V. Chen, J. Hartman, N. J. Sucher, and S. A. Lipton, *Neuron* **8,** 1087 (1992).

93. G. Weiss, G. Werner-Felmayer, E. R. Werner, K. Grunewald, H. Wachter, and M. W. Hentze, *J. Exp. Med.* **180,** 969 (1994).

Index